Fine- and Coarse-Grain
Reconfigurable Computing

Fine- and Coarse-Grain Reconfigurable Computing

Stamatis Vassiliadis
Editor

Technical University Delft, The Netherlands

Dimitrios Soudris
Editor

Democritus University of Thrace, Greece

Foreword by Yale Patt, Jim Smith and Mateo Valero

Springer

Editor

Dimitrios Soudris
Democritus University of Thrace,
Greece

Additional materials to this book can be downloaded from http://extras.springer.com

ISBN 978-94-077-9814-4 ISBN 978-1-4020-6505-7 (eBook)
DOI 10.1007/978-1-4020-6505-7

Printed on acid-free paper.

9 8 7 6 5 4 3 2 1

springer.com

To Stamatis who commenced
the Last Journey so early

Καλό Ταξίδι Δάσκαλε

Foreword

Stamatis Vassiliadis was born in the small village of Manolates, on the Mediterranean island of Samos, Greece on July 19th, 1951. He passed away on April 7th, 2007 in Pijnacker, in the Netherlands. In between, he led a very remarkable life. As a child, he walked many kilometers through the mountains to reach his school and would study at night using the light of an oil lamp; as a grown up he became one of the recognized scientific world leaders in computer architecture. For those of you who have chosen to read this book and are not familiar with the scientific achievements of Stamatis Vassiliadis, we have provided this very short snapshot.

Stamatis Vassiliadis did his doctoral study at Politechnico di Milano. As a result, Italy has always had a very special place in his heart. It was there he learned a lot about science, but also about good food, friendships and many other aspects of life. After graduation Stamatis moved to the USA and worked for IBM at the Advanced Workstations and Systems laboratory in Austin, Texas, the Mid-Hudson Valley laboratory in Poughkeepsie, New York, and the Glendale laboratory in Endicott, New York. At IBM he was involved in a number of projects in computer design, computer organization, and computer architecture. He held leadership positions in many advanced research projects. During his time at IBM, he was awarded 73 USA patents, ranking him as the top all time IBM inventor. For his accomplishments, he received numerous awards including 24 Publication Awards, 15 Invention Achievement Awards and an Outstanding Innovation Award for Engineering/Scientific Hardware Design in 1989. While working for IBM Stamatis also served on the ECE faculties of Cornell University, Ithaca, NY and the State University of New York (SUNY), Binghamton, NY.

In 1995, he returned to Europe as the Chaired Professor specializing in computer architecture at TU Delft in the Netherlands. He developed the Computer Engineering laboratory there, which is today one of the strongest groups in the field with more than 50 PhD students from many different countries. Stamatis was an IEEE Fellow, an ACM Fellow, and a member of the Royal Dutch Academy of science (KNAW).

It is impossible to describe all of Stamatis' scientific contributions in this short introduction. His work has inspired many scientists and continues to be the basis for many patents and industrial products. For example, at IBM decades ago, he was a pioneer in micro-operation fusing, a technique that is only recently seeing the light of day in products of both Intel and AMD. He called this concept "compounding."

It was the main idea behind the IBM SCISM project. Unfortunately, it was too many years ahead of its time and was never completed. More recently, in Delft, Stamatis was among the first to recognize the potential of reconfigurable computing and proposed the MOLEN reconfigurable microcoded architecture (described in this book in many details), and the programming paradigm that would allow rapid development of computing systems, ranging from tiny embedded systems to supercomputers. This contribution is being used today in many European Union funded projects. It is also undergoing today broad industrial evaluation by relevant companies.

Stamatis Vassiliadis was very proud of Samos, the island of his birth, a small piece of Greek land that produced many great scientists during its long history. He loved Samos very deeply and returned there every summer. In addition, he crafted the SAMOS International Symposium as an opportunity for students and colleagues to enjoy intensive technical interchange, while sampling fully the vibrancy of his beloved Mediterranean Sea. This year, 2007, marks the seventh SAMOS conference. All who have attended at least one of them will always remember it as a great experience, so different from any other scientific event.

Stamatis was a very industrious and clever person; he loved his job and the opportunities it provided. The devotion to his work was a constant factor that characterized all of his life. Even after being very ill in bed, he continued his work in his Computer Engineering laboratory and was making plans for the SAMOS 2007 International Symposium. He hated mediocrity, he never understood people who did not do their job in the best possible way.

At the same time, he was not only about work. He liked to combine business and pleasure, and he certainly achieved it, passionate about work and passionate about life. He wanted everyone he came in contact with to give his best to his job, but also not lose sight about having fun. He liked people and people liked him. Sometimes he would switch from a kind of "nfant terrible" attitude to the most serious, collaborator in a split second. This was his particular way of dealing with long, tedious administrative issues. Stamatis was for many of us the "Happy Warrior" in our field. He was a very optimistic, positive person who showed great courage until the end. We will always remember him as a most valued colleague and friend.

Yale Patt, Professor at the University of Austin, in Texas
Jim Smith, Professor at the University of Wisconsin, in Madison
Mateo Valero, Professor at the Technical University of Catalonia, in Barcelona

Introduction

Due to the programmability features, reconfigurable technology offers design flexibility which is supported by quite mature commercial design flows. The epoch of reconfigurable computing started with the traditional FPGAs. Moreover, FPGA architecture characteristics and capabilities changed and improved significantly the last two decades, from a simple homogeneous architecture with logic modules and horizontal and vertical interconnections to FPGA platforms (e.g. Virtex-4 logic family), which include except logic and routing, microprocessors, block RAMs etc. In other words, the FPGA architecture changed gradually from a homogeneous and regular architecture to a heterogeneous (or piece-wise homogeneous) and piece-wise regular architecture. The platform-based design allows to designer to build a customized FPGA architecture, using specific blocks, depending on the application domain requirements. The platform-based strategy changed the FPGAs role from a "general-purpose" machine to an "application-domain" machine, closing the gap with ASIC solutions. Furthermore, the need for additional performance through the acceleration of computationally-intensive parts from complex applications can be satisfied by the coarse-gain reconfigurable architectures. In coarse grain reconfigurable hardware some flexibility is traded-off for a potentially higher degree of optimisation in terms of area and power and the ability to reconfigure the platform at a rate, which is significantly faster than the changes of mode observed by a user of the application (not possible in most conventional FPGAs).

The book consists of two parts each of which has different goals and audience. In particular, the first part includes two contributions, which provide a very detailed survey of existing fine (or FPGA)- and coarse-grain reconfigurable architectures and software-support design flows both from academia and industry. Both chapters can be considered as tutorial-like chapters. The second part includes five contributions with specific research results from AMDREL project (FPGA architecture), MORPHOSYS, MOLEN, ADRES and DRESC projects (coarse-grain architectures). The last chapter provides a taxonomy of field-programmable custom computing machines with emphasis on microcode formalism.

This book is accompanied by a CD, which includes additional material useful for the interested reader to go further in the design of FPGA and coarse-gain architectures. In particular, the CD contains, among others, public-domain software tools and a number of assignments about: (i) the MEANDER design framework for FPGA architectures (http://vlsi.ee.duth.gr/amdrel) and (ii) the MOLEN reconfigurable processor and programming paradigm (http://ce.et.tudelft.nl/MOLEN). Moreover, the first two chapters about FPGA and coarse-grain reconfigurable architectures and the solved/unsolved assignments will assist both the instructor to organize some lectures and assignments of semester course about reconfigurable computing, and the student to get deeper in many concepts of FPGA and course-grain architectures. For instance, a student can design an FPGA architecture with his/her own specifications, employing elementary modules (e.g. flip-flops, LUT), which is not available from the commercial tools.

The authors of the book chapters together with the editors would like to use this opportunity to thank many people, i.e. colleagues, M.Sc. and Ph.D. students, whose dedication and industry during the projects execution led to the introduction of novel scientific results and implementation of innovative reconfigurable systems.

Dimitrios Soudris would like to thank his parents for being a constant source of moral support and for firmly imbibing into him from a very young age that *perseverantia omnia vincit* – it is this perseverance that kept him going. This book is dedicated to them.

We finally hope that the reader (instructor, engineer, student, etc) will find the book useful and constructive and enjoyable, and that the technical material presented will contribute to the continued progress in the reconfigurable architectures field.

Delft, January 2007 Stamatis Vassiliadis and Dimitrios Soudris

Contents

Contributors

Nader Bagherzadeh
Dpt. Electrical Engineering
and Computer Science
of the University
of California Irvine (USA)
Email: nader@uci.edu

Mladen Berekovic
IMEC vzw, Kapeldreef 75,
3001 Leuven, BELGIUM,
Email: berekovi@imec.be

Sorin Cotofana
Delft University of Technology,
Computer Engineering Department,
Postbus 5031, 2600 GA Delft,
The Netherlands,
Email: S.D.Cotofana@ewi.tudelft.nl

Milagros Fernandez
Dpto. Arquitectura de Computadoresy
Automatica of the Universida
Complutense de Madrid (SPAIN)
Email: mila45@dacya.ucm.es

Roman Hermida
Dpto. Arquitectura de Computadores y
Automatica of the Universida
Complutense de Madrid (SPAIN)
Email: rhermida@dacya.ucm.es

Vassilios Kalenteridis
Electronics and Computers Div.,
Department of Physics,
Aristotle University
of Thessaloniki, 54006
Thessaloniki, Greece
Email: vkale@skiathos.physics.auth.gr

George Koutroumpezis
VLSI Design and Testing Center,
Department of Electrical
and Computer Engineering,
Democritus University of Thrace,
67100, Xanthi, Greece,
Email: gkoytroy@ee.duth.gr

Fadi Kurdahi
Dpt. Electrical Engineering
and Computer Science
of the University
of California Irvine (USA)
Email: Kurdahi@uci.edu

Georgi Kuzmanov
Delft University of Technology,
Computer Engineering Department,
Postbus 5031, 2600 GA Delft,
The Netherlands
Email: G.Kuzmanov@ewi.tudelft.nl

Rafael Maestre
Qualcomm
San Diego, California, 92130
United States
Email: rmaestre@qualcomm.com

Bennet Mei
IMEC vzw, Kapeldreef 75,
3001 Leuven, BELGIUM,
Email: bennet@imec.be

Jean-Yves Mignolet
IMEC vzw, Kapeldreef 75,
3001 Leuven, BELGIUM,
Email: mignolet@imec.be

Spyros Nikolaidis
Electronics and Computers Div.,
Department of Physics,
Aristotle University of Thessaloniki,
54006 Thessaloniki, Greece,
Email: snikolaid@physics.auth.gr

Ioannis Pappas
Electronics and Computers Div.,
Department of Physics,
Aristotle University of Thessaloniki,
54006 Thessaloniki, Greece
Email: ilpap@auth.gr

Harikleia Pournara
Electronics and Computers Div.,
Department of Physics,
Aristotle University of Thessaloniki,
54006 Thessaloniki, Greece
Email: hpour@skiathos.physics.auth.gr

Marcos Sanchez-Elez
Dpto. Arquitectura de Computadores y
Automatica of the Universida
Complutense de Madrid (SPAIN)
Email: marcos@fis.ucm.es

Mihai Sima
University of Victoria,
Department of Electrical
and Computer Engineering,
P.O. Box 3055 Stn CSC, Victoria,
B.C. V8W 3P6, Canada
Email: msima@ece.uvic.ca

Kostantinos Siozios
VLSI Design and Testing Center,
Department of Electrical
and Computer Engineering,
Democritus University of Thrace,
67100, Xanthi, Greece,
Email: ksiop@ee.duth.gr

Stylianos Siskos
Electronics and Computers Div.,
Department of Physics,
Aristotle University of Thessaloniki,
54006 Thessaloniki, Greece,
Email: siskos@physics.auth.gr

Dimitrios Soudris
VLSI Design and Testing Center,
Department of Electrical
and Computer Engineering,
Democritus University of Thrace,
67100, Xanthi, Greece,
Email: dsoudris@ee.duth.gr,
dimitrios.soudris@gmail.com

Konstantinos Tatas
VLSI Design and Testing Center,
Department of Electrical
and Computer Engineering,
Democritus University of Thrace,
67100, Xanthi, Greece,
Email: ktatas@ee.duth.gr

George Theodoridis
Physics Department,
Aristotle University
of Thessaloniki, Greece,
Email: theodor@physics.auth.gr

Nikos Vasiliadis
Electronics and Computers Div.,
Department of Physics,
Aristotle University of Thessaloniki,
54006 Thessaloniki, Greece,
Email: nivas@skiathos.physics.auth.gr

Stamatis Vassiliadis
Delft University of Technology,
Computer Engineering Department,
Postbus 5031, 2600
GA Delft, The Netherlands
Email: S.Vassiliadis@ewi.tudelft.nl

List of Abbreviation Terms

ADRES	Architecture for Dynamically Reconfigurable Embedded Systems
ADSS	Application Domain-Specific System
ASICs	Application Specific Integrated Circuits
ASIP	Application – Specific Integrated Processor
CCU	Custom Configured Unit
CDFGs	Control Data Flow Graphs
CGRUs	Coarse-Grain Reconfigurable Units
CISC	Complex Instruction Set Computers
CLB	Configurable Logic Block
CPI	Cycles Per Instruction
DDRG	Data Dependency Reuse Graph
DMA	Direct Memory Access
DRESC	Dynamically Reconfigurable Embedded System Compiler
EDP	Energy Delay Product
FCCM	Field –Programmable Custom Computing Machines
FPGAs	Field-Programmable Gate Arrays
GPP	General Purpose Processor
HDL	Hardware Description Language
LUT	Look-Up Table
PE	Processing Element
PISC	Polymorphic Instruction Set Computers
PLD	Programmable Logic Devices
PNG	Portable Network Graphics
PoCR	Pipeline of Computing Resources
PZE	Potential Zero Execution
RC	Reconfigurable Computing
RFU	Reconfigurable Functional Unit
RISC	Reduced Instruction Set Computers
RPU	Reconfigurable Processing Unit
SAD	Sum of Absolute Differences
SB	Switch Box
SIMD	Single Instruction Multiple Data

SoCR	Sea of Computing Resources
SRAM	Static Random Access Memory
SW/HW	Software-Hardware
TP	Tile Processor
VLIW	Very Long Instruction Word

Part I

Chapter 1
A Survey of Existing Fine-Grain Reconfigurable Architectures and CAD tools[1]

K. Tatas, K. Siozios, and D. Soudris

Abstract This chapter contains an introduction to FPGA technology that includes architecture, power consumption and configuration models, as well as an extensive survey of existing fine-grain reconfigurable architectures that have emerged from both academia and industry. All aspects of the architectures, including logic block structure, interconnect, and configuration methods are discussed. Qualitative and quantitative comparisons in terms of testability, technology portability, design flow completeness and configuration type are shown. Additionally, the implementation techniques and CAD tools (synthesizers, LUT-mapping tools and placement and routing tools) that have been developed to facilitate the implementation of a system in reconfigurable hardware by the industry (both by FPGA manufacturers and third-party EDA tool vendors) and academia are described.

1.1 Introduction

The domain of Field-Programmable Gate Arrays (FPGAs) is an increasingly popular technology, which allows circuit designers to produce application-specific chips bypassing the time-consuming fabrication process. A FPGA can be seen as a set of reconfigurable blocks that communicate through reconfigurable interconnect. By using the appropriate configuration, FPGAs can, in principle, implement any digital circuit as long as their available resources (logic blocks and interconnect) are adequate.

A FPGA can be programmed to solve a problem at hand in a spatial fashion. The goal of reconfigurable architectures is to achieve implementation efficiency approaching that of specialized logic, while providing the silicon reusability of general-purpose processors.

The main components and features of an FPGA are:

- The logic block architecture
- The interconnect architecture

[1] This work was partially supported by the project IST-34793-AMDREL which is funded by the E.C.

S. Vassiliadis, D. Soudris (eds.), *Fine- and Coarse-Grain Reconfigurable Computing.*

- The programming technology
- The power dissipation
- The reconfiguration model

As mentioned earlier, FPGAs can be visualized as programmable logic embedded in programmable interconnect. All FPGAs are composed of three fundamental components: logic blocks, I/O blocks and programmable routing. A circuit is implemented in an FPGA by programming each logic block to implement a small portion of the logic required by the circuit, and each of the I/O blocks to act as either an input pad or an output pad, as required by the circuit. The programmable routing is configured to make all the necessary connections between logic blocks and from logic blocks to I/O blocks. The functional complexity of logic blocks can vary from simple two-input Boolean operations to larger, complex, multi-bit arithmetic operations. The choice of the logic block granularity is dependent on the target application domain.

The programming technology determines the method of storing the configuration information, and comes in different flavors. It has a strong impact on the area and performance of the array. The main programming technologies are: Static Random Access Memory (SRAM) [1], antifuse [2], and non-volatile technologies. The choice of the programming technology is based on the computation environment in which the FPGA is used.

The general model of an FPGA is shown in Fig. 1.1. The logic cell usually consists of lookup tables (LUTs), carry logic, flip-flops, and programmable multiplexers. The multiplexers are utilized to form data-paths inside the logic cell and to connect the logic cells with the interconnection resources.

When FPGAs were first introduced in the mid 1980s they were viewed as a technology for replacing standard gate arrays for certain applications. In these first-generation systems, a single configuration was created for the FPGA, and this configuration was the only one loaded into the FPGA. A second generation soon followed, with FPGAs that could use multiple configurations, but reconfiguration was done relatively infrequently. In such systems, the time to reconfigure the FPGA

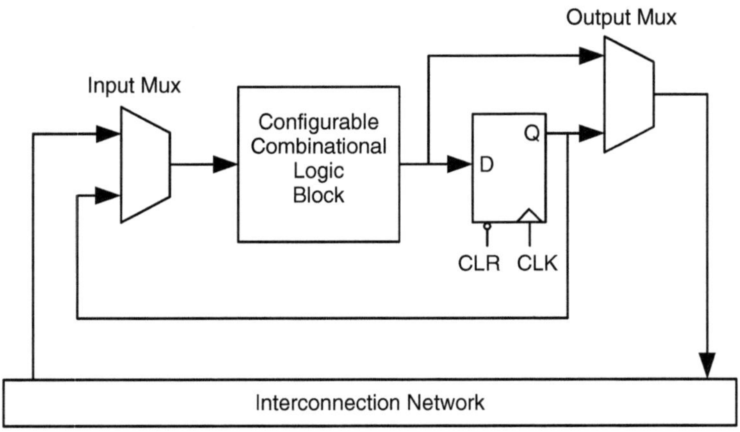

Fig. 1.1 FPGA model

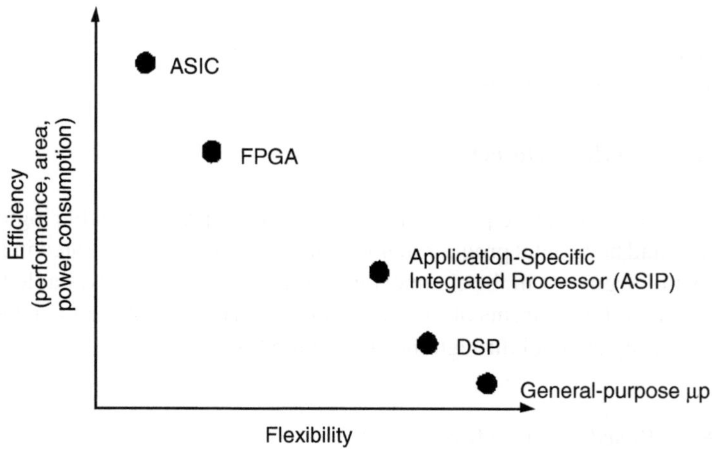

Fig. 1.2 Comparison between implementation platforms

was of little concern. Nowadays, applications demand short reconfiguration time and so a new generation of FPGAs was developed that could support many types of reconfiguration methods, depending on the application-specific needs.

Figure 1.2 shows a graphic comparison of implementation technologies in terms of efficiency (performance, area and power consumption) versus flexibility. It can be seen that FPGAs are an important implementation option since they bridge the gap between ASICs and microprocessors.

The next part of this chapter describes the Field-Programmable Gate Array (FPGA) architecture, examining alternative interconnect architectures, logic block architectures, programming technologies, power dissipation and reconfiguration models. Then follows a description of the available fine-grain reconfigurable architectures, both commercial and academic. Section 1.5 presents the available CAD tools used for programming FPGAs, also both commercial and academic (public domain).

1.2 FPGA Architecture

1.2.1 Interconnect Architecture (Routing Resources)

The FPGA interconnect architecture is realized using switches that can be programmed to realize different connections. The method of providing connectivity between the logic blocks has a strong impact on the characteristics of the FPGA architecture. The arrangement of the logic and interconnect resources can be broadly classified into six groups:

- Island style
- Row-based

- Sea-of-gates
- Hierarchical
- One-dimensional structures

1.2.1.1 Island Style Architecture

The island style architecture [3] consists of an array of programmable logic blocks with vertical and horizontal programmable routing channels as illustrated in Fig. 1.3. The number of segments in the channel determines the resources available for routing. This is quantified in terms of the channel width. The pins of the logic block can access the routing channel through the connection box.

1.2.1.2 Row-Based Architecture

As the name implies, this architecture has logic blocks arranged in rows with horizontal routing channels between successive rows. The row-based architecture [4] is shown in Fig. 1.4. The routing tracks within the channel are divided into one or more segments. The length of the segments can vary from the width of a module pair to the full length of the channel. The segments can be connected at the ends using programmable switches to increase their length. Other tracks run vertically through the logic blocks. They provide connections between the horizontal routing

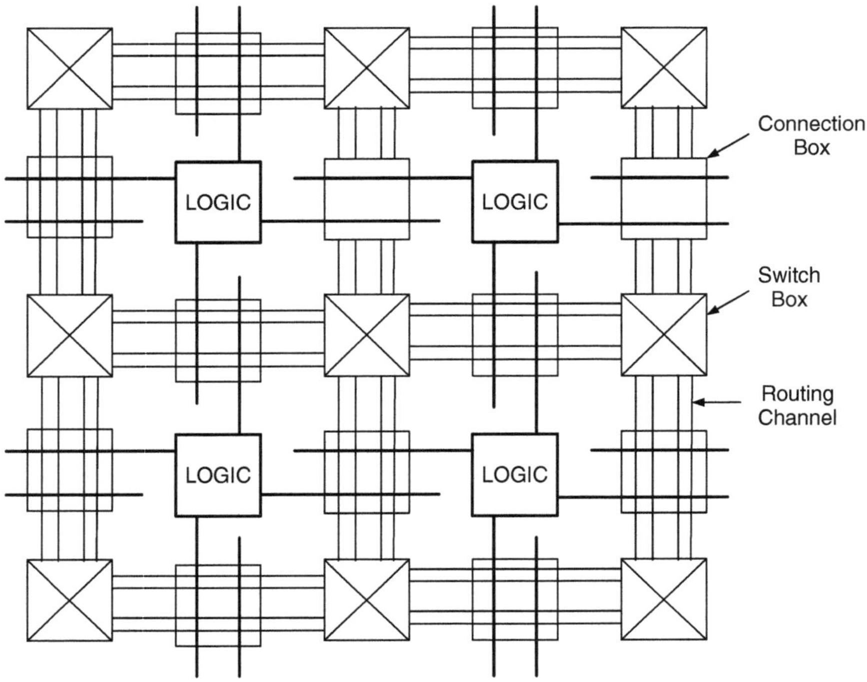

Fig. 1.3 Island style architecture

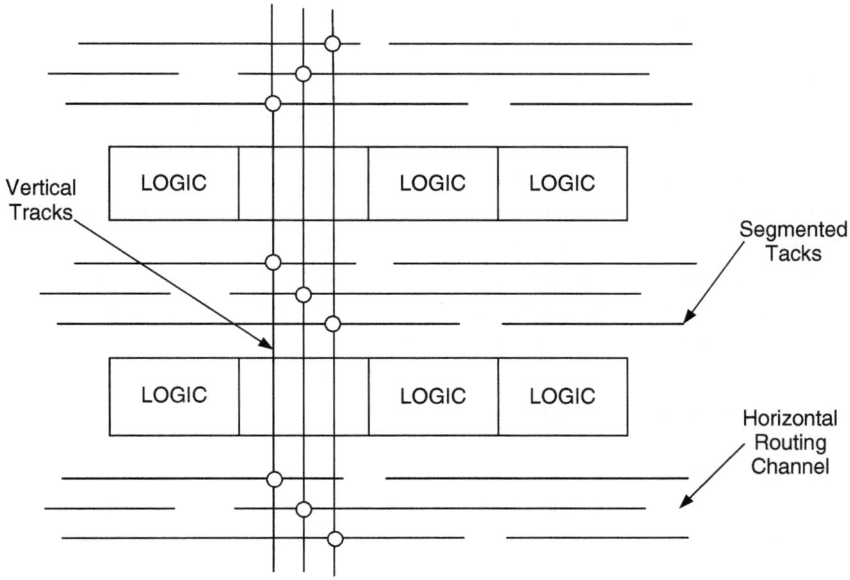

Fig. 1.4 Row-based architecture

channel and the vertical routing segments. The length of the wiring segments in the channel is determined by tradeoffs involving the number of tracks, the resistance of the routing switches, and the capacitance of the segments.

1.2.1.3 Sea-of-Gates Architecture

The sea-of-gates architecture [5], as shown in Fig. 1.5, consists of fine-grain logic blocks covering the entire floor of the device. Connectivity is realized using dedicated neighbor-to-neighbor routes that are usually faster than general routing resources. Usually the architecture also uses some general routing resources to realize longer connections.

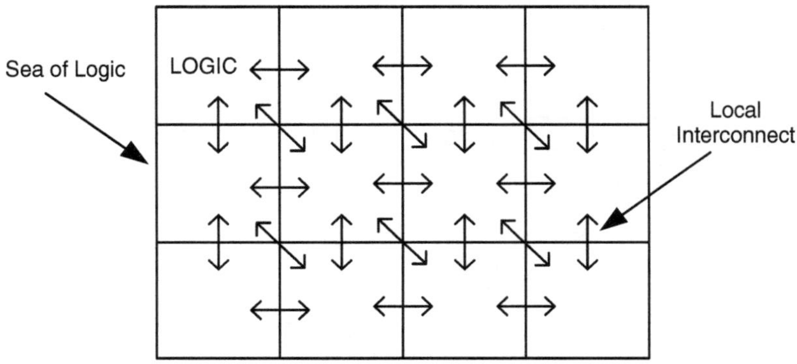

Fig. 1.5 Sea-of-gates architecture

1.2.1.4 Hierarchical Architecture

Most logic designs exhibit locality of connections, which imply a hierarchy in the placement and routing of the connections between the logic blocks. The hierarchical FPGA architecture [6] tries to exploit this feature to provide smaller routing delays and a more predictable timing behavior. This architecture is created by connecting logic blocks into clusters. These clusters are recursively connected to form a hierarchical structure. Figure 1.6 illustrates a possible architecture. The speed of the network is determined by the number of routing switches it has to pass through. The hierarchical structure reduces the number of switches in series for long connections and can hence potentially run at a higher speed.

1.2.1.5 One-Dimensional Structures

Most current FPGAs are of the two-dimensional variety. This allows for a great deal of flexibility, as any signal can be routed on a nearly arbitrary path. However, providing this level of routing, flexibility requires a great deal of routing area. It also complicates the placement and routing software, as the software must consider a very large number of possibilities. One solution is to use a more one-dimensional style of architecture [7], as shown in Fig. 1.7. Here placement is restricted along one axis. With a more limited set of choices, the placement can be performed much more quickly. Routing is also simplified, because it is generally along a single dimension as well, with the other dimension generally only used for calculations requiring a shift operation. One drawback of one-dimensional routing is that if there are not

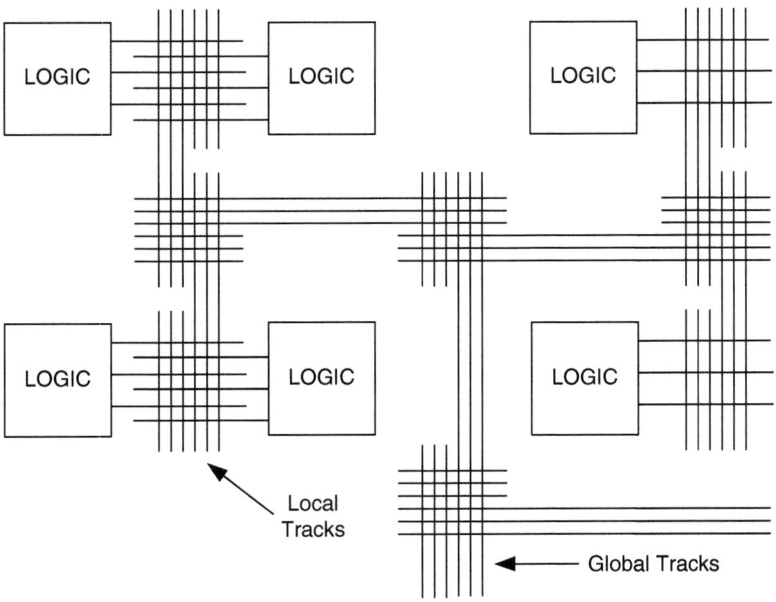

Fig. 1.6 Hierarchical architecture

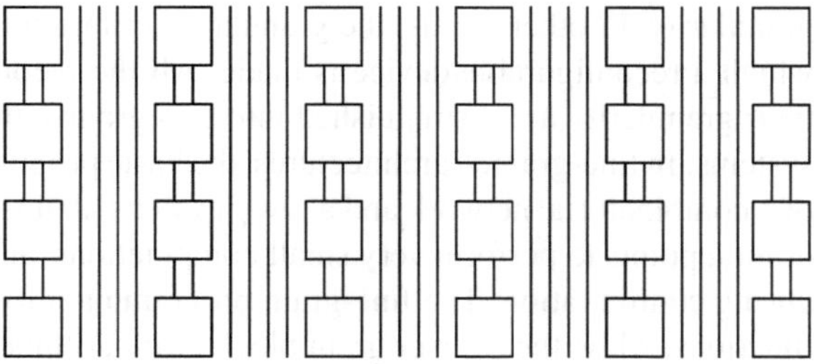

Fig. 1.7 One-dimensional structure

enough routing resources for a specific area of a mapped circuit, then the routing of the whole circuit becomes actually more difficult than on a two-dimensional array that provides more alternatives.

It should be noted that contemporary FPGAs often employ combinations of the above interconnect schemes, as will be seen in following sections. For example, an FPGA may employ nearest-neighbor connections, and at the same time longer horizontal and vertical tracks to communicate with distant logic blocks.

1.2.2 Logic Block Architecture

The configurable logic block (CLB) [3], is responsible for implementing the gate-level functionality required for each application. The logic block is defined by its internal structure and granularity. The structure defines the different kinds of logic that can be implemented in the block, while the granularity defines the maximum wordlength of the implemented functions. The functionality of the logic block is obtained by controlling the connectivity of some basic logic gates or by using LUTs and has a direct impact on the routing resources. As the functional capability increases, the amount of logic that can be packed into it increases.

A collection of CLBs, known as logic cluster, is described with the following four parameters:

- The size of (number of inputs to) a LUT.
- The number of CLBs in a cluster.
- The number of inputs to the cluster for use as inputs by the LUTs.
- The number of clock inputs to a cluster (for use by the registers).

The advantage of using a k-input LUT (k-LUT) is that it can realize any combinational logic with k inputs. Previous work [8] that evaluated the effect of the logic block on the FPGA architecture used a k-input LUT with a single output as the logic block. This structure is better for implementing random logic functions than for datapath-like bit-slice operations.

1.2.2.1 Logic Block Granularity

Logic blocks vary in complexity from very small and simple blocks that can calculate a function of only three inputs, to structures that are essentially 4-bit ALUs. The size and complexity of the basic computing blocks is referred to as the block granularity. In other words, the granularity criterion refers to the smallest block of which a reconfigurable device is made. All the reconfigurable platforms based on their granularity are distinguished into two groups, the fine-grain and coarse-grain systems. In fine-grained architectures, the basic programmed building block consists of a combinatorial network and a few flip-flops. A fine-grain array has many configuration points to perform very small computations, and thus requires more data bits during configuration. The fine-grain programmability is more amenable to control functions, while the coarser grain blocks with arithmetic capability are more useful for datapath operations. All the reconfigurable architectures that are described in this report are considered fine-grain reconfigurable architectures.

1.2.2.2 Studies on the CLB Structure

Studies on the CLB structure have shown that the best number of inputs to use in order to improve area is between 3 and 4 [8]. Also it is possible to improve the functionality by including a D flip-flop. Moreover, for multiple output LUTs, the use of 4-input LUTs minimizes the area [8], while a 5 to 6 inputs LUT minimizes delay [9]. The use of heterogeneous logic blocks that use a combination of 4 and 6-input LUTs improves the speed by 25% [10], with no additional area penalty in comparison to exclusively using 4-input LUTs. Finally, the use of clusters with 4-inputs LUT instead of one 4-input LUT, results in an area decrease of 10% [11].

1.2.3 Programming Technology

As already mentioned, the logic and routing resources of an FPGA are uncommitted, and must be programmed (configured) to realize the required functionality. The contents of the logic block can be programmed to control the functionality of the logic block, while the routing switches can be programmed to realize the desired connections between the logic blocks. There are a number of different methods to store this program information, ranging from the volatile SRAM method [12] to the irreversible antifuse technology [13]. The area of an FPGA is dominated by the area of the programmable components. Hence, the choice of the programming technology can also affect the area of the FPGA. Another factor that has to be considered is the number of times the FPGA has to be programmed (configured). Antifuse-based FPGAs can be programmed only once, while in SRAM-based FPGAs there is no limit to the number of times the array can be reprogrammed.

1.2.3.1 SRAM

In this method of programming, the configuration is stored in SRAM cells. When the interconnect network is implemented using pass-transistors, the SRAM cells control whether the transistors are on or off. In the case of the lookup tables used in the logic block, the logic is stored in the SRAM cells. This method suffers from the fact that the storage is volatile and the configuration has to be written into the FPGA each time on power-up. For systems using SRAM-based FPGAs, an external permanent storage device is usually used. This technology requires at least five transistors per cell. Due to the relatively large size of the memory cells, the area of the FPGA is dominated by configuration storage. The SRAM method of programming offers the convenience of reusing a single device for implementing different applications by loading different configurations. This feature has made SRAM-based FPGAs popular in reconfigurable platforms, which strive to obtain performance gains by customizing the implementation of functions to the specific application.

1.2.3.2 Antifuse

The antifuse programming method [13] uses a programmable connection whose impedance changes on the application of a high voltage. In the un-programmed state, the impedance of the connection is of the order of a few giga-ohms, and can be treated as an open circuit. By applying a high voltage, a physical change called fusing occurs. This results in an impedance of a few ohms though the device, establishing a connection. This method has the advantage that the area of the programming element is in the order of the size of a Via, and therefore can achieve a significant reduction in area compared to the SRAM-programmed FPGA. This programming technique is non-volatile, and does not require external configuration storage on power-down. Unlike the SRAM based technology, errors in the design cannot be corrected, since the programming process is irreversible.

1.2.3.3 EPROM, EEPROM, and FLASH

This class of non-volatile programming technology uses the same techniques as EPROM, EEPROM and Flash memory technologies [14]. This method is based on a special transistor with two gates: a floating gate and a select gate. When a large current flows through the transistor, a charge is trapped in the floating gate that increases the threshold voltage of the transistor. Under normal operation, the programmed transistors may act as open circuits, while the other transistors can be controlled using the select gates. The charge under the floating gate will persist during power-down. The floating charge can be removed by exposing the gate to ultraviolet light in the case of EPROMs, and by electrical means in the case of EEPROMs and Flash. These techniques combine the non-volatility of antifuse with the reprogrammability of SRAM. The resistance of the routing switches is larger than that of the antifuse, while the programming is more complex and time consuming than that of the SRAM technique.

1.2.4 Power Dissipation

Today's systems have become more complex, and can take advantage of the programmability offered by Field-Programmable Gate Arrays. This environment places stress on the energy efficiency of FPGAs, which is still an issue in existing commercial architectures. Another factor that has gained importance is the power density of the integrated circuits. With the reduction in feature size the transistor count per die has increased. This has resulted in an increase of power density, and the overall power dissipation per chip. Therefore, both academia and industry have developed techniques to reduce FPGA power consumption.

1.2.4.1 Components of Power

A dramatic improvement in the energy efficiency of FPGAs is required. An understanding of the energy breakdown in an FPGA is required to enable an efficient redesign process. Figure 1.8 gives the energy breakdown of a Xilinx XC4003 FPGA over a set of benchmark netlists [15].

The majority of the power is dissipated in the interconnection network. The next major component is the clock network, while the logic block consumes only 5% of the total energy. This breakdown is not specific to the Xilinx FPGA, but is representative of most of the commercial FPGA architectures.

Another aspect of power dissipation in FPGAs is in terms of dynamic versus static power consumption as can be seen in Fig. 1.9. The contribution of static power consumption in the total power budget increases as transistor sizes decrease. However, today, dynamic power consumption is still dominant.

1.2.5 Reconfiguration Models

Traditional FPGA structures have been implemented to function in a single context, only allowing one full-chip configuration to be loaded at a time. This style of reconfiguration is too limited or slow to efficiently implement run-time reconfiguration. The most well-known reconfiguration models, which could be used in order to program an FPGA, will be described next.

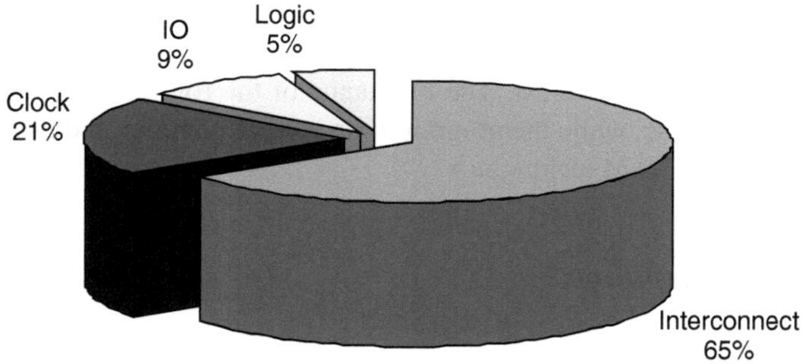

Fig. 1.8 Power breakdown in an XC4003 FPGA

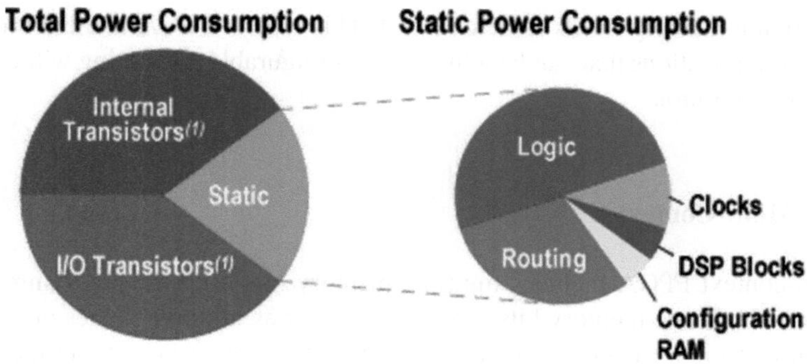

Fig. 1.9 Typical power consumption for a high-performance FPGA design [16]

1.2.5.1 Static Reconfiguration

Static reconfiguration, which is often referred to as compile-time reconfiguration, is the simplest and most common approach for implementing applications with reconfigurable logic. Static reconfiguration involves hardware changes at a relatively slow rate: hours, days, or weeks. At this strategy, each application consists of one configuration. Many of the existing reconfigurable systems are statically reconfigurable. In order to reconfigure such a system, it has to be halted while the reconfiguration is in progress and then restarted with the new program.

1.2.5.2 Dynamic Reconfiguration

On the other hand, dynamic reconfiguration [17], also known as run-time reconfiguration, uses a dynamic allocation scheme that re-allocates hardware at run-time. With this technique there is a trade-off between time and space. It can increase system performance by using highly-optimized circuits that are loaded and unloaded dynamically during the operation of the system. Dynamic reconfiguration is based on the concept of virtual hardware, which is similar to the idea of virtual memory. In this case, the physical hardware is much smaller than the sum of the resources required by all the configurations. Therefore, instead of reducing the number of configurations that are mapped, it is preferable to swap them in and out of the actual hardware, as they are needed.

1.2.5.3 Single Context

Single context FPGAs have only one configuration each time and can be programmed using a serial stream of configuration information. Because only sequential access is supported, any change to a configuration on this type of FPGA requires a complete reprogramming of the entire chip. Although this does simplify the reconfiguration hardware, it does incur a high overhead when only a small part of the configuration memory needs to be changed. This type of FPGA is therefore more suited for applications that can benefit from reconfigurable computing without run-time reconfiguration.

1.2.5.4 Multi-Context

A multi-context FPGA includes multiple memory bits for each programming bit location [18]. These memory bits can be thought of as multiple planes of configuration information, each of which can be active at a given moment, but the device can quickly switch between different planes, or contexts, of already-programmed configurations. A multi-context device can be considered as a multiplexed set of single context devices, which requires that a context be fully reprogrammed to perform any modification. This system does allow for the background loading of a context, where one plane is active and in execution, while an inactive plane is in the process of being programmed. Fast switching between contexts makes the grouping of the configurations into contexts slightly less critical, because if a configuration is on a different context than the one that is currently active, it can be activated within an order of nanoseconds, as opposed to milliseconds or longer.

1.2.5.5 Partial Reconfiguration

In some cases, configurations do not occupy the full reconfigurable hardware, or only a part of a configuration requires modification. In both of these situations, a partial reconfiguration of the array is required, rather than the full reconfiguration required by a single-context or multi-context device. In a partially reconfigurable FPGA, the underlying programming bit layer operates like a RAM device. Using addresses to specify the target location of the configuration data allows for selective reconfiguration of the array. Frequently, the undisturbed portions of the array may continue execution, allowing the overlap of computation with reconfiguration. Additionally, some applications require the updating of only a portion of a mapped circuit, while the rest should remain intact. Using this selective reconfiguration can greatly reduce the amount of configuration data that must be transferred to the FPGA.

1.2.5.6 Pipeline Reconfiguration

A modification of the partially reconfigurable FPGA design is one in which the partial reconfiguration occurs in increments of pipeline [19] stages. Each stage is configured as a whole. This is primarily used in datapath style computations, where more pipeline stages are used than can be fitted simultaneously on available hardware.

1.2.6 Run-time Reconfiguration Categories

The challenges associated with run-time reconfiguration are closely linked with the goal of reconfiguration. Therefore, it is important to consider the motivation and the different scenarios of run-time reconfiguration, which are algorithmic, architectural and functional reconfiguration. They are briefly described below.

1.2.6.1 Algorithmic Reconfiguration

The goal in algorithmic reconfiguration is to reconfigure the system with a different computational algorithm that implements the same functionality, but with different performance, accuracy, power, or resource requirements. The need for such reconfiguration arises when either the dynamics of the environment or the operational requirements change.

1.2.6.2 Architectural Reconfiguration

The goal in architectural reconfiguration is to modify the hardware topology and computation topology by reallocating resources to computations. The need for this type of reconfiguration arises in situations where some resources become unavailable either due to a fault or due to reallocation to a higher priority job, or due to a shutdown in order to minimize the power usage. For the system to keep functioning in spite of the fault the hardware topology need to be modified and the computational tasks need to be reassigned.

1.2.6.3 Functional Reconfiguration

The goal in functional reconfiguration is to execute different function on the same resources. The need for this type of reconfiguration arises in situations where a large number of different functions are to be performed on a very limited resource envelope. In such situations the resources must be time-shared across different computational tasks to maximize resource utilization and minimize redundancy.

1.2.6.4 Fast Configuration

Because run-time reconfigurable systems involve reconfiguration during program execution, the reconfiguration must be done as efficiently and as quickly as possible. This is in order to ensure that the overhead of the reconfiguration does not eclipse the benefit gained by hardware acceleration. There are a number of different tactics for reducing the configuration overhead, and they will be described below.

1.2.6.5 Configuration Prefetching

By loading a configuration into the reconfigurable logic in advance of when it is needed, it is possible to overlap the reconfiguration with useful computation. This results in a significant decrease in the reconfiguration overhead for these applications. Specifically, in systems with multiple contexts, partial run-time reconfigurability, or tightly coupled processors it is possible to load a configuration into all or part of the FPGA while other parts of the system continue computing. In this way, the reconfiguration latency is overlapped with useful computations, hiding the reconfiguration overhead. The challenge in configuration prefetching [20] is determining far enough in advance which configuration will be required next.

1.2.6.6 Configuration Compression

When multiple contexts or configurations have to be loaded in quick succession then the system's performance may not be satisfactory. In such a case, the delay incurred is minimized when the amount of data transferred from the processor to the reconfigurable hardware has to be minimized. A technique that could be used in order to compact this configuration information is configuration compression [21].

1.2.6.7 Relocation and Defragmentation in Partially Reconfigurable Systems

Partially reconfigurable systems have advantages over single context systems, but problems might occur if two partial configurations are supposed to be located at overlapping physical locations on the FPGA. A solution to this problem is to allow the final placement of the configurations to occur at run-time, allowing for run-time relocation of those configurations. By using this technique, the new configuration could be placed onto the reconfigurable hardware where it will cause minimum conflict with other needed configurations already present on the hardware. Over time, as a partially reconfigurable device loads and unloads configurations, the location of the unoccupied area on the array is likely to become fragmented, similar to what occurs in memory systems when RAM is allocated and deallocated. A configuration normally requires a continuous region of the chip, so it would have to overwrite a portion of the valid configuration in order to be placed onto the reconfigurable hardware. A system that incorporates the ability to perform defragmentation [22] of the reconfigurable array, however, would be able to consolidate the unused area by moving valid configurations to new locations.

1.2.6.8 Configuration Caching

Caching configurations [23] on an FPGA, which is similar to caching instructions or data in a general memory, is to retain the configurations on the chip so the amount of the data that needs to be transferred to the chip can be reduced. In a general-purpose computational system, caching is an important approach to hide memory latency by taking advantage of two types of locality, spatial and temporal locality. These two localities also apply to the caching of configurations on the FPGA in coupled processor-FPGA systems. The challenge in configuration caching is to determine which configurations should remain on the chip and which should be replaced when a reconfiguration occurs. An incorrect decision will fail to reduce the reconfiguration overhead and lead to a much higher reconfiguration overhead than a correct decision.

1.3 Academic Fine-Grain Reconfigurable Platforms

Some of the existing well-known academic fine-grain reconfigurable platforms are described in the next subsections. The first part of this section is about platforms that are based on fine-grain reconfigurable devices, while the second one is for stand alone reconfigurable devices. All of those architectures use one or two bits for their functions, and so they are characterized as fine-grain. At the end of this section is a summary table, where many of the previous referred systems are compared with criteria like the programmability, the reconfiguration method, the interface and the possible application domain.

1.3.1 Platforms that are Based on Fine-Grain Reconfigurable Devices

1.3.1.1 GARP

Garp [24] was developed at the University of California Berkeley. It belongs to the family of Reconfigurable Coprocessors as it integrates a reconfigurable array that has access to the processor's memory hierarchy. The reconfigurable array may be partially reconfigured as it is organized in rows. Configuration bits are included and linked as constants with ordinary C compiled programs.

In the Garp architecture, the FPGA is recast as a slave computational unit located on the same die as the processor. The reconfigurable hardware is used to speed up operations when possible, while the main processor takes care of all other computations. Figure 1.10 shows the organization of the machine at the highest level. Garp's reconfigurable hardware goes by the name of the reconfigurable array. It has been designed to fit into an ordinary processing environment, one that includes structured programs, libraries, context switches, virtual memory, and multiple users. The main thread of control through a program is managed by the processor and in fact programs never need to use the reconfigurable hardware. It is expected, however, that for certain loops or subroutines, programs will switch temporarily to the reconfigurable array to obtain a speedup. With Garp, the loading and execution of

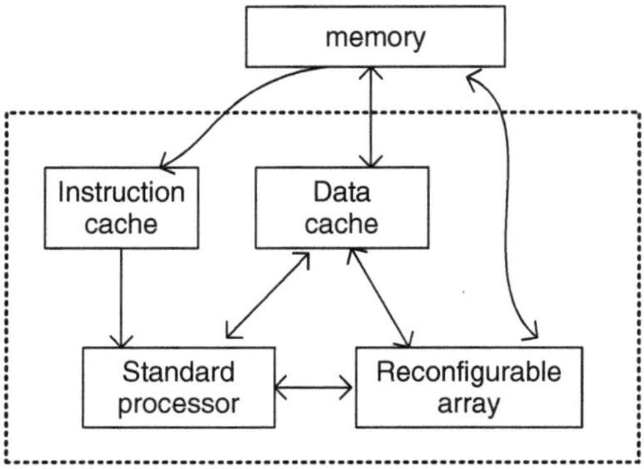

Fig. 1.10 Basic garp block diagram

configurations on the reconfigurable array is always under the control of a program running on the main processor.

Garp makes external storage accessible to the reconfigurable array by giving the array access to the standard memory hierarchy of the main processor. This also provides immediate memory consistency between array and processor. Furthermore, Garp has been defined to support strict binary compatibility among implementations, even for its reconfigurable hardware.

Garp's reconfigurable array is composed of entities called blocks. One block on each row is known as a control block. The rest of the blocks in the array are logic blocks, which correspond roughly to the CLBs of the Xilinx 4000 series. The Garp Architecture fixes the number of columns of blocks at 24, while the number of rows is implementation-specific, but can be expected to be at least 32. The architecture is defined so that the number of rows can grow in an upward-compatible fashion.

The basic "quantum" of data within the array is 2 bits. Logic blocks operate on values as 2-bit units, and all wires are arranged in pairs to transmit 2-bit quantities. Operations on data wider than 2 bits can be formed by adjoining logic blocks along a row. Construction of multi-bit adders, shifters, and other major functions is aided by hardware invoked through special logic block modes.

1.3.1.2 OneChip

The OneChip [25] architecture combines a fixed-logic processor core with reconfigurable logic resources. Typically, OneChip is useful for two types of applications. The first one is the embedded controller type problems requiring custom glue logic interfaces, while the other one is for application-specific accelerators utilizing customized computation hardware. Using the programmable components of this architecture, the performance of speed-critical applications can be improved by customizing OneChip's execution units, or flexibility can be added to the glue logic interfaces of embedded controller applications. OneChip eliminates the shortcomings of other custom compute machines by tightly integrating its reconfigurable resources into a MIPS-like processor.

1.3.1.3 Chimaera

The Chimaera [26] [27] prototype system integrates a small and fast reconfigurable functional unit (RFU) into the pipeline of an aggressive, dynamically-scheduled superscalar processor. The RFU is a small and fast field-programmable gate array-like device that can implement application specific operations. The Chimaera system is capable of collapsing a set of instructions into RFU operations, converting control-flow into RFU operations, and supporting a more powerful fine-grain data-parallel model than that supported by current multimedia extension instruction sets (for integer operations). The RFU is capable of performing computations that use up to 9 input registers and produce a single register result and it is tightly integrated with the processor core to allow fast operation (in contrast to typical FPGAs which are build as discrete components and that are relatively slow).

The Chimaera architecture, shown in Fig. 1.11, comprises the following components: the reconfigurable array (RA), the shadow register file (SRF), the execution control unit (ECU), and the configuration control and caching unit (CCCU). The RA is where operations are executed. The ECU decodes the incoming instruction stream and directs execution. The ECU communicates with the control logic of the host processor for coordinating execution of RFU operations. The CCCU is responsible for loading and caching configuration data. Finally, the SRF provides input data to the RA for manipulation.

In the core of the RFU lies the RA. The RA is a collection of programmable logic blocks organized as interconnected rows. Each row contains a number of

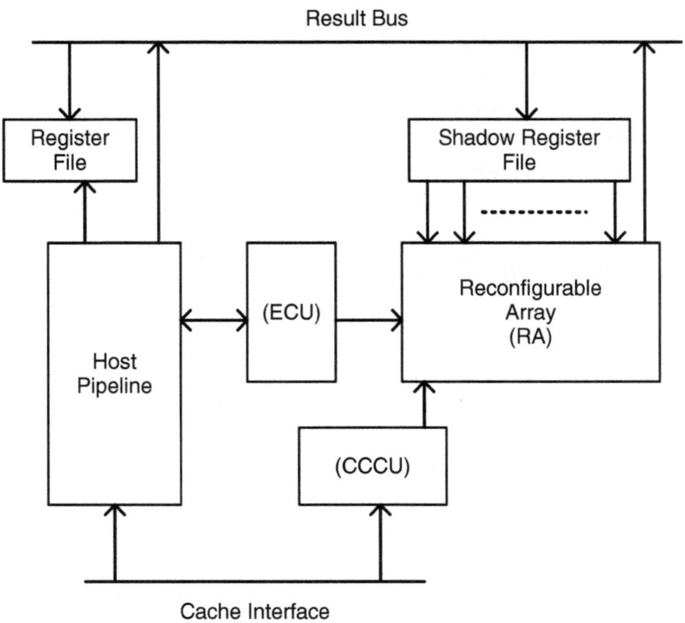

Fig. 1.11 Overview of the chimaera architecture

logic blocks, one per bit in the largest supported register data type. The logic block can be configured as a 4-LUT, two 3-LUTs, or a 3-LUT and a carry computation. Across a single row, all logic blocks share a fast-carry logic that is used to implement fast addition and subtraction operations. By using this organization, arithmetic operations such as addition, subtraction, comparison, and parity can be supported very efficiently. The routing structure of Chimaera is also optimized for such operations.

1.3.1.4 Pleiades

The Pleiades processor [28] combines an on-chip microprocessor with an array of heterogeneous programmable computational units of different granularities, which are called satellite processors, connected by a reconfigurable interconnect network. The microprocessor supports the control-intensive components of the applications as well as the reconfiguration, while repetitive and regular data-intensive loops are directly mapped on the array of satellites by configuring the satellite parameters and the interconnections between them. The synchronization between the satellite processors is accomplished by a data-driven communication protocol in accordance with the data-flow nature of the computations performed in the regular data-intensive loops.

The Maia processor combines an ARM8 core with 21 satellite processors. Those processors are two MACs, two ALUs, eight address generators, eight embedded memories (4512×16 bit, 4 $1K \times 16$ bit), and an embedded low-energy FPGA array [29]. The embedded ARM8 is optimized for low-energy operation. Both the dual-stage pipelined MAC and the ALU can be configured to handle a range of operations. The address generators and embedded memories are distributed to supply multiple parallel data streams to the computational elements. The embedded FPGA supports a 4×8 array of 5-input 3-output CLBs, optimized for arithmetic operations and data-flow control functions. It contains 3 levels of interconnect hierarchy, superimposing nearest-neighbor, mesh and tree architectures.

1.3.2 Stand Alone Fine-Grain Reconfigurable Devices

1.3.2.1 DPGA

Dynamically Programmable Gate Arrays (DPGAs) [30] differ from traditional FPGAs by providing on-chip memory for multiple array personalities. The configuration memory resources are replicated to contain several configurations for the fixed computing and interconnect resources. In effect, DPGA contains an on-chip cache of array configurations and exploits high, local on-chip bandwidth to allow reconfiguration to occur rapidly, on the order of nanoseconds instead of milliseconds. Loading a new configuration from off-chip is still limited by low off-chip bandwidth. However, the multiple contexts on the DPGA allow the array to operate on one context while other contexts are being reloaded.

The DPGA architecture consists of array elements. Each array element is a conventional 4-input LUT. Small collections of array elements are grouped together into subarrays, and these subarrays are then tiled to compose the entire array. Crossbars between the subarrays serve as inter-subarray routing connections. A single, 2-bit, global context identifier is distributed throughout the array to select the configuration for use. Additionally, programming lines are distributed to read and write configurations to memories.

The basic memory primitive is a 4×32 bit DRAM array that provides four context configurations for both the LUT and interconnect network.

1.3.2.2 Triptych

The Triptych FPGA [31], [32] matches the physical structure of the routing architecture to the fan-in/fan-out nature of the structure of digital logic by using short connections to the nearest neighbors. Segmented routing channels are used between the columns to provide for nets with fan-out greater than one. This routing architecture does not allow the arbitrary point-to-point routing available in general FPGA structures. The logic block implements logical functions using a multiplexer-based three-input lookup table followed by a master-slave D-latch and can also be used for routing. Initial results show potential implementation efficiencies in terms of area using this structure.

1.3.2.3 Montage

The Montage FPGA [32] [33] is a version of the Triptych architecture, which is modified to support asynchronous circuits and interfacing separately clocked synchronous circuits. This is achieved by the addition of an arbiter unit and a clocking scheme that allows two possible clocks or makes latches transparent.

Triptych and Montage are FPGAs designed with integrated routing and logic, and achieve higher densities than current commercial FPGAs. Both FPGAs share the same overall routing structure. The Routing and Logic Block (RLB), as shown in Fig. 1.12 consists of 3 multiplexers for the inputs, a functional unit, 3 multiplexers for the outputs, and tri-state drivers for the segmented channels. In Triptych, the functional unit is a 3-input LUT, with an optional D-latch on its output.

1.3.2.4 UTFPGA1

The work at the University of Toronto resulted in the implementation of an architecture (UTFPGA1) using three cascaded four-input logic blocks and segmented routing. UTFPGA1 [34] used information from previous architectural studies, but there was very little transistor-level optimization (for speed), and little time was spent on layout optimization. This was a first attempt that provided some insight into the problems faced in the design and layout of an FPGA.

The general architecture of UTFPGA1 is shown in Fig. 1.13. The logic block (L) contains the functionality of the circuit while the connection boxes (C) con-

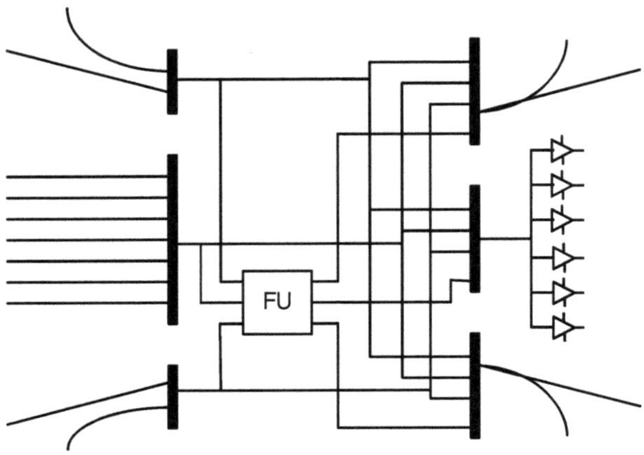

Fig. 1.12 Routing and Logic Block (RLB)

nect the logic block pins into the neighboring channel. The switch box (S) makes connections between adjacent horizontal and vertical channel segments. Connections to the I/O pads are done through I/O blocks (I), which connect to the routing channels. Configuration is done by programming static memory configured as shift registers. They have designed a single tile that contains one logic block, two connection boxes and one switch box. This tile can then be arrayed to any size. The logic block contains three cascaded four-to-one lookup tables. This configuration was chosen because results [24] have shown that significant gains in

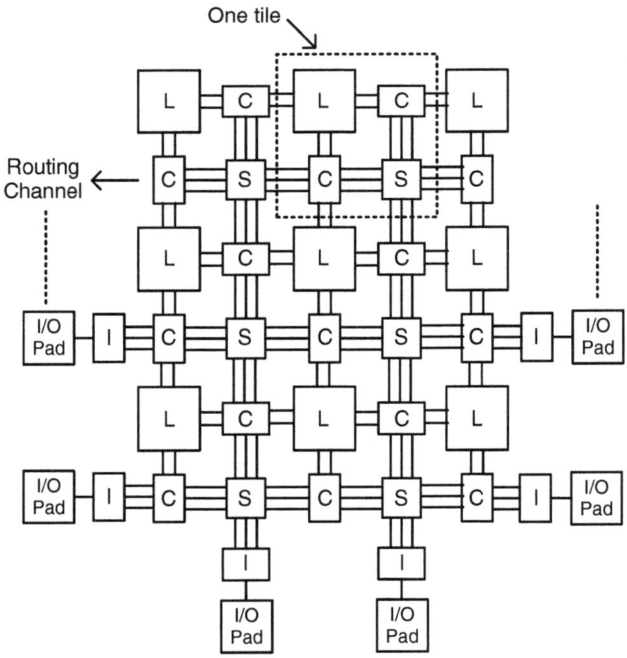

Fig. 1.13 General architecture of UTFPGA1

optimizing for delay can be achieved by having some hardwired connections between logic blocks. The block also contains a resettable D flip-flop. The routing architecture has tracks segmented into lengths of one, two, and three tiles. Such architecture provides fast paths for longer connections, improving FPGA performance.

1.3.2.5 LP_PGA

LP_PGA [35] is an energy efficient FPGA architecture. Significant reduction in the energy consumption is achieved by tackling both circuit design and architecture optimization issues concurrently. A hybrid interconnect structure incorporating Nearest Neighbor Connections, Symmetric Mesh Architecture, and Hierarchical connectivity is used. The interconnect energy is also reduced by employing low-swing circuit techniques. These techniques have been employed to design and fabricate an FPGA. Preliminary analysis shows energy improvement of more than an order of magnitude when compared to existing commercial architectures.

1.3.2.6 LP_PGA II

The LP_PGA II [36], is a stand-alone FPGA of 256 logic blocks with an equivalent logic capacity of 512 4-input LUTs. At this paragraph the implementation is described at the different components of the FPGA (logic block, connection boxes, interconnect levels, and the configuration architecture). The LP_PGA II was designed in a 0.25 μm CMOS process from STMicroelectronics.

Configurable Logic Block

The LP_PGA II CLB is illustrated in Fig. 1.14. It is implemented as a cluster of 3-input LUTs. This clustering technique makes it possible to combine the results of the four 3-input LUTs in various ways to simultaneously realize up to three different functions in a logic block. The combination of the results of the 3-input LUTs is realized using multiplexers that can be programmed at configuration time. All the outputs of the logic block can be registered if required. The flip-flops are double-edge-triggered to reduce the clock activity on the clock distribution network for a given data-throughput.

Interconnect Architecture

Three interconnect levels are used in the LP_PGA II, the nearest neighbor connection (Level-0), the mesh architecture (Level-1), and the inverse clustered tree (Level-2). The Level-0 connections provide connections between adjacent logic blocks (Sect. 1.3). Each output pin connects to one input pin of the eight immediate neighbors. The routing overhead of having eight separate lines to each input pin

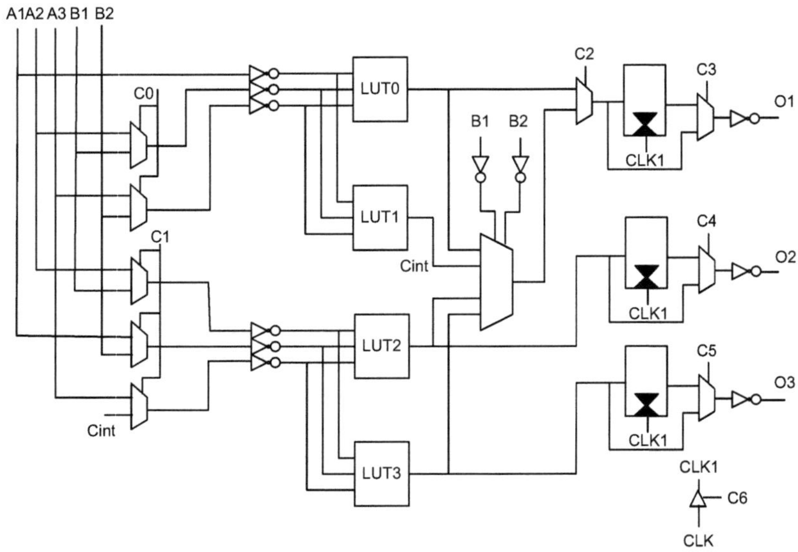

Fig. 1.14 LP PGA II Logic block architecture

from the output pins of the neighbors is quite high. The overhead can be reduced if multiple pins share the same interconnect line.

The mesh architecture (Level-1) is realized with a channel width of five. The pins of the logic block are uniformly distributed on all sides of the logic block. The pins of the logic block can access all tracks in the corresponding routing channel. The switch box allows connections between each routing segment in a given channel and the corresponding segments in the other three routing channels.

The Level-2 network provides connection between logic blocks that are farther apart on the array. The long connection can be accessed through the Mesh structure. Two tracks in each routing channel are connected using the Level-2 network. The routing through the different levels of the Level-2 network is realized using the 3-transistor routing switch.

During the physical implementation, the Level-2 network contributes a significant amount to the area. Area minimization can be achieved by recognizing that the higher levels of the network can be discarded without any significant penalty to the routability.

The routing resources account for approximately 49% of the total area of the device. As the size of the array increases, the fraction of the total area used by the routing will also increase. This is because the increase in the array size necessitates an increase in the routing resources required for each tile to ensure successful routing. The logic block contributes only 9% to the total tile area.

Configuration Architecture

The configuration method used in the LP_PGA II architecture is that of a random access technique. This makes it possible to selectively program the resources in the FPGA, without having to program the entire array each time.

Implementation

Three prototype FPGAs were built. The first prototype, LP_PGA II, was an array of sixty-four logic blocks. The purpose of this chip was to verify the architectural and circuit techniques aimed at reducing the execution energy. The second prototype was an embedded version of LP_PGA II. The array was used as an accelerator in a digital signal processor for voice band processing. Data obtained from the embedded FPGA verified the applicability of an FPGA in an energy-sensitive platform. This implementation also brought into focus the overhead associated with frequent reconfiguration of the FPGA. The last prototype, LP_PGA II incorporated the improvements to reduce the configuration energy. Measured data from the prototypes demonstrate five times to twenty-two times improvement in execution energy over comparable commercial architectures.

1.3.2.7 3D-FPGA

3D-FPGA [37] is a dynamically reconfigurable field-programmable gate array (FPGA). The architecture was developed using a methodology that examines different architectural parameters and how they affect different performance criteria such as speed, area, and reconfiguration time. The block diagram of the 3-D FPGA is shown in Fig. 1.15. The resulting architecture has high performance while the requirement of balancing the areas of its constituent layers is satisfied.

The architecture consists of three layers: the routing and logic block (RLB) layer, the routing layer (RL), and the memory layer (ML). The RLB layer is responsible for implementing logic functions and for performing limited routing. Since it is well known that, for practical applications, most nets are short, it is decided to implement in the RLB layer the portion of the routing structure that will be used for routing short nets. The remaining part of the routing structure is implemented in the RL that is formed by connecting multiple switch boxes in a mesh array structure. The memory layer is used to store configuration bits for both the RLB and routing layers. The number of configuration bits stored in this layer is determined by the size of the RLB and routing layers. The main goal is to achieve a balance between the FPGAs constituent layers. Figure 1.16 presents the internal structure of the functional unit.

A dynamically reconfigurable FPGA must provide means of communicating intermediate results between different configuration instantiations. The proposed

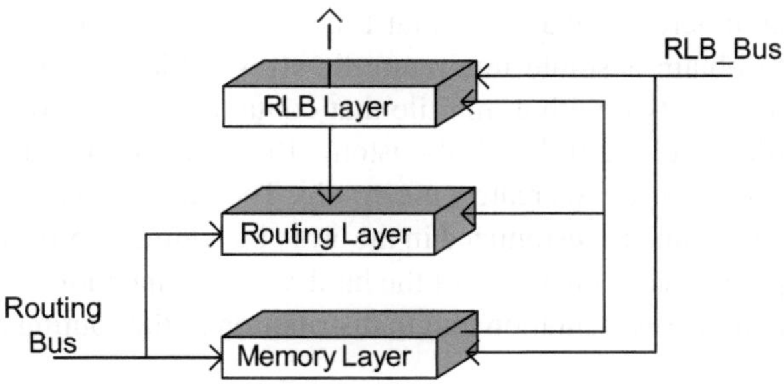

Fig. 1.15 Block diagram of the 3-D FPGA

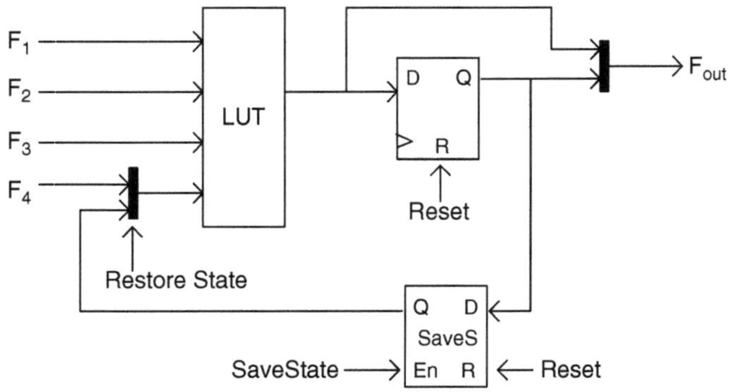

Fig. 1.16 Internal Structure of the functional unit

FPGA allows direct communication between any two configuration instantiations. The SaveState register is provided in order to allow the present state to be saved for subsequent processing. The current state can be loaded into the register when the SaveState signal is enabled. The value of the SaveS register can be retrieved by any configuration instantiation by appropriately setting the value of the RestoreState signal without disturbing the operation of the RLB during the intermediate configuration instantiations. The restored value can be used as one of the inputs into the LUT. The RLBs are organized into clusters. A cluster is formed by a square array of RLBs. The size of the cluster will be determined in Section V-B. Each cluster is associated with a cluster memory block and a switch box in the routing layer. The cluster memory block can be used to store either input data or intermediate results. The size of this cluster memory is dependent upon the mismatch between the areas of the FPGA constituent's layers.

1.3.2.8 LEGO

The LEGO [38] (Logic that's Erasable and Greatly Optimized) FPGA basic block is a four-input LUT. The designers' objective was focused on achieving a high-speed design, while keeping in mind the area tradeoffs. The most critical issues are the design of the switches and minimizing the capacitance of the routing network. The results have shown that the LEGO design compared favorably with existing commercial FPGA's of that time. Also, instead of using full-custom hand layout to obtain absolute minimum die sizes, which is both labor and time intensive, a design style with a minitile that contains a portion of the components in the logic tile, resulting in less full-custom effort, was proposed. The minitile is replicated in a 4×4 array to create a macro tile. The minitile is optimized for layout density and speed, and is customized in the array by adding appropriate vias. This technique also permits easy changing of the hard-wired connections in the logic block architecture and the segmentation length distribution in the routing architecture.

Table 1.1 Comparisons of fine-grain academic architectures

System	Granularity	Programmability	Reconfiguration	Interface	Computing model	Application domain
Splash	Fine-grain	Multiple context (for interconnect)	Static	Remote	Uniprocessor	Complex bit-oriented computations
Splash2						
DECPeRLe-1	Fine-grain	Single context	Static	Remote	Uniprocessor	Complex bit-oriented computations
Garp	Fine-grain	Multiple context	Static	Local	Uniprocessor	Bit-level image processing, cryptography
OneChip	Fine-grain	Single context	Static	Local	Uniprocessor	Embedded controllers, application accelerators
Chimaera	Fine-grain	Single context	Static	Local	Uniprocessor	Bit-level computations
DISC	Fine-grain	Single context	Dynamic	Local	Uniprocessor	General purpose
DPGA	Fine-grain	Multiple context	Dynamic	Remote	Uniprocessor	Bit-level computations

1.3.3 Summary

Table 1.1 provides the main features for some of the above described fine-grain reconfigurable architectures in terms of their programmability, the reconfiguration method, the interface and the possible application domain.

1.4 Commercial Fine-Grain Reconfigurable Platforms

1.4.1 Xilinx

In this subsection the Spartan-3, Virtex-4 and Virtex-5 families of FPGAs will be described. Besides the fine-grain resources and hard IP blocks (DSP, embedded processors) integrated in many Xilinx devices, a library of soft IP blocks are also available for the efficient implementation of complex systems.

1.4.1.1 Spartan −3 and Spartan −3L family of FPGAs

The Spartan™-3 family [39] of Field-Programmable Gate Arrays is specifically designed to meet the needs of high volume, cost-sensitive consumer electronic applications. The eight-member family offers densities ranging from 50,000 to five million system gates.

The Spartan-3 family builds on the earlier Spartan-IIE family by increasing the amount of logic resources, the capacity of internal RAM, the total number of I/Os, and the overall level of performance as well as by improving clock management functions.

Configurable Logic Block (CLB)

The Configurable Logic Blocks (CLBs) constitute the main logic resource for implementing synchronous as well as combinatorial circuits. Each CLB comprises four interconnected slices, as shown in Fig. 1.17. These slices are grouped in pairs. Each pair is organized as a column with an independent carry chain.

All four slices have the following elements in common: two logic function generators, two storage elements, wide-function multiplexers, carry logic, and arithmetic gates, as shown in Fig. 1.18.

The storage element, which is programmable as either a D-type flip-flop or a level-sensitive latch, provides a means for synchronizing data to a clock signal and storing them.

1.4.2 Interconnect

There are four kinds of interconnect in the Spartan-3 family: Long lines, Hex lines, Double lines, and Direct lines.

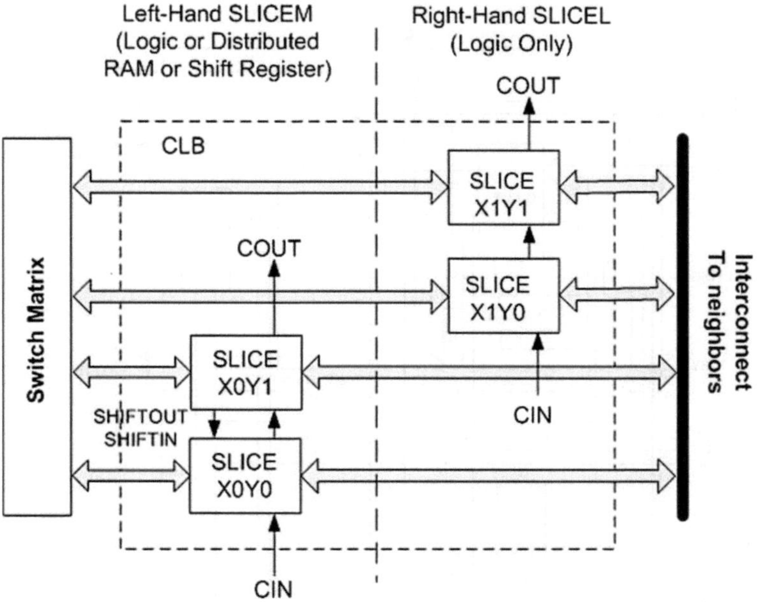

Fig. 1.17 Spartan – 3 CLB organization

- *Long lines* connect to one out of every six CLBs (see Fig. 1.19a). Because of their low capacitance, these lines are well-suited for carrying high-frequency signals with minimal loading effects (e.g. skew). Therefore, if all available Global Clock Inputs are already committed and there remain additional clock signals to be assigned, Long lines serve as a good alternative.
- *Hex lines* connect one out of every three CLBs (see Fig. 1.19b). These lines fall between Long lines and Double lines in terms of connectivity.
- *Double lines* connect to every other CLB (see Fig. 1.19c). Compared to the types of lines already discussed, Double lines provide a higher degree of flexibility when making connections.
- *Direct lines* afford any CLB direct access to neighboring CLBs (see Fig. 1.19d). These lines are most often used to conduct a signal from a "source" CLB to a Double, Hex, or Long line and then from the longer interconnect back to a Direct line accessing a "destination" CLB.

Quadrant Clock Routing

The clock routing within Virtex-3 FPGAs is quadrant-based. Each clock quadrant supports eight total clock signals. The clock lines feed the synchronous resource elements (CLBs, IOBs, block RAM, multipliers, and DCMs) within the quadrant. The top and bottom global buffers support higher clock frequencies than the left- and right-half buffers. Consequently clocks exceeding 230 MHz must use the top or bottom global buffers and, if required for the application, their associated DCMs.

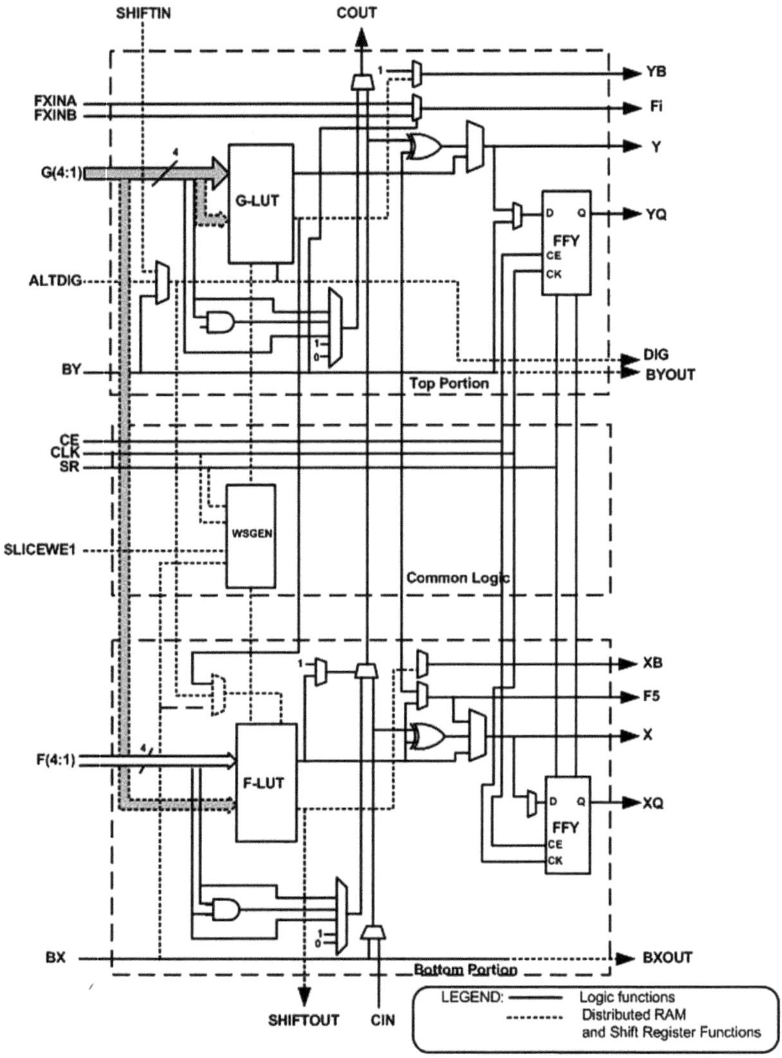

Fig. 1.18 Spartan – 3 slice

Advanced Features

Spartan-3 devices provide additional features for efficient implementation of complex systems, such as embedder RAM, embedded multipliers and Digital Clock Managers (DCMs).

Block RAM

All Spartan-3 devices support block RAM, which is organized as configurable, synchronous 18 Kb blocks. The amount of memory varies between devices from 73,728 to 1,916,928 bits.

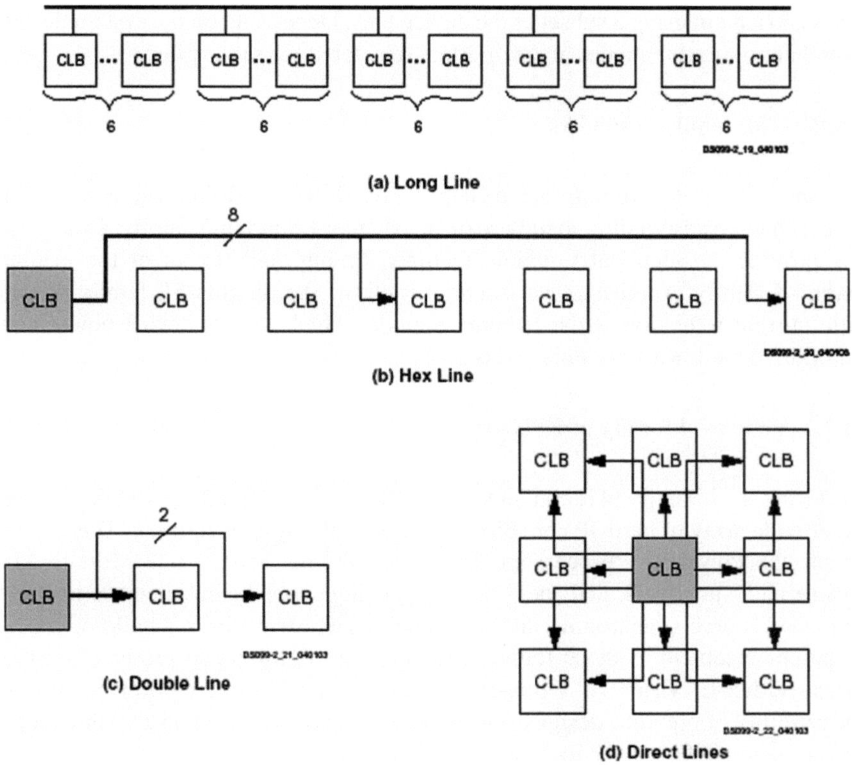

Fig. 1.19 Spartan – 3 interconnect

Dedicated Multipliers

Spartan-3 devices provide embedded multipliers that accept two 18-bit words as inputs to produce a 36-bit product.

Digital Clock Manager (DCM)

Spartan-3 devices provide flexible, complete control over clock frequency, phase shift and skew through the use of the DCM feature. To accomplish this, the DCM employs a Delay-Locked Loop (DLL), a fully digital control system that uses feedback to maintain clock signal characteristics with a high degree of precision despite normal variations in operating temperature and voltage. The DCM main functions are clock skew elimination, digital frequency synthesis and phase shifting.

Configuration

Spartan-3 devices use SRAM configuration; they are configured by loading application-specific configuration data into the internal configuration memory. Configuration is carried out using a subset of the device pins. Depending on the system design, several configuration modes are supported, selectable via mode pins.

Spartan – 3L family of FPGAs

Spartan^{TM}-3L Field-Programmable Gate Arrays (FPGAs) [40] consume less static current than corresponding members of the standard Spartan-3 family. Otherwise, they provide the identical function, features, timing, and pinout of the original Spartan-3 family. Another power-saving benefit of the Spartan-3L family beyond static current reduction is the Hibernate mode, which lowers device power consumption to the lowest possible levels.

1.4.2.1 Virtex – 4 family of FPGAs

The Virtex-4^{TM} Family [41] contains three actual families (platforms): LX, FX, and SX. A wide array of hard-IP core blocks complete the system solution. These cores include the PowerPC^{TM}processors, Tri-Mode Ethernet MACs, 622 Mb/s to 11.1 Gb/s serial transceivers, dedicated DSP slices, high-speed clock management circuitry, and source-synchronous interface blocks. The basic Virtex-4 building blocks are an enhancement of those found in the popular Virtex-based product families: Virtex, Virtex-E, Virtex-II, Virtex-II Pro, and Virtex-II Pro X, allowing upward compatibility of previous designs. Virtex-4 devices are produced on a 90 nm copper process, using 300 mm (12 inch) wafer technology.

Configurable Logic Blocks (CLBs)

A CLB resource is made up of four slices. Each slice is equivalent and contains:

- Two function generators (F & G)
- Two storage elements
- Arithmetic logic gates
- Large multiplexers
- Fast carry look-ahead chain
- Horizontal cascade chain

The function generators F & G are configurable as 4-input look-up tables (LUTs). Two slices in a CLB can have their LUTs configured as 16-bit shift registers, or as 16-bit distributed RAM. In addition, the two storage elements are either edge-triggered D-type flip-flops or level sensitive latches. Each CLB has internal fast interconnect and connects to a switch matrix to access general routing resources.

Advanced Features

Like the Spartan-3 family, Virtex-4 devices provide additional features for efficient implementation of complex systems, such as embedder RAM, embedded multiplier-accumulators and PLLs.

Block RAM

The Virtex – 4 block RAM resources are 18 Kb true dual-port RAM blocks, programmable from 16K × 1 to 512 × 36, in various depth and width configurations. Each port is totally synchronous and independent, offering three "read-during-write" modes. Block RAM is cascadable to implement large embedded storage blocks. Additionally, back-end pipeline registers, clock control circuitry, built-in FIFO support, and byte write enable are features supported in the Virtex-4 FPGA.

XtremeDSP Slices

The XtremeDSP slices contain a dedicated 18 × 18-bit 2's complement signed multiplier, adder logic, and a 48-bit accumulator. Each multiplier or accumulator can be used independently. These blocks are designed to implement extremely efficient and high-speed DSP applications.

Global Clocking

The DCM and global-clock multiplexer buffers provide a complete solution for designing high-speed clock networks. Up to twenty DCM blocks are available. To generate deskewed internal or external clocks, each DCM can be used to eliminate clock distribution delay. The DCM also provides 90°, 180°, and 270° phase-shifted versions of the output clocks. Fine-grained phase shifting offers higher resolution phase adjustment with fraction of the clock period increments. Flexible frequency synthesis provides a clock output frequency equal to a fractional or integer multiple of the input clock frequency.

Routing Resources

All components in Virtex-4 devices use the same interconnect scheme and the same access to the global routing matrix. Timing models are shared, greatly improving the predictability of the performance for high-speed designs.

Configuration

Virtex-4 devices are configured by loading the bitstream into internal configuration memory using one of the following modes:

- Slave-serial mode
- Master-serial mode
- Slave SelectMAP mode
- Master SelectMAP mode
- Boundary-scan mode (IEEE-1532)

Optional 256-bit AES decryption is supported on-chip (with software bitstream encryption) providing Intellectual Property security.

Implementation Technology

Virtex-4 devices are produced on a 90 nm copper process, using 300 mm (12 inch) wafer technology.

Power Consumption

Virtex-4 devices consume approximately 50 % the power of respective Virtex-II Pro devices due to:

- Static power reduction enabled by triple-oxide technology
- Dynamic power reduction enabled by reduced core voltage and capacitance

Virtex-4 FX Family additional features

There are certain blocks available only in the FX devices of the Virtex-4 family such as:

- 8–24 RocketIO Multi-Gigabit serial Transceiver
- One or Two PowerPC 405 Processor Cores
- Two or Four Tri-Mode (10/100/1000 Mb/s) Ethernet Media Access Control (MAC) Cores

1.4.2.2 Virtex-5 Family

The Virtex™ -5 family [42] provides the newest most powerful features among Xilinx FPGAs. The Virtex-5 LX platform contains many hard-IP system-level blocks, including powerful 36-Kb block RAM/FIFOs, second generation 25 × 18 DSP slices, SelectIO™ technology with built-in digitally-controlled impedance, ChipSync™ source-synchronous interface blocks, enhanced clock management tiles with integrated Digital Clock Managers (DCM) and phase-locked-loop (PLL) clock generators, and advanced configuration options. The Virtex-5 family of FPGAs is built on a 65 nm copper process technology.

ExpressFabric

Featuring 6-input Look-up Tables (LUT), ExpressFabric technology allows LUTs to be configured as either 6-input or dual-output 5-input generators. Functions such as 256 bits of distributed RAM, 128-bit long shift registers and 8-input functions within a single Configurable Logic Block (CLB) can be implemented.

Interconnect Network

The Virtex-5 family uses diagonally symmetric interconnects to minimize the number of interconnects required from CLB to CLB, to realize major performance improvements.

Advanced Features

DCM and PLLs

Virtex –5 devices provide:

- Digital Clock Manager (DCM) blocks for zero delay buffering, frequency synthesis, and clock phase shifting
- PLL blocks for input jitter filtering, zero delay buffering, frequency synthesis, and phase-matched clock division

DSP48E slices

The 550 MHz DSP48E slices available in all Virtex™-5 family members accelerate algorithms and enable higher levels of DSP integration and lower power consumption than previous-generation Virtex devices:

- The DSP48E slice supports over 40 dynamically controlled operating modes including; multiplier, multiplier-accumulator, multiplier-adder/subtractor, three input adder, barrel shifter, wide bus multiplexers, wide counters, and comparators.
- DSP48E slices enable efficient adder-chain architectures for implementing high-performance filters and complex math efficiently.

Embedded RAM

Virtex™ -5 FPGAs offer up to 10 Mbits of flexible embedded Block RAM. Each Virtex-5 memory block stores up to 36 Kbs of data and can be configured as either two independent 18 Kb Block RAM, or 36 Kb Block RAM. Block RAM can be configured as dual-port RAM or as FIFO and offers 64-bit error checking and correct (ECC) to improve system reliability.

Configuration

In addition to configuration with Xilinx Platform FLASH devices, Virtex-5 FPGAs offer new low-cost options, including SPI flash memory and parallel flash memory. Virtex-5 devices support partial reconfiguration.

Virtex-5 FPGAs protect designs with AES (Advanced Encryption Standard) technology which includes software-based bitstream encryption and on-chip bitstream decryption logic using dedicated memory to store the 256-bit encryption key. The encryption key and encrypted bitstream are generated using Xilinx ISE software. During configuration, the Virtex-5 device decrypts the incoming bitstream. The encryption key is stored internally in dedicated RAM. Backup is effected by a small, externally-connected battery (typical life 20+ years). The encryption key cannot be read out of the device and any attempt to remove the Virtex-5 FPGA and decapsulate the package for probing results in the instant loss of the encryption key and programming data.

Implementation technology

Virtex-5 devices are produced using a 65 nm, 12-layer metal process.

Power Consumption

Virtex-5 devices use triple-oxide technology for reducing static power consumption. Their 1.0V Core Voltage and 65 nm implementation process leads also to dynamic power consumption reduction in comparison to Virtex-4 devices.

1.4.3 ALTERA

The Cyclone, Cyclone II, Stratix/Stratix GX, and Stratix II/Stratix II GX FPGA families are described in this subsection.

1.4.3.1 Cyclone Family

The Cyclone™ field-programmable gate array family [42] is based on a 1.5-V, 0.13-µm, all-layer copper SRAM process, with densities up to 20,060 logic elements (LEs) and up to 288 Kbs of RAM. Their features include phaselocked loops (PLLs) for clocking and a dedicated double data rate (DDR) interface to meet DDR SDRAM and fast cycle RAM (FCRAM) memory. Cyclone devices support various I/O standards, including LVDS at data rates up to 640 megabits per second (Mbps), and 66- and 33 MHz, 64- and 32-bit peripheral component interconnect (PCI), for interfacing with and supporting ASSP and ASIC devices. Support for multiple intellectual property (IP) cores further extends the designer's capabilities for the implementation of complex systems on the Cyclone platform.

Logic Array Blocks and Logic Elements

The logic array consists of Logic Array Blocks (LAB)s, with 10 Logic Elements (LEs) in each LAB. A LE is a small unit of logic providing efficient implementation of user logic functions. LABs are grouped into rows and columns across the device.

Logic Array Blocks

More specifically, each LAB consists of 10 LEs, LE carry chains, LAB control signals, a local interconnect, look-up table (LUT) chain, and register chain connection lines. The local interconnect transfers signals between LEs in the same LAB. LUT chain connections transfer the output of one LE's LUT to the adjacent LE for fast sequential LUT connections within the same LAB. Register chain connections transfer the output of one LE's register to the adjacent LE's register within an LAB. The Quartus® II Compiler places associated logic within an LAB or adjacent LABs, allowing the use of local, LUT chain, and register chain connections for performance and area efficiency.

LAB Interconnects

The LAB local interconnect can drive LEs within the same LAB. The LAB local interconnect is driven by column and row interconnects and LE outputs within the same LAB. Neighboring LABs, PLLs, and M4K RAM blocks from the left and right can also drive an LAB's local interconnect through direct link connection. The direct link connection feature minimizes the use of row and column interconnects, providing higher performance and flexibility. Each LE can drive 30 other LEs through fast local and direct link interconnects.

Logic Elements

The smallest unit of logic in the Cyclone architecture, the LE (Fig. 1.20), is compact and provides advanced features with efficient logic utilization. Each LE contains a four-input LUT, which is a function generator that can implement any function of four variables. In addition, each LE contains a programmable register and carry chain with carry select capability. A single LE also supports dynamic single bit addition or subtraction mode selectable by an LAB-wide control signal. Each LE drives all types of interconnects: local, row, column, LUT chain, register chain, and direct link interconnects.

The Cyclone LE can operate in one of the following modes: i) normal mode and ii) dynamic arithmetic mode. Each mode uses LE resources differently. In each mode, eight available inputs to the LE, the four data inputs from the LAB local interconnect, carry-in0 and carry-in1 from the previous LE, the LAB carry-in from

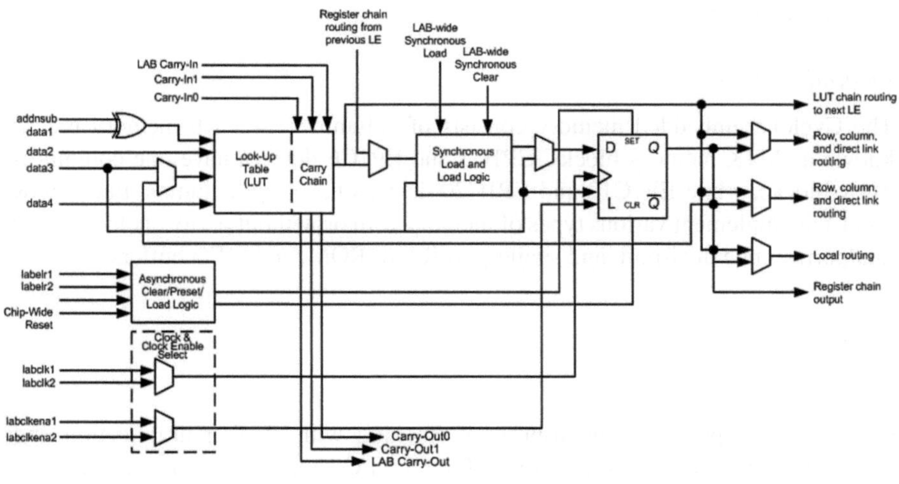

Fig. 1.20 Cyclone Logic Element (LE)

the previous carry-chain LAB, and the register chain connection are directed to different destinations to implement the desired logic function. LAB-wide signals provide clock, asynchronous clear, asynchronous preset/load, synchronous clear, synchronous load, and clock enable control for the register. These LAB-wide signals are available in both LE modes. The addnsub control signal is allowed in arithmetic mode.

MultiTrack Interconnect

In the Cyclone architecture, connections between LEs, memory blocks, and device I/O pins are provided by the MultiTrack interconnect structure with DirectDrive technology. The MultiTrack interconnect consists of continuous, performance-optimized routing lines of different speeds used for inter- and intra-design block connectivity. The Quartus II Compiler automatically places critical design paths on faster interconnects to improve design performance. DirectDrive technology is a deterministic routing technology that ensures identical routing resource usage for any function regardless of placement within the device. The MultiTrack interconnect and DirectDrive technology simplify the integration stage of block-based designing by eliminating the re-optimization cycles that typically follow design changes and additions. The MultiTrack interconnect consists of row and column interconnects that span fixed distances. A routing structure with fixed length resources for all devices allows predictable and repeatable performance when migrating through different device densities. Dedicated row interconnects route signals to and from LABs, PLLs, and memory blocks within the same row.

Advanced Features

Embedded Memory

The Cyclone embedded memory consists of columns of 4.5 Kb memory blocks known as M4K memory blocks. EP1C3 and EP1C6 devices have one column of M4K blocks, while EP1C12 and EP1C20 devices have two columns. Each M4K block can implement various types of memory with or without parity, including true dualport, simple dual-port, and single-port RAM, ROM, and FIFO buffers.

PLLs

Cyclone PLLs provide general-purpose clocking with clock multiplication and phase shifting as well as outputs for differential I/O support. Cyclone devices contain two PLLs, except for the EP1C3 device, which contains one PLL.

External RAM Interfacing

Cyclone devices support DDR SDRAM and FCRAM interfaces at up to 133 MHz through dedicated circuitry.

Configuration

Designers can load the configuration data for a Cyclone device with one of three configuration schemes chosen on the basis of the target application. Designers can use a configuration device, intelligent controller, or the JTAG port to configure a Cyclone device. A configuration device can automatically configure a Cyclone device at system power-up.

Implementation Technology

The Cyclone field-programmable gate array family is based on a 1.5-V, 0.13-μm, all-layer copper SRAM process.

1.4.3.2 Cyclone II

Cyclone II [44] FPGAs benefit from using TSMC's 90 nm low-k dielectric process to extend the Cyclone FPGA density range to 68,416 logic elements (LEs) and provide up to 622 usable I/O pins and up to 1.1 Mbits of embedded memory.

The I/O, logic array blocks/logic element and interconnect architecture of the Cyclone II device family are similar to the Cyclone family respective features. Embedded RAM is also included. A significant addition to the Cyclone features is the addition of embedded multiplier blocks for the efficient implementation of digital signal processing functions.

Embedded Multipliers

Cyclone II devices have embedded multiplier blocks optimized for multiplier-intensive digital signal processing (DSP) functions, such as finite impulse response (FIR) filters, fast Fourier transform (FFT) functions, and discrete cosine transform (DCT) functions. Each embedded multiplier can be used in one of two basic operational modes, depending on the application needs: i) as a single 18-bit multiplier, or ii) one or two independent 9-bit multipliers.

1.4.3.3 Stratix II and Stratix II GX

The Stratix II and Stratix II GX [45] FPGA families are based on a 1.2-V, 90 nm, all-layer copper SRAM process and offer up to 9 Mbits of on-chip, TriMatrix™ memory for demanding, memory intensive applications and have up to 96 DSP blocks with up to 384 (18-bit × 18-bit) multipliers for efficient implementation of high performance filters and other DSP functions. Various high-speed external memory interfaces are supported, including double data rate (DDR) SDRAM and DDR2 SDRAM, RLDRAM II, quad data rate (QDR) II SRAM, and single data rate (SDR) SDRAM. Stratix II devices support various I/O standards along with support for 1-gigabit per second (Gbps) source synchronous signaling with DPA circuitry. Stratix II devices offer a complete clock management solution with internal clock frequency of up to 550 MHz and up to 12 phase-locked loops (PLLs). Stratix II devices include the ability to decrypt a configuration bitstream using the Advanced Encryption Standard (AES) algorithm to protect designs.

Logic Array Blocks

Each Stratix II LAB consists of eight Adaptive Logic Modules (ALMs), carry chains, shared arithmetic chains, LAB control signals, local interconnect, and register chain connection lines. The local interconnect transfers signals between ALMs in the same LAB. Register chain connections transfer the output of an ALM register to the adjacent ALM register in a LAB.

Adaptive Logic Modules

The basic building block of logic in the Stratix II architecture, the Adaptive Logic Module (ALM), contains a variety of look-up table (LUT)-based resources that can be divided between two adaptive LUTs (ALUTs) as can be seen in Fig. 1.21. With up to eight inputs to the two ALUTs, one ALM can implement various combinations of two functions. This adaptability allows the ALM to be completely backward-compatible with four-input LUT architectures. One ALM can also implement any function of up to six inputs and certain seven-input functions.

In addition to the adaptive LUT-based resources, each ALM contains two programmable registers, two dedicated full adders, a carry chain, a shared arithmetic chain, and a register chain. Through these dedicated resources, the ALM can efficiently implement various arithmetic functions and shift registers.

The Stratix II ALM can operate in one of the following modes:

- Normal mode
- Extended LUT mode
- Arithmetic mode
- Shared arithmetic mode

Each mode uses ALM resources differently. In each mode, eleven available inputs to the ALM-the eight data inputs from the LAB local interconnect; carry-in from the previous ALM or LAB; the shared arithmetic chain connection from the previous ALM or LAB; and the register chain connection-are directed to different destinations to implement the desired logic function. LAB-wide signals provide clock, asynchronous clear, asynchronous preset/load, synchronous clear,

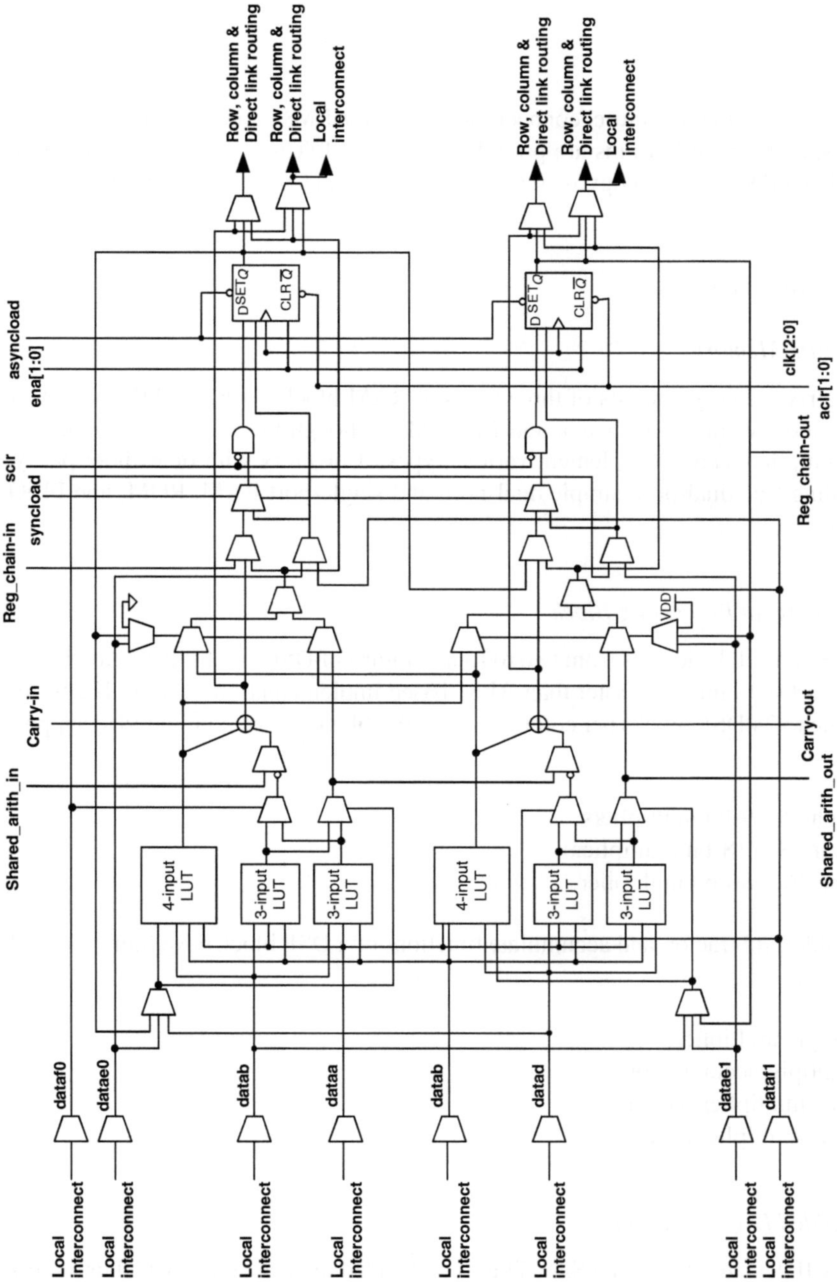

Fig. 1.21 Stratix II Adaptive Logic Module (ALM)

synchronous load, and clock enable control for the register. These LAB-wide signals are available in all ALM modes.

MultiTrack Interconnect

In the Stratix II architecture, connections between ALMs, TriMatrix memory, DSP blocks, and device I/O pins are provided by the MultiTrack interconnect structure with DirectDrive technology seen also in Cyclone and Cyclone II devices.

Advanced Features

TriMatrix Memory

TriMatrix memory consists of three types of RAM blocks: M512, M4K (as in Cyclone and Cyclone II devices), and M-RAM. Although these memory blocks are different, they can all implement various types of memory with or without parity, including true dual-port, simple dual-port, and single-port RAM, ROM, and FIFO buffers.

Digital Signal Processing Block

Each Stratix II device has from two to four columns of DSP blocks to efficiently implement DSP functions faster than ALM-based implementations. Stratix II devices have up to 24 DSP blocks per column. Each DSP block can be configured to support up to:

- Eight 9×9-bit multipliers
- Four 18×18-bit multipliers
- One 36×36-bit multiplier

The adder, subtractor, and accumulate functions of a DSP block have four modes of operation:

- Simple multiplier
- Multiply-accumulator
- Two-multipliers adder
- Four-multipliers adder

Embedded Logic Analyzer

Stratix II devices feature the SignalTap II embedded logic analyzer, which monitors design operation over a period of time through the embedded JTAG circuitry. The designer can analyze internal logic at speed without bringing internal signals to the I/O pins.

Configuration

The logic, circuitry, and interconnects in the Stratix II architecture are configured with CMOS SRAM elements. Stratix II devices are configured at system power-up with data stored in an Altera configuration device or provided by an external controller. They can be configured using the fast passive parallel (FPP), active serial (AS), passive serial (PS), passive parallel asynchronous (PPA), and JTAG configuration schemes.

The Stratix II device's optimized interface allows microprocessors to configure it serially or in parallel, and synchronously or asynchronously. The interface also enables microprocessors to treat Stratix II devices as memory and configure them by writing to a virtual memory location.

Implementation Technology

The Stratix II FPGA family is based on a 1.2-V, 90 nm, all-layer copper SRAM process.

Power Consumption

Stratix II FPGAs use a variety of techniques in order to reduce static and dynamic power consumption including:

- *Increased Vt (Threshold Voltage)* which reduces static power consumption at the cost of transistor performance, therefore used only in non-critical paths.
- *Increased Transistor Length* which also reduces static power consumption at the cost of transistor performance.
- *Architectural Changes: 4-Input LUT to 7-Input Variable Adaptive Logic Module (ALM)* which reduces the active power (dynamic power) consumption by minimizing the amount of interconnect and total silicon also positively affecting static power consumption at a lesser extent.
- *Low-K Dielectric Process* which reduces dynamic power approximately by 10%.
- *Lower I/O Pin Capacitance* which reduces the I/O power consumption and therefore total dynamic power.
- *Power Effiecient Clocking Structure* which reduces dynamic power by shutting down parts of the clock network.

Stratix II GX

The Stratix® II GX family of devices is Altera's third generation of FPGAs to combine high-speed serial transceivers with a scalable, highperformance logic array. Stratix II GX devices include 4 to 20 high-speed transceiver channels, each incorporating clock/data recovery unit (CRU) technology and embedded SERDES capability at data rates of up to 6.375 gigabits per second (Gbps). The transceivers are grouped into four-channel transceiver blocks, and are designed for low power consumption and small die size. The Stratix II GX FPGA technology is built upon the Stratix II architecture, and offers a 1.2-V logic array with the logic element, interconnect, embedded RAM and DSP blocks offered by the Stratix II family. Stratix II GX devices have somewhat fewer logic resources than the respective Stratix II devices due to the space occupied by the tranceivers.

1.4.3.4 Stratix and Stratix GX

The Stratix and Stratix GX families [46] are based on a 1.5-V, 0.13-µm, all-layer copper SRAM process, with densities up to 114,140 logic elements (LEs) and up to 10 Mbits of RAM. Stratix devices offer up to 28 digital signal processing (DSP) blocks with up to 224 (9-bit × 9-bit) embedded multipliers, optimized for DSP applications that enable efficient implementation of high-performance filters and multipliers. Stratix devices support various I/O standards and also offer a complete clock management solution with its hierarchical clock structure with up to 420 MHz performance and up to 12 phase-locked loops (PLLs).

Logic Array Blocks

Each LAB consists of 10 LEs, LE carry chains, LAB control signals, local interconnect, LUT chain, and register chain connection lines. The local interconnect transfers signals between LEs in the same LAB. LUT chain connections transfer the output of one LE's LUT to the adjacent LE for fast sequential LUT connections within the same LAB. Register chain connections transfer the output of one LE's register to the adjacent LE's register within an LAB. The Quartus II Compiler places associated logic within an LAB or adjacent LABs, allowing the use of local, LUT chain, and register chain connections for performance and area efficiency.

LAB Interconnects

The LAB local interconnect can drive LEs within the same LAB. The LAB local interconnect is driven by column and row interconnects and LE outputs within the same LAB. Neighboring LABs, M512 RAM blocks, M4K RAM blocks, or DSP blocks from the left and right can also drive an LAB's local interconnect through the direct link connection. The direct link connection feature minimizes the use of row and column interconnects, providing higher performance and flexibility. Each LE can drive 30 other LEs through fast local and direct link interconnects.

Logic Elements

The smallest unit of logic in the Stratix architecture, the LE, is compact and provides advanced features with efficient logic utilization. Each LE contains a four-input LUT, which is a function generator that can implement any function of four

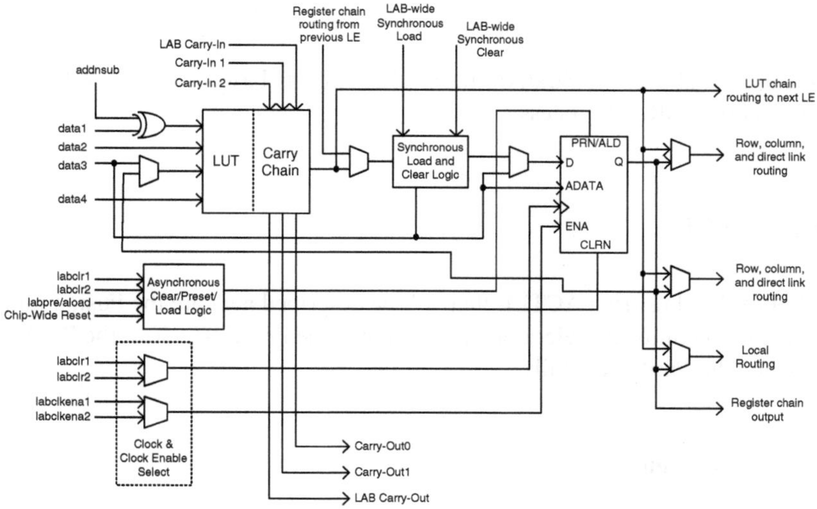

Fig. 1.22 Stratix logic element

variables. In addition, each LE contains a programmable register and carry chain with carry select capability. A single LE also supports dynamic single bit addition or subtraction mode selectable by an LAB-wide control signal. Each LE drives all types of interconnects: local, row, column, LUT chain, register chain, and direct link interconnects. The Stratix logic element schematic is shown in Fig. 1.22.

Each LE has three outputs that drive the local, row, and column routing resources. The LUT or register output can drive these three outputs independently. Two LE outputs drive column or row and direct link routing connections and one drives local interconnect resources. This allows the LUT to drive one output while the register drives another output. This feature, called register packing, improves device utilization because the device can use the register and the LUT for unrelated functions. Another special packing mode allows the register output to feed back into the LUT of the same LE so that the register is packed with its own fan-out LUT. This provides another mechanism for improved fitting. The LE can also drive out registered and unregistered versions of the LUT output.

MultiTrack Interconnect

In the Stratix architecture, connections between LEs, TriMatrix memory, DSP blocks, and device I/O pins are provided by the MultiTrack interconnect structure with DirectDrive technology available in Stratix II/Stratix II GX, Cyclone and Cyclone II devices.

TriMatrix Memory

TriMatrix memory consists of the same three types of RAM blocks (M512, M4K, and MRAM blocks) seen in Stratix II devices

Implementation Technology

As mentioned above, the Stratix/Stratix GX family is based on a 1.5-V, 0.13-μm, all-layer copper SRAM process.

1.4.4 ACTEL

The FPGA families from ACTEL that will be described next are the Fusion family, the ProASIC3 family, Axcelerator, the eX family, the ProASIC 500K, the ProASIC-PLUS, and the VariCore family.

1.4.4.1 Fusion Family

The Actel Fusion family [47], based on the highly successful ProASIC® 3 and ProASIC3E Flash FPGA architecture, has been designed as a high-performance, programmable, mixed-signal platform. For that purpose, Fusion devices combine an advanced Flash FPGA core with Flash memory blocks and analog peripherals.

VersaTiles

The Fusion core consists of VersaTiles, which are also used in the successful Actel ProASIC3 family. The Fusion VersaTile can be configured as one of the following:

- All three-input logic functions—LUT-3 equivalent
- Latch with clear or set
- D-flip-flop with clear or set and optional enable

Advanced Features

Embedded memories

Fusion devices provide three types of embedded memory:

Flash memory blocks: The Flash memory available in each Fusion device is composed of 1 to 4 Flash blocks, each 2 Mbits in density. Each block operates independently with a dedicated Flash controller and interface. Fusion devices support two methods of external access to the Flash memory blocks: i) a serial interface that features a built-in JTAG-compliant port and ii) a soft parallel interface. FPGA logic or an on-chip soft microprocessor can access Flash memory through the parallel interface.

In addition to the Flash blocks, Actel Fusion devices have 1 kb of user-accessible, nonvolatile FlashROM on-chip. The FlashROM is organized as 8 × 128-bit pages. The FlashROM can be used in diverse system applications:

- Internet protocol addressing (wireless or fixed)
- System calibration settings

- Device serialization and/or inventory control
- Subscription-based business models (for example, set-top boxes)
- Secure key storage for secure communications algorithms
- Asset management/tracking
- Date stamping
- Version management

The FlashROM can be programmed (erased and written) via the JTAG programming interface, and its contents can be read back either through the JTAG programming interface or via direct FPGA core addressing.

SRAM and FIFO

Fusion devices have embedded SRAM blocks along the north and south sides of the device. Each SRAM block is 4,608 bits in size. Available memory configurations are 256 × 18, 512 × 9, 1k × 4, 2k × 2, and 4k × 1 bits. The individual blocks have independent read and write ports that can be configured with different bit widths on each port. In addition, every SRAM block has an embedded FIFO control unit. The control unit allows the SRAM block to be configured as a synchronous FIFO (with the appropriate flags and counters) without using additional core VersaTiles.

Clocking Resources

Each member of the Fusion family contains six blocks of Clock Conditioning Circuitry (CCC). In the two larger family members, two of these CCCs also include a PLL; the smaller devices support one PLL. The inputs of the CCC blocks are accessible from the FPGA core or from one of several I/O inputs with dedicated CCC block connections.

The CCC block has the following key features:

- Wide input frequency range (fIN_CCC) = 1.5 MHz to 350 MHz
- Output frequency range (fOUT_CCC) = 0.75 MHz to 350 MHz
- Clock phase adjustment via programmable and fixed delays
- Clock skew minimization (PLL)
- Clock frequency synthesis (PLL)

In addition to the CCC and PLL support described above, there are on-chip oscillators as well as a comprehensive global clock distribution network. The integrated RC oscillator generates a 100 MHz clock. It is used internally to provide a known clock source to the Flash memory read and write control. It can also be used as a source for the PLLs. The crystal oscillator supports the following operating modes:

- Crystal (32.768 kHz to 20 MHz)
- Ceramic (500 kHz to 8 MHz)
- RC (32.768 kHz to 4 MHz)

Analog Components

Fusion devices include built-in analog peripherals such as a configurable 32:1 input analog multiplexer (MUX), up to 10 independent metal-oxide semiconductor field-effect transistor (MOSFET) gate driver outputs, and a configurable Analog to Digital Converter (ADC). The ADC supports 8-, 10-, and 12-bit modes of operation with a cumulative sample rate up to 600 k samples per second (ksps), differential nonlinearity (DNL) < 1.0 LSB, and Total Unadjusted Error (TUE) of ± 4 LSB in 10-bit mode.

Configuration

Fusion devices, once configured using their internal Flash memory do not require reconfiguration at power-up like SRAM-configuration devices.

Implementation Technology

The Fusion family is based on a 130 nm, 7-Layer Metal, Flash-Based CMOS Process.

1.4.4.2 ProASIC3 and ProASICPLUS Families

The ProASIC3 and ProASICPLUS families of FPGAs [48, 49] are older generations of Flash-based FPGAs with many of the features provided by the Fusion family such as on-chip FlashROM, Versa tiles, Segmented, Hierarchical Routing, PLLs and embedded SRAM.

1.4.4.3 Axcelerator Family

Actel's Axcelerator FPGA family [50] offers high performance at densities of up to two million equivalent system gates. Based upon Actel's new AX architecture, Axcelerator has several system-level features such as embedded SRAM (with complete FIFO control logic), PLLs, segmentable clocks, chip-wide highway routing, PerPin FIFOs, and carry logic.

1.4.4.4 VariCore

VariCore IP blocks [51] are embedded, reprogrammable "soft hardware" cores designed for use in ASIC and ASSP SoC applications. The available VariCore embedded programmable gate array (EPGA) blocks have been designed in 0.18 micron CMOS SRAM technology.

1.4.5 Atmel

Here are described the FPGAs that are available from ATMEL. These are the AT40K, AT40KLV, and AT6000 families.

1.4.5.1 AT40K/AT40KLV FPGA family

The AT40K/AT40KLV [52] is a family of fully PCI-compliant, SRAM-based FPGAs with distributed 10 ns programmable synchronous/asynchronous, dual-port/single-port SRAM, 8 global clocks, Cache Logic ability (partially or fully reconfigurable without loss of data), automatic component generators, and range in size from 5,000 to 50,000 usable gates.

The AT40K/AT40KLV is designed to quickly implement high-performance, large gate count designs through the use of synthesis and schematic-based tools. Atmel's design tools provide seamless integration with industry standard tools such as Synplicity, ModelSim, Exemplar and Viewlogic. The AT40K/AT40KLV can be used as a coprocessor for high-speed (DSP/processorbased) designs by implementing a variety of computation intensive, arithmetic functions. These include adaptive finite impulse response (FIR) filters, fast Fourier transforms (FFT), convolvers, interpolators and discrete-cosine transforms (DCT) that are required for video compression and decompression, encryption, convolution and other multimedia applications.

Cell architecture

The AT40K/AT40KLV FPGA core cell (Fig. 1.23) is a highly configurable logic block based around two 3-input LUTs (8 × 1 ROM), which can be combined to produce one 4-input LUT. This means that any core cell can implement two functions of 3 inputs or one function of 4 inputs. There is a Set/Reset D flip-flop in every cell, the output of which may be tristated and fed back internally within the core cell. There is also a 2-to-1 multiplexer in every cell, and an upstream AND gate in the

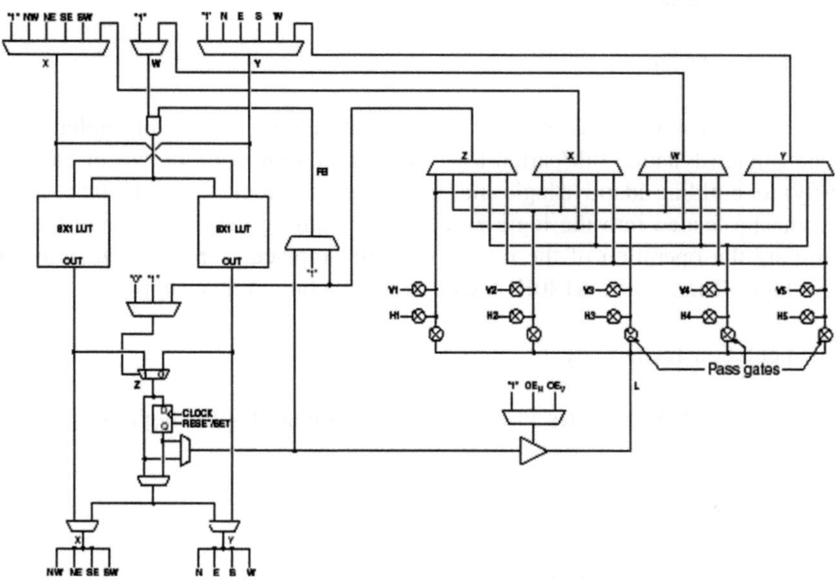

Fig. 1.23 The AT40K/AT40KLV FPGA core cell

"front end" of the cell. This AND gate is an important feature in the implementation of efficient array multipliers.

SRAM

The AT40K/AT40KLV FPGA offers a patented distributed 10 ns SRAM capability where the RAM can be used without losing logic resources. Multiple independent, synchronous or asynchronous, dual-port or single-port RAM functions (FIFO, scratch pad, etc.) can be created using Atmel's macro generator tool.

Array and Vector Multipliers

The AT40K/AT40KLV's patented 8-sided core cell with direct horizontal, vertical and diagonal cell-to-cell connections implements fast array multipliers without using any busing resources.

Automatic Component Generators

The AT40K/AT40KLV FPGA family is capable of implementing user-defined, automatically generated, macros in multiple designs; speed and functionality are unaffected by the macro orientation or density of the target device. The Automatic Component Generators work seamlessly with industry standard schematic and synthesis tools to create the fastest, most efficient designs available. The patented AT40K/AT40KLV series architecture employs a symmetrical grid of small yet powerful cells connected to a flexible busing network.

Devices range in size from 5,000 to 50,000 usable gates in the family, and have 256 to 2,304 registers.

Cache Logic Design

The AT40K/AT40KLV, AT6000 and FPSLIC families are capable of implementing Cache Logic (dynamic full/partial logic reconfiguration, without loss of data, on-the-fly) for building adaptive logic and systems. As new logic functions are required, they can be loaded into the logic cache without losing the data already there or disrupting the operation of the rest of the chip; replacing or complementing the active logic. The AT40K/AT40KLV can act as a reconfigurable coprocessor.

Implementation Technology

The AT40K/AT40KLV series FPGAs utilize a reliable 0.6μm single-poly, CMOS process.

1.4.5.2 AT6000 FPGA Family

AT6000 Series [53] SRAM-based Field-Programmable Gate Arrays (FPGAs) are ideal for use as reconfigurable coprocessors and implementing compute-intensive logic. Supporting system speeds greater than 100 MHz and using a typical operating current of 15 to 170 mA, AT6000 Series devices are ideal for high-speed, compute-intensive designs. These FPGAs are designed to implement Cache Logic, which provides the user with the ability to implement adaptive hardware and perform hardware acceleration. The patented AT6000 Series architecture employs a symmetrical grid of small yet powerful cells connected to a flexible busing network.

Devices range in size from 4,000 to 30,000 usable gates, and 1024 to 6400 registers. Pin locations are consistent throughout the AT6000 Series for easy design migration. High-I/O versions are available for the lower gate count devices. AT6000 Series FPGAs utilize a reliable 0.6 μm single-poly, double-metal CMOS process Multiple design entry methods are supported. The Atmel architecture was developed to provide the highest levels of performance, functional density and design flexibility in an FPGA. The cells in the Atmel array are small, very efficient and contain the most important and most commonly used logic and wiring functions. The cell's small size leads to arrays with large numbers of cells, greatly multiplying the functionality in each cell. A simple, high-speed busing network provides fast, efficient communication over medium and long distances.

Symmetrical Array

At the heart of the Atmel architecture is a symmetrical array of identical cells. The array is continuous and completely uninterrupted from one edge to the other, except for bus repeaters spaced every eight cells. In addition to logic and storage, cells can also be used as wires to connect functions together over short distances and are useful for routing in tight spaces.

Cell Structure

The Atmel cell is simple and small and yet can be programmed to perform all the logic and wiring functions needed to implement any digital circuit. Its four sides are functionally identical, so each cell is completely symmetrical. The Atmel AT6000 Series cell structure is shown in Fig. 1.24.

In addition to the four local-bus connections, a cell receives two inputs and provides two outputs to each of its North (N), South (S), East (E) and West (W) neighbors. These inputs and outputs are divided into two classes: "A" and "B". There is an A input and a B input from each neighboring cell and an A output and a B output driving all four neighbors. Between cells, an A output is always connected to an A input and a B output to a B input. Within the cell, the four A inputs and the four B inputs enter two separate, independently configurable multiplexers. Cell flexibility is enhanced by allowing each multiplexer to select also the logical constant "1". The two multiplexer outputs enter the two upstream AND gates.

Logic States

The Atmel cell implements a rich and powerful set of logic functions, stemming from 44 logical cell states which permutate into 72 physical states. Some states use

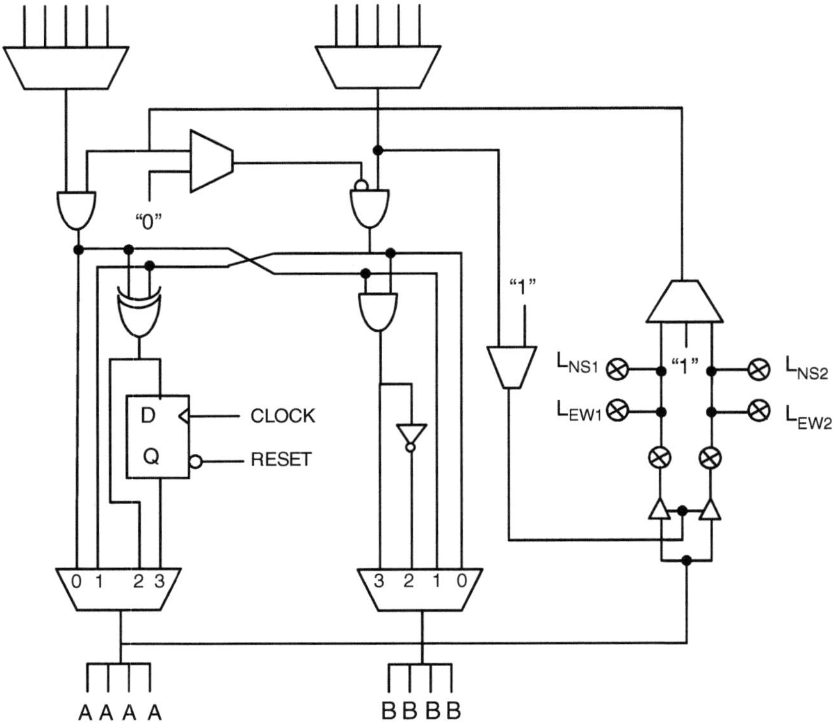

Fig. 1.24 The AT6000 series cell structure

both A and B inputs. Other states are created by selecting the "1" input on either or both of the input multiplexers.

1.4.6 Quicklogic

The available FPGA families from QuickLogic are the PolarPro and Eclipse II.

1.4.6.1 PolarPro Family

The PolarPro FPGA technology [54] was purposely architected to meet the interconnect and system logic requirements of power sensitive and portable applications. Through a new and innovative logic cell architecture, versatile embedded memory with built-in FIFO control logic and advanced clock management control units, the PolarPro architecture is synthesis friendly and logic mapping efficient.

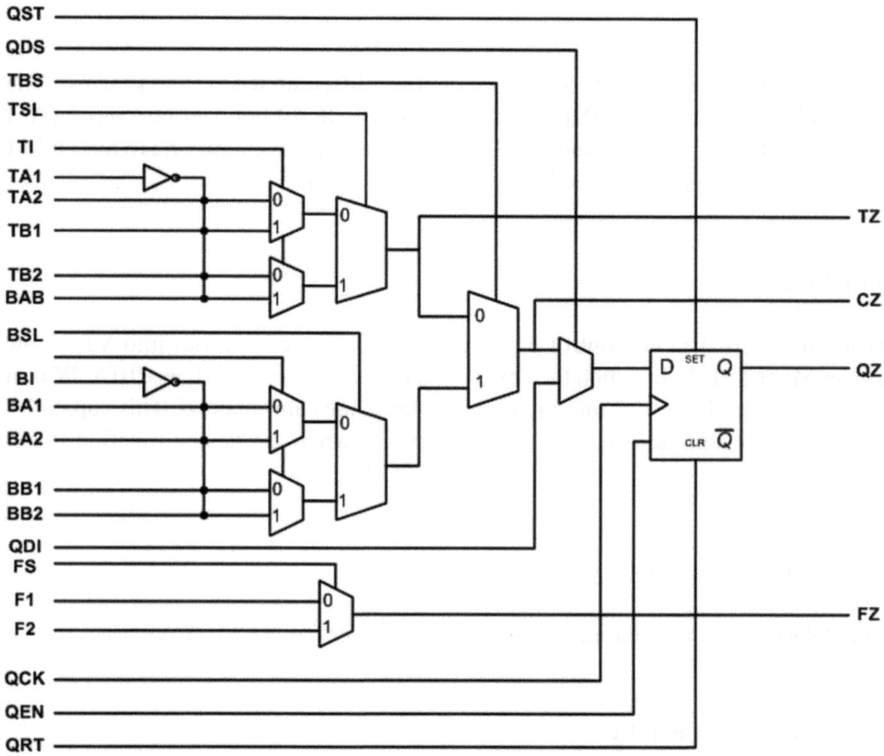

Fig. 1.25 PolarPro logic cell

Programmable Logic Architectural Overview

The QuickLogic PolarPro logic cell structure presented in Fig. 1.25 is a single register, multiplexer-based logic cell. It is designed for wide fan-in and multiple, simultaneous output functions. The cell has a high fan-in, fits a wide range of functions with up to 24 simultaneous inputs (including register control lines), and four outputs (three combinatorial and one registered). The high logic capacity and fan-in of the logic cell accommodates many user functions with a single level of logic delay.

The QuickLogic PolarPro logic cell can implement:

- Two independent 3-input functions
- Any 4-input function
- 8 to 1 mux function
- Independent 2 to 1 mux function
- Single dedicated register with clock enable, active high set and reset signals
- Direct input selection to the register, which allows combinatorial and register logic to be used separately
- Combinatorial logic that can also be configured as an edge-triggered master-slave D flip-flop

RAM Modules

The PolarPro family of devices include two different RAM block sizes. The QL1P075, QL1P100, QL1P200, and QL1P300 have 4-kilobit (4,608 bits) RAM blocks, while the QL1P600 and QL1P1000 devices have 8-kilobit (9,216 bits) RAM blocks. The devices include embedded FIFO controllers.

VLP Mode

From an external input control pin, the FPGA device can be put into Very Low Power Mode (VLP), in which the device will typically draw less than 10 μA. Within the VLP mode, I/O states and internal register values are retained. This capability provides an instant ability to save battery power when the device function is not needed.

Implementation Technology

The PolarPro family is based on a 0.18 μm, six layer metal CMOS process.

1.4.6.2 Eclipse II Family

Logic Cell

The Eclipse II [55] logic cell structure presented in Fig. 1.26 is a dual register, multiplexer-based logic cell. It is designed for wide fan-in and multiple, simultaneous output functions. Both registers share CLK, SET, and RESET inputs. The second register has a two-to-one multiplexer controlling its input. The register can be loaded from the NZ output or directly from a dedicated input.

The complete logic cell consists of two six-input AND gates, four two-input AND gates, seven two-to-one multiplexers, and two D flip-flops with asynchronous SET and RESET controls. The cell has a fan-in of 30 (including register control lines), fits a wide range of functions with up to 17 simultaneous inputs, and has six outputs (four combinatorial and two registered). The high logic capacity and fan-in of the logic cell accommodates many user functions with a single level of logic delay while othe architectures require two or more levels of delay.

RAM Modules

The Eclipse II Product Family includes up to 24 dual-port 2,304-bit RAM modules for implementing RAM, ROM, and FIFO functions. Each module is user-configurable into two different block organizations and can be cascaded horizontally to increase their effective width, or vertically to increase their effective depth.

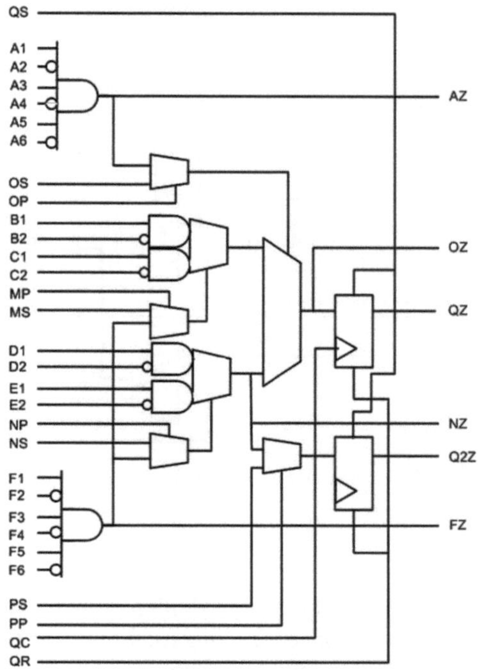

Fig. 1.26 Eclipse II logic cell

Embedded Computational Unit (ECU)

By embedding a dynamically reconfigurable computational unit, the Eclipse II device can address various arithmetic functions efficiently. This approach offers greater performance and utilization than traditional programmable logic implementations. The embedded block is implemented at the transistor level as shown in Fig. 1.27.

Programmable Logic Routing

Eclipse II devices are engineered with six types of routing resources as follows: short (sometimes called segmented) wires, dual wires, quad wires, express wires, distributed networks, and default wires. Short wires span the length of one logic cell, always in the vertical direction. Dual wires run horizontally and span the length of two logic cells. Short and dual wires are predominantly used for local connections. Default wires supply VCC and GND (Logic '1' and Logic '0') to each column of logic cells.

Quad wires have passive link interconnect elements every fourth logic cell. As a result, these wires are typically used to implement intermediate length or medium fan-out nets.

Express lines run the length of the device, uninterrupted. Each of these lines has a higher capacitance than a quad, dual, or short wire, but less capacitance than shorter wires connected to run the length of the device. The resistance will also be lower

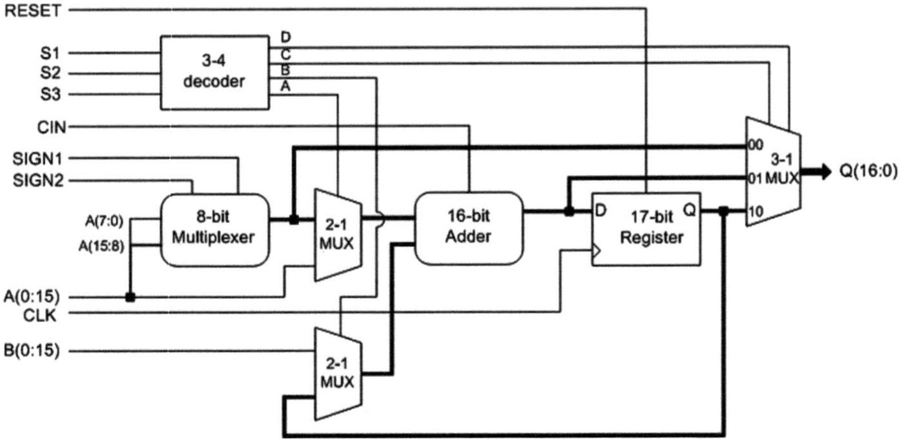

Fig. 1.27 Eclipse II embedded computational unit

because the express wires don't require the use of pass links. Express wires provide higher performance for long routes or high fan-out nets.

Distributed networks span the programmable logic and are driven by quad-net buffers.

PLLs

The QL8325 and QL8250 devices contain four PLLs, the remaining Eclipse II devices do not contain PLLs. There is one PLL located in each quadrant of the FPGA. QuickLogic PLLs compensate for the additional delay created by the clock tree itself, as previously noted, by subtracting the clock tree delay through the feedback path.

Low Power Mode

Quiescent power consumption of all Eclipse II devices can be reduced significantly by de-activating the charge pumps inside the architecture. By applying 3.3 V to the VPUMP pin, the internal charge pump is deactivated —this effectively reduces the static and dynamic power consumption of the device. The Eclipse II device is fully functional and operational in the Low Power mode. Users who have a 3.3 V supply available in their system should take advantage of this low power feature by tying the VPUMP pin to 3.3 V. Otherwise, if a 3.3 V supply is not available, this pin should be tied to ground.

Implementation Technology

The Eclipse II family is based on a 0.18 μm, six layer metal CMOS process.

1.4.7 Lattice

The LatticeECP2 and XP families are described next.

1.4.7.1 LatticeECP2 family

The LatticeECP2 [56] family of FPGAs, apart from the "traditional" FPGA fabric (logic blocks and interconnect) provides a number of features for implementing complex systems such as embedded RAM, DSP blocks, PLLs and DLLs.

Logic blocks

There are two kinds of logic blocks in LatticeECP2 devices, the Programmable Functional Unit (PFU) and Programmable Functional Unit without RAM (PFF). The PFU contains the building blocks for logic, arithmetic, RAM and ROM functions. The PFF block contains building blocks for logic, arithmetic and ROM functions. Both PFU and PFF blocks are optimized for flexibility allowing complex designs to be implemented quickly and efficiently. Logic Blocks are arranged in a two-dimensional array. Only one type of block is used per row. Each PFU block consists of four interconnected slices. Each slice (Fig. 1.28) has up to four potential modes of operation: Logic, Ripple, RAM and ROM.

Embedded RAM

The LatticeECP2 family of devices contain up to two rows of sysMEM EBR blocks. sysMEM EBRs are large dedicated 18K fast memory blocks. Each sysMEM block can be configured in variety of depths and widths of RAM or ROM. In addition, LatticeECP2 devices contain up to two rows of DSP Blocks. Each DSP block has multipliers and adder/accumulators, which are the building blocks for complex signal processing capabilities.

PLLs

Other blocks provided include PLLs, DLLs and configuration functions. The LatticeECP2 architecture provides two General PLLs (GPLL) and up to four Standard PLLs (SPLL) per device. In addition, each LatticeECP2 family member provides two DLLs per device. The GPLLs and DLLs blocks are located in pairs at the end

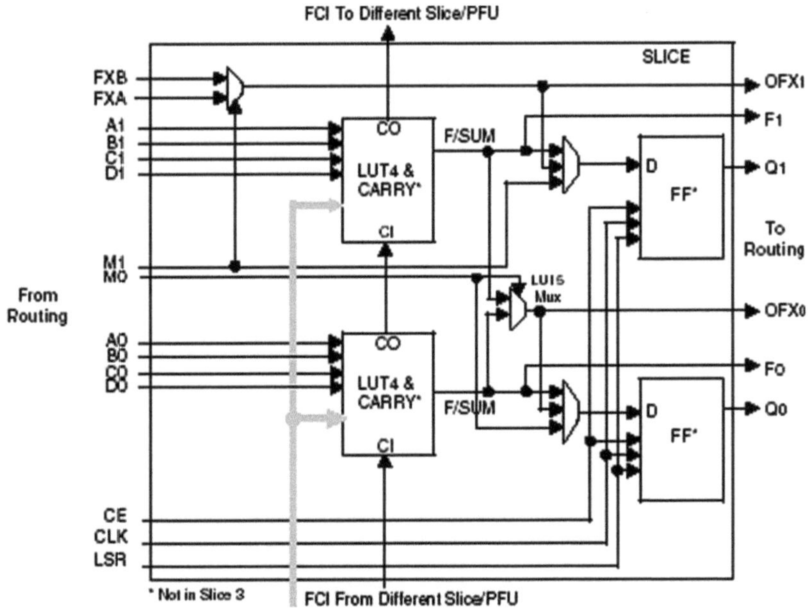

Fig. 1.28 LatticeECP2 slice

of the bottom-most EBR row; the DLL block located towards the edge of the device. The SPLL blocks are located at the end of the other EBR/DSP rows.

sysDSP blocks

The sysDSP block in the LatticeECP2 family supports four functional elements in three 9, 18 and 36 data path widths. The user selects a function element for a DSP block and then selects the width and type (signed/unsigned) of its operands. The operands in the LatticeECP2 family sysDSP Blocks can be either signed or unsigned but not mixed within a function element. Similarly, the operand widths cannot be mixed within a block. In LatticeECP2 family of devices the DSP elements can be concatenated.

The resources in each sysDSP block can be configured to support the following four elements:

- MULT (Multiply)
- MAC (Multiply, Accumulate)
- MULTADD (Multiply, Addition/Subtraction)
- MULTADDSUM (Multiply, Addition/Subtraction, Accumulate)

Configuration

LatticeECP2 devices use SRAM configuration with enhanced configuration features such as:

- **Decryption Support:** LatticeECP2 devices provide on-chip, non-volatile key storage to support decryption of a 128-bit AES encrypted bitstream, securing designs and deterring design piracy.
- **TransFR (Transparent Field Reconfiguration):** TransFR I/O (TFR) is a technology feature that allows users to update their logic in the field without interrupting system operationby allowing I/O states to be frozen during device configuration. Thus the device can be field updated with a minimum of system disruption and downtime.
- **Dual Boot Image Support:** Dual boot images are supported for applications requiring reliable remote updates of configuration data for the system FPGA. After the system is running with a basic configuration, a new boot image can be downloaded remotely and stored in a separate location in the configuration storage device. Any time after the update the LatticeECP2 can be re-booted from this new configuration file. If there is a problem such as corrupt data during down or incorrect version number with this new boot image, the LatticeECP2 device can revert back to the original backup configuration and try again. All this can be done without power cycling the system.

1.4.7.2 LatticeXP Family

The LatticeXP family [57] has a lot of features in common with the LatticeECP2 family such as logic block (PFU/PFFs), interconnect and clock networks, embedded RAM and DLLs. The LatticeXP family, however lacks sysDSP blocks and additionally features a non-volatile configuration memory.

Sleep Mode

The LatticeXP "C" devices (VCC = 1.8/2.5/3.3V) have a sleep mode that allows standby current to be reduced by up to three orders of magnitude during periods of system inactivity. Entry and exit to Sleep Mode is controlled by the SLEEPN pin. During Sleep Mode, the FPGA logic is non-operational, registers and EBR contents are not maintained and I/Os are tri-stated. Do not enter Sleep Mode during device programming or configuration operation. In Sleep Mode, power supplies can be maintained in their normal operating range, eliminating the need for external switching of power supplies. Table 2–9 compares the characteristics of Normal, Off and Sleep Modes.

Configuration

LatticeXP devices include a nonvolatile memory that is programmed in configuration mode. On power up, the configuration data is transferred from the Non-volatile Memory Blocks to the configuration SRAM. With this technology, expensive external configuration memories are not required and designs are secured from unauthorized read-back. This transfer of data from non-volatile memory to configuration

SRAM via wide busses happens in microseconds, providing an "instant-on" capability that allows easy interfacing in many applications.

Security

The LatticeXP devices contain security bits that, when set, prevent the readback of the SRAM configuration and non-volatile memory spaces. Once set, the only way to clear security bits is to erase the memory space.

Internal Logic Analyzer Capability (ispTRACY)

All LatticeXP devices support an internal logic analyzer diagnostic feature. The diagnostic features provide capabilities similar to an external logic analyzer, such as programmable event and trigger condition and deep trace memory. This feature is enabled by Lattice's ispTRACY. The ispTRACY utility is added into the user design at compile time.

1.4.8 Summary

Table 1.2 summarizes some of the main characteristics about the FPGAs that have been described previously at this section. The comparison of the FPGAs is based on the technology maturity, the design flow, the technology implementation, the technology portability, the available data-sheet information and their testability.

Tables 1.3 to 1.5 provide a number of quantitative and qualitative comparisons among the commercial FPGA described in the previous sections in terms of features, technology and circuit-level low power techniques.

1.5 Academic Software Tools for Designing Fine-Grain Platforms

1.5.1 Introduction

An efficient FPGA platform, will not lead to optimal application implementation without an equally efficient set of CAD tools to map the target application on the resources of the device, verify its functionality and timing constraints, and finally, produce the configuration bitstream.

A typical programmable logic design involves three steps:

- Design entry
- Design implementation
- Design verification

Table 1.2 Comparison between some the most well-known FPGAs

FPGA	Technology maturity	Design flow	Technology implementation	Technology portability	Data-sheet information	Testability
Altera	Chips and development board available	Complete design tools, Third party EDA tools support	Standard SRAM process	Firm, HardCopy devices can be used to transfer from PLD to ASIC	Complete	JTAG and PC trace debugging, graphical view of floor planning
Xilinx	Chips and development board available	Third party EDA tools support	Standard SRAM process	Firm	Complete	JTAG debugging environment
Atmel	Chips available	ASIC design flow	Standard SRAM FPGA and RISC micro-controller, standard peripherals	No	Enough	Co-verification environment, Source-level debugging
Actel	Chips available	Compatible design flow with ASIC, third party EDA tools support	Standard CMOS SRAM technology	Yes, leading silicon foundries support	Enough	Built-in self test interface

Table 1.3 Comparison among commercial devices

Device family	Embedded RAM	Interconnect	Technology	Configuration
Spartan – 3	72–1872 Kb	4-type Segmented	90 nm	SRAM
Virtex – 4	864–9,936 Kb	4-type Segmented	90 nm	SRAM
Virtex – 5	Upto 10 Mb	Diagonal	65 nm	SRAM
Cyclone	Upto 288 Kb	3-type Segmented	130 nm	SRAM
Cyclone II	Upto1.1 Mb	3-type Segmented	90 nm	SRAM
Stratix II	Upto 9 Mb	3-type Segmented	90 nm	SRAM
Stratix		3-type Segmented	130 nm	SRAM
PolarPro	36–198 Kb		180 nm	
Eclipse II	9.2–55.3 Kb		180 nm	
Fusion	27–270 Kb		130 nm	Flash
ProASIC3	Upto 144 Kb		130 nm	Flash
LatticeECP2	55–1,032 Kb			SRAM
LatticeXP	54–396 Kb			Non-volatile & SRAM

Table 1.4 Features of commercial devices

Device family	Multipliers	PLL/DCM	Max flip-flops	Max user I/O
Spartan – 3	4–104 18×18	2 or 4	1728–74880	124–784
Virtex – 4	32–192 18×18	4–20		320–896
Virtex – 5	32–192 18×18	2/4–6/12	19200–207,360	400–1200
Cyclone	N/A	1–2	2,910–20,060	104–301
Cyclone II	13 – 150 18×18 or 26–300 9×9	2 or 4	4,608–68,416	158–622
Stratix II	96–768 9×9 or 48–384 18×18 or 12–96 36×36	6–12	12,480–143,520	366–1170
Stratix II GX	128–504 9×9 or 64–252 18×18 or 16–63 36×36	4–8	27,104–106,032	361–734
Stratix	48–96 9×9 or 24–48 18×18 or 6–12 36×36	6–10	21,140–64,940	426–726
Stratix GX	48–112 9×9 or 24–56 18×18 or 6–14 36×36	4 or 8	21,140–82,500	362–624
Fusion	N/A	1–2	2,304–38,400	75–252 digital, 20–40 analog
PolarPro	N/A	1–2	512–7,680	168–652
Eclipse II	N/A	Upto 4	532–4,002	92–310
LatticeECP2	24–172 9×9 or 12–88 18×18 or 3–22 36×36		12,000–136,000	192–628
LatticeXP	N/A	2–4	12,000–79,000	136–340

Table 1.5 Technology-level low-power techniques in 90 and 65 nm FPGAs

Device family	Increased Vt	Increased transistor length	Triple oxide
Virtex – 4			YES
Virtex – 5			YES
Stratix II	YES	YES	
Stratix II GX	YES	YES	

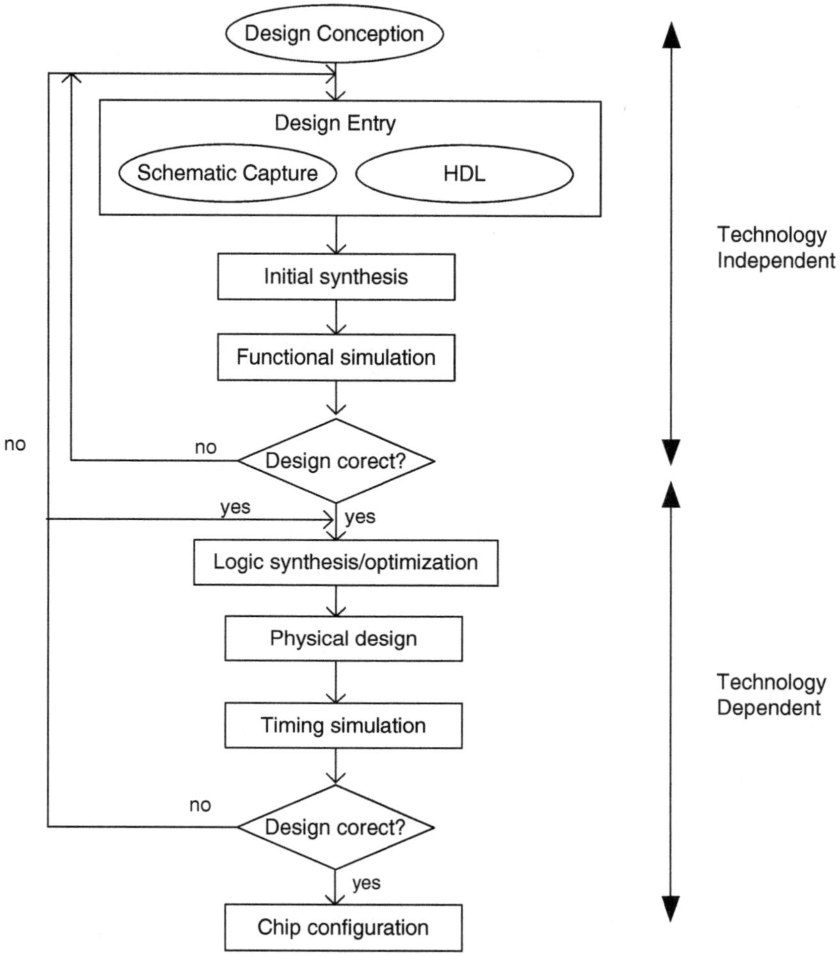

Fig. 1.29 Traditional design synthesis approach and the modeling approach

All three steps, which are shown in Fig. 1.29, are described briefly below.

1.5.1.1 Design Entry

A variety of tools are available to accomplish the design entry step. Some designers prefer to use their favorite schematic entry package while others prefer to specify their design using a hardware description language such as Verilog, VHDL, or ABEL. Others prefer to mix both schematic and language-based entry in the same design. There has been an on-going battle as to which method is best. Traditionally, schematic-based tools provided experienced designers more control over the physical placement and partitioning of logic on the device. However, this extra tailoring took time. Likewise, language-based tools allowed quick design entry but often at the cost of lower performance or density. Synthesis for language-based designs has significantly improved in the last few years, especially for FPGA design. In either case, learning the architecture and the tool helps you to create a better design. Technology-ignorant design is quite possible, but at the expense of density and performance.

1.5.1.2 Design Implementation

After the design is entered and synthesized, it is ready for implementation on the target device. The first step involves converting the design into the format supported internally by the tools. Most implementation tools read "standard" netlist formats and the translation process is usually automatic. Once translated, the tools perform a design rule check and optimization on the incoming netlist. Then the software partitions the designs into the logic blocks available on the device. Partitioning is an important step for FPGAs, as good partitioning results in higher routing completion and better performance for FPGAs. After that, the implementation software searches for the best location to place the logic block among all of the possibilities. The primary goal is to reduce the amount of routing resources required and to maximize system performance. This is a compute-intensive operation for FPGA tools. The implementation software monitors routing length and routing track congestion while placing the blocks. In some systems, the implementation software also tracks the absolute path delays in order to meet user-specified timing constraints. Overall, the process mimics printed circuit board place and route. When the placement and routing process is complete, the software creates the binary programming file used to configure the device. In large or complex applications, the software may not be able to successfully place and route the design. Some packages allow the software to try different options or to run more iterations in an attempt to obtain a fully-routed design. Also, some vendors supply floor-planning tools to aid in physical layout. Layout is especially important for larger FPGAs because some tools have problems recognizing design structure. A good floor-planning tool allows the designer to convey this structure to the place and route software.

1.5.1.3 Verification

Design verification occurs at various levels and steps throughout the design. There are a few fundamental types of verification as applied to programmable logic. Functional simulation is performed in conjunction with design entry, but before place and route, to verify correct logic functionality. Full timing simulation must wait until after the place and route step. While simulation is always recommended, programmable logic usually does not require exhaustive timing stimulation like gate arrays. In a gate array, full timing simulation is important because the devices are mask-programmed and therefore not changeable. In a gate array, you can not afford to find a mistake at the silicon level. One successful technique for programmable logic design is to functionally simulate the design to guarantee proper functionality, verify the timing using a static timing calculator, and then verify complete functionality by testing the design in the system. Programmable logic devices have a distinct advantage over gate arrays. Changes are practically free. With in-system programmable (ISP) devices, such as SRAM based FPGAs, changes are possible even while the parts are mounted in the system. Using in-system verification techniques, the design is verified at full speed, with all the other hardware and software. Creating timing simulation vectors to match these conditions would be extremely difficult and time consuming. Some of the device vendors supply additional in-system debugging capabilities.

1.5.2 Public Domain Tools

This section describes the available public domain CAD tools that support a range of architectures. Those tools are open-source, which means their source code is publicly available in order to make any changes targeting improvement of their functionality. The main providers of these tools are UCLA and the Toronto FPGA Research Group.

1.5.2.1 Tools from UCLA

The available CAD Tools from UCLA can be used for interconnection, technology mapping and as a multilayer router. Those tools are:

TRIO

TRIO [58] stands for Tree, Repeater, and Interconnect Optimization. It includes many optimization engines in order to perform Routing-tree construction, Buffer (repeater) insertion, Device and wire sizing, and Spacing. TRIO uses two types of models to compute the device delay and also two types of interconnect capacitance models.

RASP_SYN

RASP_SYN [59] is a LUT-based FPGA technology mapping package and is the synthesis core of the UCLA RASP System. It uses a lot of mapping algorithms, some of them are the Depth minimization, Depth optimal, Optimal mapping with retiming, Area-delay tradeoff, FPGA resynthesis, Simultaneous area delay minimization, Mapping for FPGAs with embedded memory blocks for area minimization while maintaining the delay, Delay optimal mapping for heterogenous FPGAs, Delay-oriented mapping for heterogenous FPGAs with bounded resources, Performance-driven mapping for PLA with area/delay trade-offs, Simultaneous logic decomposition with technology mapping. The first step of the entire flow of the

RASP_SYN package involves gate decomposition, in order to obtain a K-bounded circuit, where K is the fan-in limit of LUTs of the target architecture. Then, generic LUT mapping is run, and post-processing mainly for area reduction. Finally, architecture specific mapping takes place.

IPEM

IPEM [60] is another tool from UCLA which provides a set of procedures that estimate interconnect performance under various performance optimization algorithms for deep submicron technology. Since it adopts several models derived from corresponding interconnection optimization algorithms, IPEM is fast and accurate. Also it has the advantage that it is user-friendly thanks to its ANSI C interface and library. The output of this tool produces considerable interconnect optimization in logic level synthesis, as well as interconnect planning.

MINOTAUR

The next available tool from UCLA is MINOTAUR [61] which is a performance-driven multilayer general area router. It utilizes current high-performance interconnect optimization results in order to obtain interconnect structures with strict address delay and signal integrity requirements. In addition to that, the tool considers global congestion by routing all layers simultaneously, and places no restriction on the layers a route may use. Moreover it combines the freedom and flexibility of maze routing solutions with the global optimization abilities of the iterative deletion method.

FPGAEVA

FpgaEva [62] is a heterogeneous FPGA evaluation tool that incorporates a set of architecture evaluation related features into a user friendly Java interface. This tool uses the state-of-the-art mapping algorithms and supports user-specified circuit models like area/delay of LUTs of different size, while it allows the user to compare multiple architectures. In addition to that, fpgaEva has the advantage that it is written in Java and so, the remote evaluation mode allows the user to run it from any computer.

1.5.2.2 Tools from Toronto FPGA Research Group

Apart from the available tools from UCLA there are also CAD tools from the Toronto FPGA Research Group. Those tools can be used for variable serial data width arithmetic module generation, for placement, routing and for technology mapping. A briefly description of the characteristics of those tools is following.

PSAC-Gen

The first tool from this group is PSAC-Gen [63] which stands for Parametrizeable Serial Arithmetic Core Generator. It is a tool that allows design and implementation of bit-serial and digit-serial arithmetic circuits using simple arithmetic expressions. In other words, it is used to easily generate a wide variety of arithmetic circuits involving addition, subtraction, and multiplication. The PSAC-Gen takes as input an arithmetic circuit description and creates a set of VHDL files that describe the circuit.

Edif2Blif

EDIF is an industry-standard file format that allows EDA tools to communicate with each other, including the ability to transfer netlists, timing parameters, graphical representations, and any other data the vendors wish. The Edif2Blif tool [64] converts netlists from the industry standard Electronic Data Interchange Format (EDIF) to the academic Berkeley Logic Interchange Format (BLIF).

SEGA

SEGA [65] was developed as a tool to evaluate routing algorithms and architectures for array-based Field-Programmable Gate Arrays. It was written in a modular fashion to permit flexibility between modifying the routing algorithm and representing the routing architecture. Both SEGA and CGE solve the detailed routing resource allocation problem for array-based FPGAs, but SEGA has improved upon CGE in that it considers the speed-performance of the routed circuit an important goal (instead of just routability).

PGARoute

PGARoute [66] is a global router for symmetric FPGAs. In order to make the placement, it uses the Xaltor program. When the PGARoute finishes its work, it prints out the number of logic blocks it used in the longest and in the shortest row.

Transmogrifier C

Transmogrifier C [67] is a compiler for a simple hardware description language. It takes a program written in a restricted subset of the C programming language, and produces a netlist for a sequential circuit that implements the program in a Xilinx XC4000 series FPGA.

Chortle

The next available tool from the Toronto FPGA Research Group is Chortle [68], which is used to map a Boolean network into a circuit of lookup tables. During this mapping, it attempts to minimize the number of lookup tables required to implement the Boolean network.

VPR and T-VPACK

VPR [69] is a placement and routing tool for array-based FPGAs that was developed from the Toronto FPGA Research Group. VPR was developed to allow circuits to be placed and routed on a wide variety of FPGAs. It is used for perform placement and either global routing or combined global and detailed routing. Although this tool was initially developed for island-style FPGAs, it can also be used with row-based FPGAs. The cost function that is used in this tool is the "linear congestion cost" while the router is based on the Pathfinder negotiated congestion algorithm.

Figure 1.30 summarizes the CAD flow with the VPR tool.

First, a system for sequential circuit analysis (SIS) is used to perform technology-independent logic optimization on a circuit. Next this circuit is technology-mapped by FlowMap into four-input look-up tables (4-LUTs) and registers. The Flowpack post-proccessing algorithm is then run to further optimize the mapping and reduce the number of LUTs required. VPack packs 4-LUTs and registers together into the logic blocks. The netlist of logic blocks and a description of the FPGA global routing architecture are then read into the placement and routing tool. VPR first places the circuit, and then repeatedly globally routes (or attempts to route) the circuit with different number of tracks in each channel, or channel capacities. VPR performs a binary search on the channel capacities, increasing them after a failed routing and reducing them after a successful one, until it finds the minimum number of tracks required for the circuit to globally route successfully on a given global routing architecture.

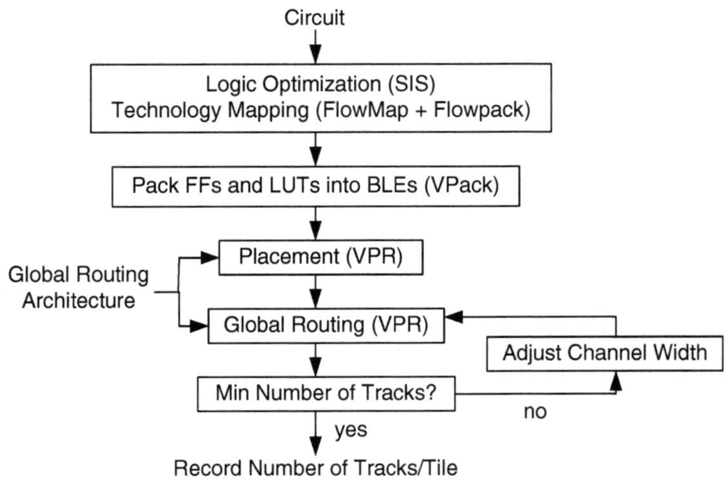

Fig. 1.30 The CAD flow with the VPR tool

Power Model (VPR)

Power Model [70] is built on top of the original VRP CAD tool. Figure 1.31 shows the VPR framework with Power Model, which is part of the area and delay model. An activity estimator is used to estimate the switching frequencies of all nodes in the circuit. In the current implementation, the activity estimator and the power model are not used to guide the placement and routing. It estimates the power consumption only after placement and routing has occurred.

Power Model includes terms for dynamic power, short-circuit, and leakage power. It is flexible enough to target FPGAs with different LUT sizes, different interconnect strategies (segment length, switch block type, connection flexibility), different cluster sizes (for a hierarchical FPGA), and different process technologies.

1.6 Commercial Software Tools for Designing Fine-Grain Platforms

In this section are described some of the most well-known commercial software tools for designing fine-grain reconfigurable platforms. These tools are sorted alphabetically and they are grouped by the vendor company that produces them.

1.6.1 Actel

1.6.1.1 Development Software

- Libero v2.2 Integrated Design Environment (IDE): Actel's Libero v2.2 IDE [71] offers best in class tools from such EDA powerhouses as Mentor Graphics, SynaptiCAD, Synplicity, and custom developed tools from Actel integrated into a single design package. It includes also Actel's "Designer" software,. Designer

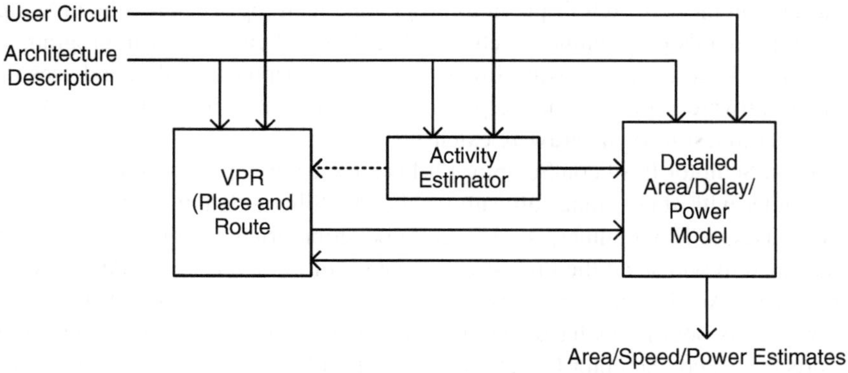

Fig. 1.31 Framework with power model

offers premier backend design support for physical implementation. Libero IDE supports all currently released Actel devices and is available in three flavors: Libero Silver, Libero Gold, and Libero Platinum. Some of the Libero's IDE Software features are the powerful design management and flow control environment, the easy schematic and HDL design, the VHDL or Verilog Behavioral, post-synthesis and post-layout simulation capability, the VHDL / Verilog synthesis, and the physical implementation with place and route.

- Actel Designer R1-2002 Software: The Actel Designer [72] offers an easy to use and flexible solution for all Actel FPGA devices. It gives designers the flexibility to plug and play with other third party tools. Advanced place-and-route algorithms accommodate the needs of today's increasingly complex design and density requirements. The architecture expertise are built into the tools to create the most optimized design. The Actel Designer software interface offers both automated and manual flows, with the push-button flow achieving the optimal solution in the shortest cycle. User driven tools like ChipEdit, PinEdit, and Timing Constraint Editor give expert users maximum flexibility to drive the place-and-route tools to achieve the timing required. The Actel Designer software supports all the established EDA standards like Verilog/VHDL/EDIF netlist formats. I/O handling tools like I/O-Attribute Editor and PinEdit enable designers to assign different attributes including capacitance, slew, pin, and hot swap capabilities to individual I/Os. Actel's highly efficient place and route algorithms allow designers to assign package pins locations during the design development phase with confidence that the design will place and route as specified. Silicon Explorer enables the user to debug the design in real time by probing internal nodes for viewing while the design is running at full speed.

1.6.1.2 Programming/Configuration

- Silicon Sculptor II: Silicon Sculptor II [73] is a robust, compact, single device programmer with stand alone software for the PC. Designed to allow concurrent programming of multiple units from the same PC, with speeds equivalent to, or faster than those of Actel's previous programmers. It replaces the Silicon Sculptor I as Actel's programmer of choice. The Silicon Sculptor II can program all Actel packages, it works with Silicon Sculptor I adapter modules, and uses the same software as the Silicon Sculptor I. In addition to that, it could allow self-test in order to test its own hardware extensively.
- Silicon Sculptor I: Silicon Sculptor [74] is a robust, compact, single device programmer with stand alone software for the PC. Silicon Sculptor 6X Concurrent Actel Device Programmer, is a six site production oriented device programmer designed to withstand the high stress demands of high volume production environments. Actel no longer offers the Silicon Sculptor I and Silicon Sculptor 6X for sale, as both items have been discontinued. On the other hand, Actel supports the Silicon Sculptor I and Silicon Sculptor 6X by continuing to release new software that allows the programming of new Actel devices.

1.6.1.3 Verification and Debug

- Silicon Explorer II: Actel's antifuse FPGAs contain ActionProbe circuitry that provides built-in, no-cost access to every node in a design, enabling 100% real-time observation and analysis of a device's internal logic nodes without design iteration. Silicon Explorer II [75] is an easy to use integrated verification and logic analysis tool for the PC, accesses the probe circuitry that allows designers to complete the design verification process at their desks.

1.6.2 Cadence

FPGA HDL design, synthesis, and verification are more demanding than ever due to today's complex system-on-programmable-chips (SoPC). There is a need for tools and solutions to proficiently manage complex FPGA designs, to dramatically increase design efficiencies, and to significantly reduce system costs and development time. Cadence provides the tools and solutions to achieve all that. It provides exclusive transaction-level verification capabilities that can handle HDL schematics-including component-level and block-based decomposition-along with algorithmic entry, mixed-language, and mixed-signal simulation.

1.6.2.1 Signal Processing Worksystem (SPW)

The Cadence Signal Processing Worksystem (SPW) [76] starts by building your design with pre-authored library blocks. Additionally, it is possible to simulate the design and analyze the results by easily integrating C, C++, or SystemC code or MATLAB models. From there take the design to application-specific integrated circuit (ASIC) or field-programmable gate array (FPGA) implementation by describing the hardware architectures using VHDL, Verilog, SystemC, or graphical-based blocks, and verify and debug it together with previously-generated testbenches. The generation of register transfer level (RTL) allows targeting an unparalleled efficient datapath synthesis step.

1.6.2.2 Cadence FPGA Verification

The Cadence NC-Sim simulation family [77] is a verification solution for high-end FPGA design. The native compiled simulator offers the freedom to transparently mix VHDL and Verilog. This makes Cadence NC-Sim a very flexible and adaptable simulator, allowing seamless integration into today's complex FPGA design flows.

1.6.2.3 ORCAD Capture

With its fast, universal design entry capabilities, Orcad Capture [78] schematic entry has quickly become one of the world's favorite design entry tools. From designing a new analog circuit, revising schematic diagrams on an existing PCB, or drafting a block diagram of HDL modules, Orcad Capture provides everything you need to complete and verify the designs quickly.

1.6.2.4 Cadence Verilog Desktop

The Cadence Verilog Desktop [79] brings the quality and reliability of the Cadence NC-Verilog simulator to every desktop. Built on technology from NC-Verilog, the Verilog Desktop is ideal for engineering teams that want to leverage the performance and capacity created to validate multimillion gate ASIC designs. Its unique debug features make Verilog Desktop a perfect fit for FPGA and CPLD development and verification. It comes complete with the SimVision Graphical Analysis Environment, and the Signalscan waveform display tool.

1.6.3 Mentor Graphics

The available tools from Mentor Graphics are described in this subsection.

1.6.3.1 Integrated FPGA Design Flow

- FPGA Advantage: FPGA Advantage [80] provides a complete and seamless integration of design creation, management, simulation and synthesis, empowering the FPGA designer to have a faster path from concept to implementation.

1.6.3.2 HDL Design

- HDL Designer: HDL Designer [81] is a complete design and management solution that includes all the point tools of the HDL Designer Series. It allows to standardize on a toolset that can be used to share designs and designers. HDL visualization and creation tools, along with automatic documentation features, foster a consistent style of HDL for improved design reuse, so it can fully leverage existing IP.
- Debug Detective: Debug Detective [82] takes debugging of HDL designs to the next level. As a snap-on to ModelSim it renders on-the-fly graphical and tabular views of HDL source code to aid understanding and control, and delivers interactive debug and analysis between these views and the ModelSim user interface.

This combination enables faster debug and improved productivity of the HDL design.

- HDL Detective: HDL Detective [83] allows you to understand, visualize and navigate complex designs without forcing you to change the design methodology. Its fully automated documentation and communication features provide a push-button process for reusing HDL designs and commercial IP, so it is possible to visualize the current state of any design. HDL Detective also automatically generates documentation for newly developed HDL. By translating HDL to diagrammatic representations, the time it takes to understand an unfamiliar design can be reduced dramatically.
- HDL Author: HDL Author [84] integrates all the design management features of HDL Pilot, and adds best-in-class text-based and graphics-based editors to provide a comprehensive environment for design creation, reuse and management. To accommodate the fullest range of design preferences, HDL Author is available in three flavors that give the flexibility to design systems using pure HDL source code, pure graphics, or a combination of both.
  - HDL Author Text provides absolute control over all aspects of the design process. It includes a Block Editor and an Interface-Based Design (IBD) editor for writing code directly, creating documentation, following a reuse methodology, and integrating blocks from multiple locations.
  - HDL Author Graphics allows intuitive design, using diagrams from which HDL is automatically generated and documentation is implicitly available. It includes a Block Editor, State Machine Editor, Flow Chart Editor and Truth Table Editor for creating a design and documentation using a graphical methodology that's ideally suited to designers or organizations that are migrating to HDL methodologies.
  - HDL Author Pro includes all the above features in a single, economical solution that provides complete creative control.
- HDL Pilot: HDL Pilot [85] is a comprehensive environment for managing HDL designs and data from start to finish. It provides an easy-to-use cockpit from which designers can launch common tools for developing complex Verilog, VHDL and mixed-HDL designs. HDL Pilot automatically and incrementally imports and analyzes HDL files to simplify design navigation, and introduces a simple but effective GUI for the use of version control. Common operations such as data compilation for simulation and synthesis are performed automatically. And HDL Pilot can be easily customized to recognize different data types and tools.

1.6.3.3 Synthesis

- Precision Synthesis: The Precision Synthesis environment [86] has a highly intuitive interface that drives the most advanced FPGA synthesis technology available, delivering correct results without iterations. Timing constraints,

coupled with state-of-the-art timing analysis, guide optimization when and where it's needed most, achieving excellent results for even the most aggressive designs.
- LeonardoSpectrum: With one synthesis environment, it is possible to create PLDs, FPGAs, or ASICs in VHDL or Verilog. LeonardoSpectrum [87] from Mentor Graphics combines push-button ease of use with the powerful control and optimization features associated with workstation-based ASIC tools. Users faced with design challenges can access advanced synthesis controls within LeonardoSpectrum's exclusive PowerTabs. In addition, the powerful debugging features and exclusive five-way cross-probing in LeonardoInsight accelerate the analysis of synthesis results. Finally, Leonardo can be also used for HDL synthesis on FPGAs.

1.6.3.4 Simulation

- ModelSim [88] is one of the most popular and widely used VHDL and mixed-VHDL/Verilog simulators and the fastest-growing Verilog simulator. ModelSim products are uniquely architected using technology such as Optimized Direct Compile for faster compile times and simulation performance, Single Kernel Simulation (SKS) and Tcl/Tk for greater levels of openness and faster debugging. Exclusive to ModelSim, these innovations result in leading compiler/simulator performance, complete freedom to mix VHDL and Verilog and the unmatched ability to customize the simulator. In addition, with each ModelSim license, designers enjoy Model Technology's ease of use, debugging support, robust quality and technical support.

1.6.4 QuickLogic Development Software

QuickLogic provides support for Windows, Unix, and Web-Based comprehensive design environment ranging from schematic and HDL-based design entry, HDL language editors and tutorials, logic synthesis place and route, timing analysis, and simulation support [89].

1.6.5 Synplicity

- Synplify: The Synplify synthesis solution [90] is a high-performance, sophisticated logic synthesis engine that utilizes proprietary Behavior Extracting Synthesis Technology (B.E.S.T.) to deliver fast, highly efficient FPGA and CPLD designs. The Synplify product takes Verilog and VHDL Hardware Description Languages as input and outputs an optimized netlist in most popular FPGA vendor formats.
- Synplify Pro: Synplify Pro software [91] extends the capability of the Synplify solution to meet the needs of today's complex, high density designs. Team design,

integration of IP, complex project management, graphical FSM debugging, testability and other features are included in the Synplify Pro solution.

- Synplify Premier [92]:
- HDL Analyst: HDL Analyst [93] adds to Synplify the ability to create an RTL block diagram of the design from the HDL source code. A post-mapped schematic diagram is also created that displays timing information for critical paths. Bi-directional cross-probing between all three design views allows to instantly understand exactly what the HDL code produced while dramatically improving debug time.
- Amplify Physical Optimizer: The Amplify Physical Optimizer [94] product is the first and only physical synthesis tool designed specifically for programmable logic designers. By performing simultaneous placement and logic optimization, the Amplify product has demonstrated an average of over 21% performance improvement and over 45% improvement in some cases when compared with logic synthesis alone. Now the Amplify product includes Total Optimization Physical Synthesis (TOPS) technology. This boosts performance further and also reduces design iterations through highly accurate timing estimations. The Amplify Physical Optimizer product was created for programmable logic designers utilizing Altera and Xilinx devices, and who need to converge on aggressive timing goals as quickly as possible. RT Level physical constraints, along with standard timing constraints, are provided to the Amplify product's highly innovative new physical synthesis algorithms, resulting in superior circuit performance in a fraction of the time normally required by traditional methodologies.
- Certify SC: A new member of Synplicity's Certify [95] verification synthesis software family, the Certify SC software is a tool aimed at ASIC and intellectual property (IP) prototyping on a single FPGA, and providing advanced hardware debug capabilities to FPGA designers. Introducing new features targeted at ASIC conversion and debug access, including integration with Xilinx ChipScope debugging tools, the Certify SC software is designed to enable ASIC designers to either prototype IP or portions of ASIC designs onto high-density FPGAs. Additionally, FPGA designers can now take advantage of the advanced debug insertion features of the Certify product as an upgrade option to the Synplify Pro advanced FPGA synthesis solution.

1.6.6 Synopsys

- FPGA Compiler II: By leveraging Synopsys expertise in multimillion-gate ASIC synthesis technology and applying this expertise to FPGA architecture-specific synthesis, FPGA Compiler II [96] provides traditional FPGA or ASIC-like design flows that precisely meet the needs of programmable logic designers while at the same time utilizing an intuitive GUI or scripting mode for design realization.

1.6.7 Altera

1.6.7.1 Quartus II

The Quartus II [97] software provides a complete flow (Fig. 1.32) for creating high-performance system-on-a-programmable-chip (SOPC) designs. It integrates design, synthesis, place-and-route, and verification into a seamless environment, including interfaces to third-party EDA tools.

The standard Quartus II compilation flow consists of the following essential modules:

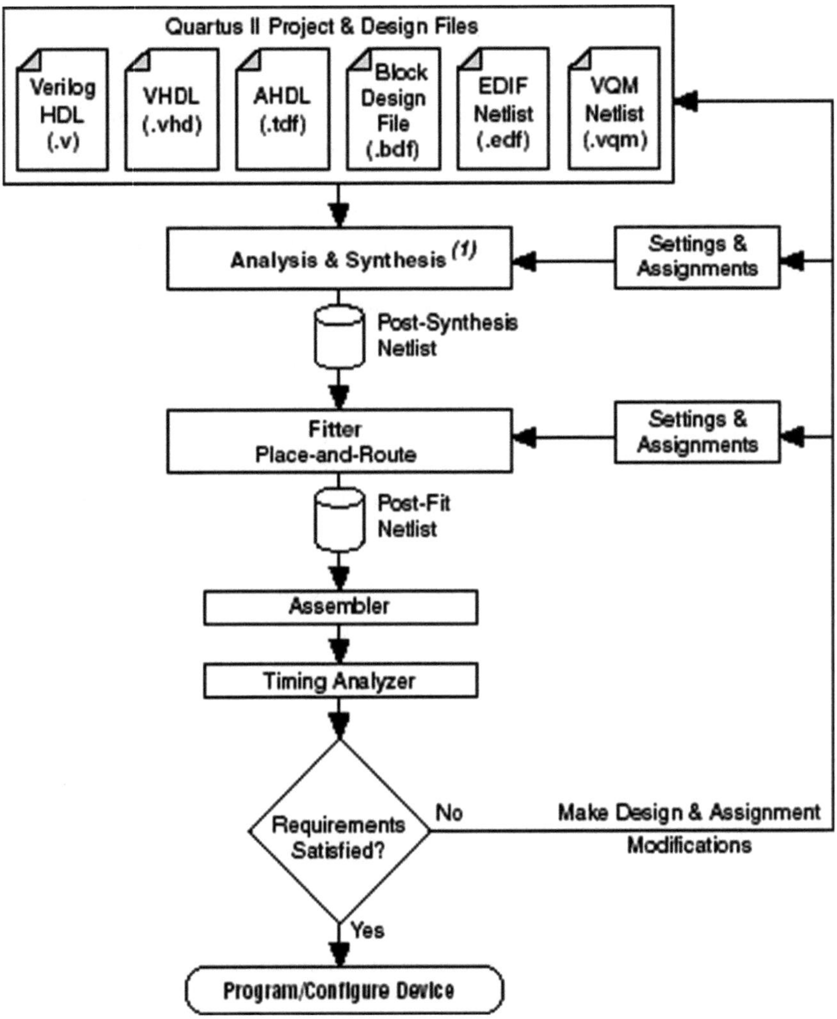

Fig. 1.32 Quartus II standard design flow

- **Analysis & Synthesis**—performs logic synthesis to minimize the design logic and performs technology mapping to implement the design logic using device resources such as logic elements. This stage also generates the project database that integrates the design files (including netlists from third-party synthesis tools).
- **Fitter**—places and routes the logic of a design into a device.
- **Assembler**—converts the Fitter's device, logic, and pin assignments into programming files for the device.
- **Timing Analyzer**—analyzes and validates the timing performance of all the logic in a design.

LogicLock Block-Based Design

LogicLock block-based design is a design methodology available through the Quartus II software. With the LogicLock design flow, the Quartus II software is a programmable logic device (PLD) design software which includes block-based design methodologies as a standard feature, helping to increase designer productivity and shorten design and verification cycles. The LogicLock design flow provides the capability to design and implement each design module independently. Designers can integrate each module into a top-level project while preserving the performance of each module during integration. The LogicLock flow shortens design and verification cycles because each module is optimized only once.

The Quartus II software supports both VHDL and Verilog hardware description language (HDL) text and graphical based design entry methods and combining the two methods in the same project. Using the Quartus II block design editor, top-level design information can be edited in graphical format and converted to VHDL or Verilog for use in third-party synthesis and simulation flows.

NativeLink integration facilitates the inter-operation and seamless transfer of information between the Quartus II software and other EDA tools. It allows third-party synthesis tools to map primitives directly to Altera device primitives. Because primitives are mapped directly, the synthesis tool has control over how the design is mapped to the device. Direct mapping shortens compile times and eliminates the need for extra library mapping translations that could limit performance gains provided by the third-party synthesis tool. The NativeLink flow allows designers to use the Quartus II software pre-place-and-route estimates in third-party EDA tools to optimize synthesis strategies. The Quartus II software can pass post-place-and-route timing information to third-party EDA simulation and timing analysis tools, addressing chip-level and board-level verification issues.

The Quartus II software allows designers to develop and run scripts in the industry-standard tool command language (Tcl). The use of Tcl scripts in the Quartus II software automates compilation flows and makes assignments, automates complex simulation test benches, and creates custom interfaces to third-party tools.

Quartus II Synthesis

The Quartus II design software includes integrated VHDL and Verilog hardware description language (HDL) synthesis technology and NativeLink integration to third-party synthesis software from Mentor Graphics, Synopsys, and Synplicity. Through these close partnerships, Altera offers synthesis support for all its latest device families and support for the latest Quartus II software features in industry-leading third-party synthesis software.

Place & Route

The PowerFit place-and-route technology in the Quartus II design software uses the designer's timing specifications to perform optimal logic mapping and placement. The timing-driven router algorithms in the Quartus II software intelligently prioritize which routing resources are used for each of the design's critical timing paths. Critical timing paths are optimized first to help achieve timing closure faster and deliver faster performance (f_{MAX}). The Quartus II software supports the latest Altera device architectures such as the families described previously in this chapter. This cutting-edge place-and-route technology provides Quartus II software users with superior performance and productivity, including the fastest compile times in the industry. The Quartus II software versions 2.0 and later also include the fast fit compilation option for up to 50% faster compile times.

Quartus II Verification & Simulation

Design verification can be the longest process in developing high-performance system-on-a-programmable-chip (SOPC) designs. Using the Quartus II design software the verification times could be reduced because this high-performance software includes a suite of integrated verification tools which integrate with the latest third-party verification products. The Quartus II Verification Solutions is shown in Table 1.6.

Quartus II Web Edition Software

The Quartus II Web Edition software is an entry-level version of the Quartus II design software supporting selected Altera devices. With PowerFit place-and-route technology, Quartus II Web Edition software lets to experience the performance and compile time benefits of the Quartus II software.

The Quartus II Web Edition software includes a complete environment for programmable logic device (PLD) design including schematic- and text-based design entry, HDL synthesis, place-and-route, verification, and programming.

1.6.8 Xilinx

1.6.8.1 Xilinx ISE

Xilinx ISE [98] is an integrated design environment (tool flow) that supports both Xilinx and third-party tools for the implementation of digital systems on Xilinx devices. The Xilinx flow is similar to other academic and industrial tool flows.

Table 1.6 Quartus II verification solutions

Verification method	Description	Quartus II software support or subscription support	Third-Party support
Design rule checking	Checks designs before synthesis and fitting for coding styles that could cause synthesis, simulation, or design migration problems	Quartus II software to HardCopy device migration design rule checking	Atrenta: SpyGlass Synopsys: Leda
Functional verification	Checks if a design meets functional requirements before fitting	ModelSim-Altera software	Cadence: NC-Verilog, NC-VHDL Mentor Graphics: ModelSim Tool Synopsys: VCS, Scirrocco
Testbench generation	Reduces amount of hand-generated test vectors	-Waveform-to-testbench converter-Testbench template generator	
Static timing analysis	Analyzes, debugs, and validates a design's performance after fitting	Quartus II software static timing analyzer	Synopsys: PrimeTime
Timing simulation	Performs a detailed gate-level timing simulation after fitting	-Quartus II software simulator -ModelSim-Altera software	Cadence: NC-Verilog, NC-VHDL Mentor Graphics: ModelSim Synopsys: VCS, Scirrocco
Hardware/ Software co-simulation	Quickly simulates interaction between PLD hardware, embedded processor, memory, and peripherals	ModelSim-Altera software	ModelSim
In-system verification	Reports behavior of internal nodes in-system and at system speeds	-Quartus II SignalTap II logic analyzer-Quartus II SignalProbe feature	Bridges to silicon

(*Continued*)

Verification method	Description	Quartus II software support or subscription support	Third-Party support
Board-Level timing analysis	Verifies PLD and entire board meets system timing requirements		Innoveda: Blast Mentor Graphics: Tau
Signal integrity analysis & EMC	Verifies that high speed I/O signals will be transmitted reliably and within EMC guidelines	Quartus II software design-specific IBIS model generation	Cadence: SpectraQuest Innoveda: XTK, Hyperlynx Mentor Graphics: Interconnectix
Formal verification	Identifies differences between source register transfer level (RTL) net lists and post place-and-route net lists without the user creating any test vectors		Synopsys: FormalityVerplex: Conformal LEC
Power estimation	Estimates the power consumption of your device using your design's operating characteristics	-Quartus II software simulator-ModelSim-Altera software	Mentor Graphics: ModelSim

Table 1.7 Xilinx ISE features and supported tools

Design Entry	Schematic Editor
	HDL Editor
	State Diagram Editor
	Xilinx CORE Generator™ System™
	RTL & Technology Viewers
	PACE (**P**inout & **A**rea **C**onstraint **E**ditor)
	Architecture Wizards
	3rd Party RTL Checker Support
	Xilinx System Generator for DSP
Embedded system design	Embedded Design Kit (EDK)
Synthesis	XST – Xilinx Synthesis Technology
	Mentor Graphics Leonardo Spectrum
	Mentor Graphics Precision RTL
	Mentor Graphics Precision Physical
	Synopsys DC-FPGA Compiler
	Synplify/Pro/Premier
	Synplicity Amplify Physical Synthesis
	ABEL
Implementation	FloorPlanner
	PlanAhead™
	Timing Driven Place & Route
	Incremental Design
	Timing Improvement Wizard
	Xplorer
Programming	iMPACT / System ACE™ / CableServer
Board-level integration	IBIS, STAMP, and HSPICE** models
	ELDO Models** (MGT only)
Verification	ChipScope™ Pro
	Graphical Testbench Editor
	ISE Simulator Lite
	ISE Simulator
	ModelSim® XE III Starter
	ModelSim XE III
	Static Timing Analyzer
	FPGA Editor with Probe
	ChipViewer
	XPower (Power Analysis)
	3rd Party Equivalence Checking Support
	SMARTModels for PowerPC™ and RocketIO™
	3rd Party Simulator Support

Table 1.7 shows the tools (Xilinx and 3rd party) that can be used in each step of the Xilinx flow.

Design Entry

ISE provides support for today's most popular methods for design capture including HDL and schematic entry, integration of IP cores as well as robust support for reuse of IP. ISE even includes technology called IP Builder, which allows to capture an IP and to reuse it in other designs.

ISE's Architecture Wizards allow easy access to device features like the Digital Clock Manager and Multi-Gigabit I/O technology. ISE also includes a tool called

PACE (Pinout Area Constraint Editor) which includes a front-end pin assignment editor, a design hierarchy browser, and an area constraint editor. By using PACE, designers are able to observe and describe information regarding the connectivity and resource requirements of a design, resource layout of a target FPGA, and the mapping of the design onto the FPGA via location/area.

Synthesis

Synthesis is one of the most essential steps in the design methodology. It takes the conceptual Hardware Description Language (HDL) design definition and generates the logical or physical representation for the targeted silicon device. A state of the art synthesis engine is required to produce highly optimized results with a fast compile and turnaround time. To meet this requirement, the synthesis engine needs to be tightly integrated with the physical implementation tool and have the ability to proactively meet the design timing requirements by driving the placement in the physical device. In addition, cross probing between the physical design report and the HDL design code will further enhance the turnaround time.

Xilinx ISE provides the seamless integration with the leading synthesis engines from Mentor Graphics, Synopsys, and Synplicity. It is possible to use any of the above synthesis engines. In addition, ISE includes Xilinx proprietary synthesis technology, XST. It gives the option to use multiple synthesis engines to obtain the best-optimized result of the programmable logic design.

Implementation & Configuration

Programmable logic design implementation assigns the logic created during design entry and synthesis into specific physical resources of the target device. The term "place and route" has historically been used to describe the implementation process for FPGA devices and "fitting" has been used for CPLDs. Implementation is followed by device configuration, where a bitstream is generated from the physical place and route information and downloaded into the target programmable logic device.

Verification

There are five types of verification available in Xilinx ISE:

- Functional Verification verifies syntax and functionality of a design at the DHL level.
- Gate-Level Verification allows you to directly verify your design at the RTL level after it has been generated by the Synthesis tool.
- Timing Verification is used to verify timing delay in a design ensuring timing specification is met.

- Advanced Verification offers designers different options beyond the traditional verification tools.
- Using Board Level Verification tools ensures your design performs as intended once integrated with the rest of the system.

Advanced Design Techniques

As the FPGA requirements grow, the design problems can change. High-density design environments mean multiple teams working through distributed nodes on the same project, located in different parts of the world, or across the aisle. ISE advanced design options are targeted at making the high-density design as easy to realize as the smallest glue-logic.

- Floorplanner – The Xilinx High-Level Floorplanner is a graphic planning tool that lets to map the design onto the target chip. Floorplanning can efficiently drive the high-density design process.
- Modular Design – The ability to partition a large design into individual modules. Each of those modules can then be floorplanned, designed, implemented, and then locked until the remaining modules are finished.
- Partial Reconfigurability – Partial reconfiguration is useful for applications requiring the loading of different designs into the same area of the device, or the ability to flexibly change portions of a design without having to either reset or completely reconfigure the entire device.
- Incremental Design – By first Area Mapping your design, Incremental Design makes sure that any late design changes don't force a full re-implementation of the chip. Only the area involved in the change must be re-implemented, the rest of the design stays intact.
- High-Level Languages – As design densities increase, the need for a higher-level of abstraction becomes more important. Xilinx is driving and supporting the industry standards and their supporting tools.

Board Level Integration

Xilinx understands the critical issues such as complex board layout, signal integrity, high-speed bus interface, high-performance I/O bandwidth, and electromagnetic interference for system level designers. To ease the system level designers' challenge, ISE provides support to all Xilinx leading FPGA technologies:

- System IO
- XCITE
- Digital clock management for system timing
- EMI control management for electromagnetic interference

ISE WebPACK

ISE WebPACK is a free version of ISE that supports a subset of the Virtex, Virtex-E, Virtex-2/Virtex-2 Pro and Virtex-4 as well as a subset of Spartan II/IIE and Spartan – 3/3E/3L devices.

1.7 Conclusions

This chapter included both and introduction to FPGA technology and an extensive survey of the existing fine-grain reconfigurable architectures from both academia and industry, which indicated both the strengths and limitations of fine-grain reconfigurable hardware. An important consideration in dynamically reconfigurable systems is the reconfiguration latency and power consumption. Various techniques have been employed to reduce the reconfiguration latency, such as prefetching and configuration caching. Prefetch techniques can reduce the reconfiguration latency by allowing pipelining of reconfiguration and execution operations. Prefetching requires knowing beforehand what the next configuration will be, while caching simply requires knowledge of the most common and often required reconfigurations, so they can stored in the configuration cache.

In recent years, increased density has helped integrate coarse-grain elements in FPGAs such as SRAM, dedicated arithmetic units (multipliers etc.) and DLLs, and also a great number of logic gates, making them significant alternatives to ASICs. In fact 75 per cent of the ASICs produced in 2001 could fit in a commercial FPGA, and 60 per cent of them have timing constraints that could be met in an FPGA implementation.

Although fine-grain architectures with building blocks of 1-bit are highly reconfigurable, the systems exhibit low efficiency, when it comes to more specific tasks. An example of this category is if an 8-bit adder is implemented in a fine-grain circuit, it will be inefficient compared to a reconfigurable array of 8-bit adders, when performing an addition-intensive task. In addition to that, an 8-bit adder will also occupy more space in the fine-grain implementation. On the other hand, when a system uses building blocks with more than 1-bit, for example 2-bit, it has a major advantage compared to the 1-bit building blocks. This advantage is that the system could utilize the chip area better, since it is optimized for the specific operations. However, a drawback of this approach is represented in a high overhead when synthesizing operations that are incompatible with the simplest logic block architecture.

References

1. Hsieh, H., W. Carter, J. Y. Ja, E. Cheung, S. Schreifels, C. Erickson, P. Freidin, and L. Tinkey, "Third-generation architecture boosts speed and density of field-programmable gate arrays", in Proc. Custom Integrated Circuits Conf., 1990, pp. 31.2.1–31.2.7.

2. Ahrens, M., A. El Gamal, D. Galbraith, J. Greene, and S. Kaptanoglu, "An FPGA family optimized for high densities and reduced routing delay", in Proc. Custom Integrated Circuits Conf., 1990, pp. 31.5.1–31.5.4.
3. George, V. and J. M. Rabaey, "Low-Energy FPGAs: Architecture and design", Kluwer Academic Publishers, 2001.
4. http://www.actel.com, Accelerator Series FPGAs – ACT3 Family, Actel Corporation, 1997.
5. http://www.actel.com, SX Family of High Performance FPGAs, Actel Corporation, 2001.
6. Butts M. and Batcheller J, "Method of using electronically reconfigurable logic circuits", 1991, US Patent 5,036,473.
7. Hauck S, "The roles of FPGAs in reprogrammable systems", in Proc. IEEE 86, 4, pp. 615–638, 1998.
8. Rose, J., R. J. Francis, D. Lewis, and P. Chow, "Architecture of field-programmable array: The effect of logic block functionality on area efficiency", IEEE Journal of Solid State Circuits, Vol. 25, No. 5, October 1990, pp. 1217–1225.
9. Kouloheris, J. L. and A. El Gamal, "FPGA Performance versus cell granularity", Proceedings of the IEEE custom integrated circuits conference, San Diego, California, 1991, pp. 6.2.1–6.2.4.
10. Singh, S., J. Rose, P. Chow, and D. Lewis, "The effect of logic block architecture on FPGA performance", IEEE Journal of Solid-State Circuits, Vol. 27, no. 3, March 1992, pp. 281–287.
11. He, J. and J. Rose, "Advantages of heterogeneous logic block architecture for FPGAs", Proceedings of the IEEE Custom integrated circuits conference, San Diego, California, 1993, pp. 7.4.1–7.4.5.
12. H. Hsieh, W. Carter, J. Y. Ja, E. Cheung, S. Schreifels, C. Erickson, P. Freidin, and L. Tinkey, "Third-generation architecture boosts speed and density of field-programmable gate arrays", in Proc. Custom Integrated Circuits Conf., 1990, pp. 31.2.1–31.2.7.
13. Trimberger, S. "Effects of FPGA Architecture on FPGA Routing", in Proceedings of the 32nd ACM/IEEE Design Automation Conference (DAC), San Francisco, California, USA, 1995, pp. 574–578.
14. http://aplawrence.com/Makwana/nonvolmem.html
15. Betz, V. and J. Rose, "Cluster-based logic blocks for FPGAs: Area-Efficiency vs. Input sharing and size", IEEE Custom Integrated Circuits Conference, Santa Clara, California, 1997, pp. 551–554.
16. http://www.altera.com/products/devices/stratix_II/features/Stratix II 90 nm Silicon Power Optimization.htm
17. "Reconfigurability requirements for wireless LAN products", Electronic document available at http://www.imec.be/adriatic/deliverables/ec-ist-adriatic_deliverable-D1–1.zip
18. Trimberger, S. D., Carberry, A. Johnson, and J. Wong, "A time-multiplexed FPGA", IEEE Symposium on field-programmable custom computing machines, 1997, pp. 22–28.
19. Schmit, H., "Incremental reconfiguration for pipelined applications", 5th IEEE Symposium on FPGA-Based Custom Computing Machines (FCCM '97), Napa Valley, CA, April 1997, 99 16–18.
20. Hauck, S., "Configuration prefetch for single context reconfigurable coprocessors", ACM/SIGDA International Symposium on Field-Programmable Gate Arrays, 1998.
21. Hauck, S., Z. Li, and E. Schewabe, "Configuration compression for the xilinx XC6200 FPGA", IEEE Symposium on FPGAs for Custom Computing Machines, 1998.
22. Li, K., Z. Compton, J. Cooley, S. Knol, and S. Hauck, "Configuration relocation and defragmentation for run-time reconfigurable computing", IEEE Trans., VLSI System, 2002.
23. Li, Z., K. Compton, and S. Hauck, "Configuration caching techniques for FPGA", IEEE Symposium on FPGAs for Custom Computing Machines, 2000.
24. Hauser, J. R. and J. Wawrzynek, "Garp: A MIPS Processor with a reconfigurable coprocessor", University of California, Berkeley.
25. Wittig, R. D. and P. Chow, "One chip: An FPGA Processor with reconfigurable logic".
26. Hauck, S., T. W. Fry, M. M. Holser, and J. P. Kao, "The chimaera reconfigurable functional unit", IEEE Symposium on field-programmable custom computing machines, pp. 87–96, 1997.

27. Ye, Z. A., A. Moshovos, S. Hauck, and P. Banerjee, "Chimaera: A high-performance architecture with a tightly-coupled reconfigurable function unit".
28. Zhang, H., V. Prabhu, V. George, M. Wan, M. Benes, A. Abnous, and J. M. Rabaey, "A 1V Heterogeneous reconfigurable processor IC for baseband wireless applications", ISCA 2000.
29. George, V., H. Zhang and J. Rabaey, "The design of a low energy FPGA", in Proc. of Int. Symp. on Low Power Electronics and Design (ISLPED) 1999, pp. 18.–193
30. Tau, E., D. Chen, I. Eslick, J. Brown, and A. DeHon, "A first generation DPGA implementation", FPD'95, Third canadian workshop of field-programmable devices, May 29-June 1, 1995, Montreal, Canada.
31. Ebeling, C., G. Borriello, S. A. Hauck, D. Song, and E.A. Walkup, "TRIPTYCH: A new FPGA architecture", in FPGA's, W. Moore and W. Luk, Eds. Abingdon, U.K.L Abingdon, 1991, ch 3.1, pp. 75–90.
32. Borriello, G., C. Ebeling, S. A. Hauck, and S. Burns, "The triptych FPGA architecture", IEEE Trans. VLSI Syst., Vol 3, pp. 491–500, Dec. 1995.
33. Hauck, S., G. Borriello, S. Burns, and C. Ebeling, "MONTAGE: An FPGA for synchronous and asynchronous circuits", in Proc. 2nd Int. Workshop Field-Programmable Logic Applicat., Vienna, Austria, Sept. 1992.
34. Chow, P., S. O. Seo, D. Au, T. Choy, B. Fallah, D. Lewis, C. Li, and J. Rose, "A 1.2 μm CMOS FPGA using cascaded logic blocks and segmented routing", in FPGA's W. Moore and W. Luk, Eds. Abingdon, U.K.: Abingdon, 1991, ch 3.2, pp. 91–102.
35. George, V. and J. M. Rabaey, "Low-Energy FPGAs: Architecture and design", Kluwer Academic Publishers, 2001.
36. Chiricescu, S., M. Leeser, and M. M. Vai, "Design and analysis of a dynamically reconfigurable three-dimensional FPGA", IEEE Transactions on Very Large Scale Integration (VLSI) Systems, Vol. 9, No. 1, February 2001.
37. Chow, P., S. O. Seo, J. Rose, K. Chung, G. Paez-Monzon, and I. Rahardja, "The design of a SRAM-Based field-programmable gate Array-Part I: Architecture", IEEE Transactions on Very Large Scale Integration (VLSI) Systems, Vol. 7, No. 2, June 1999.
38. Chow, P., S. O. Seo, J. Rose, K. Chung, G. Paez-Monzon, and I. Rahardja, "The design of a SRAM-Based field-programmable gate Array-Part II: Circuit design and layout", IEEE Transactions on Very Large Scale Integration (VLSI) Systems, Vol. 7, No. 3, September 1999.
39. http://www.xilinx.com/products/silicon_solutions/fpgas/virtex/virtex5/index.htm
40. http://www.xilinx.com/products/silicon_solutions/fpgas/virtex/virtex4/index.htm
41. http://www.xilinx.com/products/silicon_solutions/fpgas/spartan_series/spartan3_fpgas/index.htm
42. http://www.xilinx.com/products/silicon_solutions/fpgas/spartan_series/spartan3l_fpgas/index.htm
43. http://www.altera.com/products/devices/stratix2/st2-index.jsp
44. http://www.altera.com/products/devices/cyclone2/cy2-index.jsp
45. http://www.altera.com/products/devices/cyclone/cyc-index.jsp
46. http://www.altera.com/literature/ds/ds_stx.pdf
47. http://www.actel.com/products/fusion/
48. http://www.actel.com/products/pa3/index.aspx
49. http://www.actel.com/products/proasicplus/index.html
50. http://www.actel.com/docs/datasheets/AXDS.pdf
51. http://www.actel.com/varicore/support/docs/VariCoreEPGADS.pdf
52. http://www.actel.com/docs/datasheets/MXDS.pdf
53. http://www.atmel.com/atmel/acrobat/doc0264.pdf
54. http://www.quicklogic.com/
55. http://www.quicklogic.com/
56. http://www.lattice.com/
57. http://www.lattice.com/
58. http://ballade.cs.ucla.edu/~trio/
59. http://ballade.cs.ucla.edu/software_release/rasp/htdocs/
60. http://ballade.cs.ucla.edu/software_release/ipem/htdocs/

61. Cha, Y-J., C. S. Rim, and K. Nakajima, "A simple and effective greedy multilayer router for MCMS", Proceedings of the International Symposium on Physical Design, Napa Valley, California, United States, 1997.
62. http://cadlab.cs.ucla.edu/~xfpga/fpgaEva/index.html
63. http://www.eecg.toronto.edu/~jayar/software/psac/psac.html
64. http://www.eecg.toronto.edu/~jayar/software/edif2blif/edif2blif.html
65. Electronic document available at http://www.eecg.toronto.edu/~lemieux/sega/sega.html
66. Electronic document available at ftp://ftp.eecg.toronto.edu/pub/software/pgaroute/
67. http://www.eecg.toronto.edu/EECG/RESEARCH/tmcc/tmcc/
68. Electronic document available at ftp://ftp.eecg.toronto.edu/pub/software/Chortle/
69. Electronic document available at http://www.eecg.toronto.edu/~vaughn/vpr/vpr.html
70. Kara K., W. Poon, A. Yan, and S. J. E. Wilton, "A flexible power model for FPGAs", 12th International Conference, FPL 2002 Montpellier, France, September 2002.
71. http://www.actel.com/download/software/libero
72. http://www.actel.com/products/software/designer
73. http://www.embeddedstar.com/weblog/2006/08/22/actel-silicon-sculptor-3-fpga-tool/
74. http://www.actel.com/documents
75. http://www.actel.com/documents/SiExIIpib.pdf
76. http://www.cadence.com/company/newsroom/press_releases/pr.aspx?xml=013101_SPW
77. www.cadence.com/whitepapers/FPGA_Dev_Using_NC-Sim.pdf.
78. http://www.orcad.com
79. http://www.cadence.com/datasheets/4492C_IncisiveVerilog_DSfnl.pdf
80. http://www.mentor.com/products/fpga_pld/fpga_advantage/index.cfm
81. http://www.mentor.com/products/fpga_pld/hdl_design/hdl_designer_series/
82. http://www.embeddedstar.com/software/content/m/embedded239.html
83. http://www.mentor.com/products/fpga_pld/hdl_design/hdl_detective/
84. http://www.mentor.com/products/fpga_pld/hdl_design/hdl_author/
85. http://www.mentor.com/products/fpga_pld/news/hds2002_1_pr.cfm
86. http://www.mentor.com/products/fpga_pld/synthesis/
87. http://www.mentor.com/products/fpga_pld/synthesis/leonardo_spectrum/
88. http://www.model.com
89. http://www.quicklogic.com
90. http://www.embeddedstar.com/software/content/s/embedded382.html
91. http://www.fpgajournal.com/news_2006/04/20060411_01.htm
92. http://www.synplicity.com/products/synplifypremier/index.html
93. http://www.synplicity.com/literature/pdf/hdl_analyst_1103.pdf
94. http://www.synplicity.com/corporate/pressreleases/2003/SYB-207final.html
95. http://www.fpgajournal.com/news_2007/01/20070129_06.htm
96. http://www.synopsis.com
97. http://www.altera.com/products/software/products/quartus2/qts-index.html
98. http://www.xilinx.com

Chapter 2
A Survey of Coarse-Grain Reconfigurable Architectures and Cad Tools

Basic Definitions, Critical Design Issues and Existing Coarse-grain Reocnfigurable Systems

G. Theodoridis, D. Soudris, and S. Vassiliadis

Abstract According to the granularity of configuration, reconfigurable systems are classified in two categories, which are the fine- and coarse-grain ones. The purpose of this chapter is to study the features of coarse-grain reconfigurable systems, to examine their advantages and disadvantages, to discuss critical design issues that must be addressed during their development, and to present representative coarse-grain reconfigurable systems that have been proposed in the literature.

Key words: Coarse-grain reconfigurable systems/architectures · design issues of coarse-grain reconfigurable systems · mapping/compilation methods · reconfiguration mechanisms

2.1 Introduction

Reconfigurable systems have been introduced to fill the gap between Application Specific Circuits (ASICs) and micro-processors (μPs) aiming at meeting the multiple and diverse demands of current and future applications.

As the functionality of the employed Processing Elements (PEs) and the interconnections among PEs can be reconfigured in the field, special-purpose circuits can be implemented to satisfy the requirements of applications in terms of performance, area, and power consumption. Also, due to the inherent reconfiguration property, flexibility is offered that allows the hardware to be reused in many applications, avoiding the manufacturing cost and delay. Hence, reconfigurable systems are an attractive alternative proposal to satisfy the multiple, diverse, and rapidly-changed requirements of current and future applications with reduced cost and short time-to-market. Based on the granularity of reconfiguration, reconfigurable systems are classified in two categories, which are the fine- and the coarse-grain ones [1]–[8].

A fine-grain reconfigurable system consists of PEs and interconnections that are configured at bit-level. As the PEs implement any 1-bit logic function and rich

S. Vassiliadis, D. Soudris (eds.), *Fine- and Coarse-Grain Reconfigurable Computing.*

interconnection resources exist to realize the communication links between PEs, fine-grain systems provide high flexibility and can be used to implement theoretically any digital circuit. However, due to fine-grain configuration, these systems exhibit low/medium performance, high configuration overhead, and poor area utilization, which become pronounced when, they are used to implement processing units and datapaths that perform word-level data processing.

On the other hand, a coarse-grain reconfigurable system consists of reconfigurable PEs that implements word-level operations and special-purpose interconnections retaining enough flexibility for mapping different applications onto the system. In these systems the reconfiguration of PEs and interconnections is performed at word-level. Due to their coarse-grain granularity, when they are used to implement word-level operators and datapaths, coarse-grain reconfigurable systems offer higher performance, reduced reconfiguration overhead, better area utilization, and lower power consumption than the fine-grain ones [9].

In this chapter we are dealing with the coarse-grain reconfigurable systems. The purpose of the chapter is to study the features of these systems, to discuss their advantages and limitations, to examine the specific issues that should be addressed during their development, and to describe representative coarse-grain reconfigurable systems. The fine-grain reconfigurable systems are described in detailed manner by the Chapter 1.

The chapter is organized as follows: In Section 2.2, we examine the needs and features of modern applications and the design goals to meet the applications' needs. In Section 2.3, we present the fine- and coarse-grain reconfigurable systems and discuss their advantages and drawbacks. Section 2.4 deals with the design issues related with the development of a coarse-grain reconfigurable system, while Section 2.5 is dedicated to a design methodology for developing coarse-grain reconfigurable systems. In Section 2.6, we present representative coarse-gain reconfigurable systems. Finally, conclusions are given in Section 7

2.2 Requirements, Features, and Design Goals of Modern Applications

2.2.1 Requirements and Features of Modern Applications

Current and future applications are characterized by different features and demands, which increase the complexity of developing systems to implement them. The majority of contemporary applications, for instance DSP or multimedia ones, are characterized by the existence of computationally-intensive algorithms. Also, high speed and throughput are frequently needed since real-time applications (e.g. video conferencing) are widely-supported by modern systems. Moreover, due to the wide spread of portable devices (e.g. laptops, mobile phones), low-power consumption becomes an emergency need. In addition, electronic systems, for instance, consumer electronics may have strict size constraints, which make the silicon area a critical design issue. Consequently, *the development of special-purpose circuits/systems in needed to meet the above design requirements.*

However, apart from the circuit specialization, systems must also exhibit flexibility. As the needs of the customers change rapidly and new standards appear, systems must be flexible enough to satisfy the new requirements. Also, flexibility is required to support possible bug fixes after the system's fabrication. These can be achieved by changing (reconfiguring) the functionality of the system in the field, according to the needs of each application. In that way, the same system can be reused in many applications, its lifetime in the market increases, while the development time and cost are reduced. However, the reconfiguration of the system must be accomplished without introducing large penalties in terms of performance. Consequently, *it is demanded the development of flexible systems that can be reconfigured in the field and reused in many applications.*

Besides the above, there are additional features that should be considered and exploited, when a c*ertain application domain* is considered. Thanks to 90/10 rule, it is known that for a given application domain a small portion of each application (about 10 %) accounts for a large fraction of execution time and energy consumption (about 90 %). These computationally-intensive parts are, usually, called kernels and exhibit regularity and repetitive execution. Typical example of a kernel is the nested loops of DSP applications. Moreover, the majority of the kernels perform word-level processing on data with wordlength greater than one bit (usually 8- or 16-bit).

Kernels also exhibit similarity, which is observed in many abstraction levels. In lower abstraction levels similarity appears as common performed operations. For instance, in multimedia kernels apart from the basic logical and arithmetic operations, there are also more complex operations such as multiple-accumulate, add-compare-select, and memory addressing calculations, which are frequently appear. In higher abstraction levels, a set of functions also appear as building modules in many algorithms. Typical examples are the FFT, DCT, FIR, and IIR filters in DSP applications. Depending on the considered domain, additional features may exist such as locality of references and inherent parallelism that should be also taken into account during the development of the system.

Summarizing, the applications demand special-purpose circuits to satisfy performance, power consumption, and area constraints. They also demand flexible systems meeting the rapidly-changed requirements of customers and applications, increasing the lifetime of the system in the market, and reducing design time and cost. When a certain application domain is targeted, there are special features that must be considered and exploited. Specifically, the number of computationally-intensive kernels is small, word level processing is performed, and the computations exhibit similarity, regularity, and repetitive execution.

2.2.2 Design Goals

Concerning the two major requirements of modern and future applications, namely the circuit specialization and the flexibility, two conventional approaches exist to satisfy them, which are the ASIC- and μP-based approach. However, none of them can satisfy both these requirements optimally. Due to their special-purpose nature, ASICs offer high performance, small area, and low energy consumption, but they are not flexible enough as applications demand. On the other hand, μP-based solutions offer the maximal flexibility since the employed μP(s) can be programmed and used in many applications. Comparing ASIC- and μP-based solutions, the latter ones suffer from lower performance and higher power consumption because μP(s) are general-purpose circuits.

What actually is needed is *a trade off between flexibility and circuit specialization.* Although flexibility can be achieved via processor programming, when rigid time, or power consumption constraints have to be met, this solution is prohibitive due to the general-purpose nature of these circuits. Hence, we have to develop new systems of which we can change the functionality of the hardware in the field according to the needs of the application meeting in that way the requirements of circuitry specialization and flexibility. To achieve this we need PEs that can be reconfigured to implement a set of logical and arithmetic operations (ideally any arithmetic/logical operation). Also, we need programmable interconnections to realize the required communication channels among PEs [1], [2].

Although Field Programmable Gate Arrays (FPGAs) can be used to implement any logic function, due to their fine-grain reconfiguration (the underlying PEs and interconnections are configured at bit-level), they suffer by large reconfiguration time and routing overhead, which becomes more profound when they are used to implement word-level processing units and datapaths [4]. To build a coarse-grain unit, a number of PEs must be configured individually to implement the required functionality at bit-level, while the interconnections among the PEs, must be also programmed individually at bit-level. This increases the number of configuration signals that must be applied. Since reconfiguration is performed by downloading the values of the reconfiguration signals from the memory, the reconfiguration time increases, while large memories are demanded for storing the data of each reconfiguration. Also, as a large number of programmable switches are used for configuration purposes, the performance is reduced and the power consumption increases. Finally, FPGAs exhibit poor area utilization as in many times the area that is spent for routing is by far larger than the area used for logic [4]–[6]. We will discus the FPGAs and their advantages and shortcomings in more details in a next section.

To overcome the limitations imposed by the fine-grain reconfigurable systems, new architectures must be developed. When word-level processing is required, this can be accomplished by developing architectures that support coarse-grain reconfiguration. Such architecture consists of optimally-designed coarse-grain PEs, which perform word-level data processing and can be configured at word-level, and proper interconnections that are also configured at word-level. Due to the word-level reconfiguration, a small number of configuration bits is required resulting into a massive reduction of configuration data, memory needs, and reconfiguration time. For a coarse-grain reconfigurable unit we do not need to configure each slice of the unit individually at bit-level. Instead, using few configuration (control) bits, the functionality of the unit can be determined based on a set of predefined operations that the unit supports. The same also holds for interconnections, since they are grouped in buses and configured by a single control signal instead of using separate control signals for each wire, as it happens in fine-grain systems. Also, because few programmable switches are used for configuration purposes and the PEs are optimally-designed hardwired units; high performance, small area, and low power consumption are achieved.

The development of universal coarse-grain architecture to be used in any application is an unrealistic goal. A huge amount of PEs to execute any possible operation must be developed. Also, a reconfigurable interconnection network realizing any communication pattern between the processing units must be built. However, if we focus on a *specific application domain* and exploit its special features, the design of coarse-grain reconfigurable systems remains a challenging problem but it becomes manageable and realistic. As it was mentioned, when a certain application domain is considered, the number of computationally-intensive kernels is small and perform similar functions. Therefore, the number of PEs and the interconnections required to implement these kernels is not be so large. In addition, as we target to a specific domain, the kernels are known in advance or they can be derived after profiling representative applications of the considered domain. Also, any additional property of the domain such as the inherent parallelism and regularity, which is appeared in the dominant kernels, must be taken into account. However, as PEs and interconnections are designed for a specific application domain, only circuits and kernels/algorithms of the considered domain can be implemented optimally.

Taking into account the above, *the primary design objective is to develop application domain-specific coarse-grain reconfigurable architectures,* which achieve high performance and energy efficiency approaching those of ASICs, while retain adequate flexibility, as they can be reconfigured to implement the dominant kernels of the considered application domain. In that way, executing the computationally-intensive kernels on such architectures, we meet the requirements of circuitry specialization and flexibility for the target domain. The remaining non-computationally intensive parts of the applications may executed by a μP, which is also responsible for controlling and configuring the reconfigurable architecture.

In more details, the *goal is to develop application domain-specific coarse-grain reconfigurable systems with the following features*:

- The dominant kernels are executed by optimally-designed hardwired coarse-grain reconfigurable PEs.
- The reconfiguration of interconnections is done at word-level, while they must be flexible and rich enough to ensure the communication patterns required interconnecting the employed PEs.
- The reconfiguration of PEs and interconnections must be accomplished with the minimal time, memory requirements, and energy overhead.
- A good matching between architectural parameters and applications' properties must exist. For instance, in DSP the computationally-intensive kernels exhibit similarity, regularity, repetitive execution, and high inherent parallelism that must be considered and exploited.
- The number and type of resources (PEs and interconnections) depend on the application domain but benefits form the fact that the dominant kernels are not too many and exhibit similarity.

- A methodology for deriving such architectures supported by tools for mapping applications onto the generated architectures is required.

For the shake of completeness, we start the next section with a brief description of fine-grain reconfigurable systems and discuss their advantages and limitations. Afterwards, we will discuss the coarse-grain reconfigurable systems in details.

2.3 Features of Fine- and Coarse-Grain Reconfigurable Systems

A reconfigurable system includes a set of programmable processing units called *reconfigurable logic*, which can be reconfigured in the filed to implement logic operations or functions, and programmable interconnections called *reconfigurable fabric*. The reconfiguration is achieved by downloading from a memory a set of configuration bits called *configuration context*, which determines the functionality of reconfigurable logic and fabric. The time needed to configure the whole system is called *reconfiguration time,* while the memory required for storing the reconfiguration data called *context memory*. Both the reconfiguration time and context memory constitute the *reconfiguration overhead*.

2.3.1 Fine-Grain Reconfigurable Systems

Fine-grain reconfigurable systems are those systems that both reconfigurable logic and fabric are configured at bit-level. The FPGAs and CPLDs are the most representative fine-grain reconfigurable systems. In the following paragraphs we focus on FPGAs but the same also holds for CPLDs.

2.3.1.1 Architecture Description

A typical FPGA architecture is shown in Fig. 2.1. It is consists of a 2-D array of Computational Logic Blocks (CLBs) used to implement combinational and sequential logic. Each CLB typically contains two or four identical programmable slices. Each slice usually contains two programmable cores with few inputs (typically four inputs) that can be programmed to implement 1-bit logic function. Also, programmable interconnects surround CLBs ensuring the communication between them, while programmable I/O cells surround the array to communicate with the environment. Finally, specific I/O ports are employed to download the reconfiguration data from the context memory. Regarding with the interconnections between CLBs, either direct connections via programmable switches or a mesh structure using Switch Boxes (S-Box) can be used. Each S-Box contains a number of programmable switches (e.g. pass transistor) to realize the required interconnections between the input and output wires.

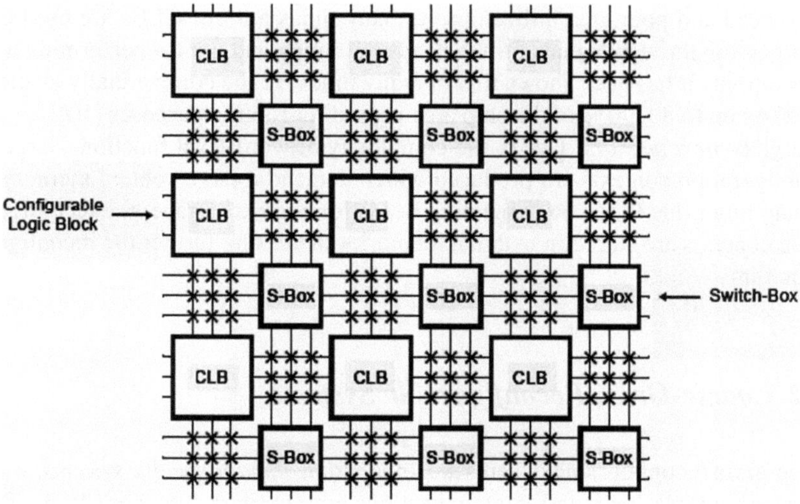

Fig. 2.1 A typical FPGA architecture

2.3.1.2 Features

Since each CLB implements any 1-bit logic function and the interconnection network provides a rich connectivity between CLBs, FPGAs can be treated as general-purpose reconfigurable circuits to implement control and datapath units. Although, some FPGA manufactures developed devices such as Virtex 4 and Stratix, which contain coarse-grain units (e.g. multipliers, memories or processor cores), they are still fine-grain and general-purpose reconfigurable devices. Also, as FPGAs are used for more two decades, mature and robust commercial CAD frameworks are developed for physical implementation of an application onto the device starting from an HDL description and ending up to placement and routing onto the device.

However, due to their fine-grain configuration and general-purpose nature, fine-grain reconfigurable systems suffer by a number of drawbacks, which become more pronounced when they are used to implement word-level units and datapaths [9]. These drawbacks are discussed in the following.

- *Low performance and high power consumption.* This happens because word-level modules are built by connecting a number of CLBs using a large number of programmable switches causing performance degradation and power consumption increase.
- *Large context and configuration time.* The configuration of CLBs and interconnections wires is performed at bit-level by applying individual configuration signals for each CLB and wire. This results in a large configuration context that have to be downloaded from the context memory and consequently in large configuration time. The large reconfiguration time may degrade performance when multiple and frequently-occurred reconfigurations are required.
- *Huge routing overhead and poor area utilization.* To build a word-level unit or datapth a large number of CLBs must be interconnected resulting in huge routing

overhead and poor area utilization. In many times a lot of CLBs are used only for passing through signals for the needs of routing and not for performing logic operations. It has been shown that in many times for the commercially available FPGAs up to 80–90 % of the chip area is used for routing purposes [10].

- *Large context memory*. Due to the complexity of word-level functions, large reconfiguration contexts are produced which demand a large context memory. In many times due to the large memory needs for context storage, the reconfiguration contexts are stored in external memories increasing further the reconfiguration time.

2.3.2 Coarse-Grain Reconfigurable Systems

Coarse-grain reconfigurable systems are *application domain-specific systems*, whose the reconfigurable logic and interconnections are configured at word-level. They consist of programmable hardwired coarse-grain PEs which support a predefined set of word-level operations, while the interconnection network is based on the needs of the circuits of the specific domain.

2.3.2.1 Architecture Description

A generic architecture of a coarse-grain reconfigurable system is illustrated in Fig. 2.2. It encompasses a set of Coarse-Grain Reconfigurable Units (CGRUs), a programmable interconnection network, a configuration memory, and a controller. The coarse-grain reconfigurable part undertakes the computationally-intensive parts of the application, while the main processor is responsible for the remaining parts. Without loss of generality, we will use this generic architecture to present the basic

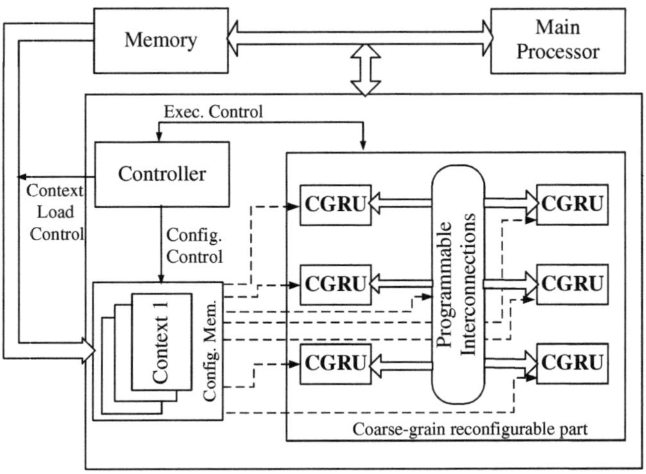

Fig. 2.2 A Generic Coarse-Grain Reconfigurable System

concepts and discuss the features of coarse-grain reconfigurable systems. Considering the target application domain and design goals, the type, number, and organization of the CGRUs, the interconnection network, the reconfiguration memory, and the controller are tailored to the domain's needs and an instantiation of the architecture is obtained.

The CGRUs and interconnections are programmed by proper configuration (control) bits that are stored in configuration memory. The configuration memory may store one or multiple configuration contexts, but each time one context is active. The controller is responsible to control the loading of configuration context from the main memory to configuration memory, to monitor the execution process of the reconfigurable hardware and to activate the reconfiguration contexts. In many cases the main processor undertakes the operations that are performed by the controller.

Concerning the interconnection network, it consists of programmable interconnections that ensure the communication among CGRUs. The wires are grouped in buses each of which is configured by a single configuration bit instead of applying individual configuration bits for each wire as it happens in fine-grain systems. The interconnection network can be realized by a crossbar, mesh or a mesh variation structure.

Regarding with processing units, each unit *is a domain-specific hardwired Coarse-Grain Reconfigurable Unit (CGRU) that executes a useful operation autonomously.*

By the term *useful operation* we mean a logical or arithmetic operation required by the considered domain. The term *autonomously* means that the CGRU can execute by itself the required operation(s). In other words, the CGRU does not need any other primitive resource for implementing the operation(s). In contrary, in fine-grain reconfigurable systems the PEs (CLBs) are treated as primitive resources because a number of them are configured and combined to implement the desired operation.

By the term *coarse-grain reconfigurable unit* we mean that the unit is configured at word level. The configuration bits are applied to configure the entire unit and not each slice individually at bit level. Theoretically, the granularity of the unit may range from 1-bit, if it is the granularity of the useful operation, to any word length. However, in practice the majority of applications perform processing on data with the word-length greater or equal to 8-bits. Consequently, the granularity of a CGRU is usually greater or equal of 8-bits.

The term *domain-specific* is referred to the functionality of CGRU. A CGRU can be designed to perform any word-level arithmetic or logical operations. As, coarse-grain reconfigurable systems target at a specific domain, the CGRU is designed having in mind the operations required by the domain. Finally, the CGRUs are physically implemented as *hardwired* units. Because they are special-purpose units developed to implement the operations of a given domain, they are usually implemented as hardwired units to improve performance, area, and power consumption.

2.3.2.2 Features

Considering the above, coarse-grain reconfigurable systems are characterized by the following features:

- *Small configuration contexts*. The CGRUs need a few configuration bits, which are order of magnitude less than those required if FPGAs were used to implement the same operations. Also, a few configuration bits are needed to establish the interconnections among CGRUs because the interconnection wires are also configured at word level
- *Reduced reconfiguration time*. Due to the small configuration context, the reconfiguration time is reduced. This permits coarse-grain reconfigurable systems to be used in applications that demand multiple and run-time reconfigurations.
- *Reduced context memory size*. Due to the reduction of configuration contexts, the context memory size reduces. This allows using on-chips memories which permits switching from one configuration to another with low configuration overhead.
- *High performance* and *low power consumption*. This stems from the hardwired implementation of CGRUs and the optimally design of interconnections for the target domain.
- *Silicon area efficiency and reduced routing overhead*. This comes from the fact that the CGRUs are optimally-designed hardwired units which are not built by combing a number of CLBs and interconnection wires resulting in reduced routing overhead and better area utilization

However, as the use of coarse-grain reconfigurable systems is new computing paradigm, new methodologies and design frameworks for design space exploration and application mapping on these systems are demanded. In the following sections we discuss the design issues related with the development of coarse-grain reconfigurable systems.

2.4 Design Issues of Coarse-Grain Reconfigurable Systems

As mentioned, the development of a reconfigurable system is characterized by a trade off between flexibility and circuit specialization. We start by defining flexibility and then we discuss issues related to flexibility. Afterwards, we study in details the design issues for developing coarse-grain reconfigurable systems.

2.4.1 Flexibility Issues

By the term flexibility, it is meant the capability of the system to adapt and respond to the new requirements of the applications implementing circuits and algorithms that were not considered during the system's development.

To address flexibility two issues should be examined. The first issue is how flexibility is measured, while the second one is how the system must be designed to achieve a certain degree of flexibility supporting future applications, functionality upgrades, and bug fixes after its fabrication. After studying these issues, we present a classification of coarse-grain reconfigurable systems according to the provided flexibility.

2.4.1.1 Flexibility Measurement

If a large enough set of circuits from a user's domain is available, the measurement of flexibility is simple. A set of representative circuits of the considered application domain is provided to the design tools, the architecture is generated, and then the flexibility of the architecture is measured by testing how many of the domain members are efficiently mapped onto that system. However, in many cases we have not enough representative circuits for this purpose. Also, as reconfigurable systems are developed to be reused for implementing future applications, we have to further examine whether the system can be used to realize new applications. Specifically, we need to examine whether some design decisions, which are proper for implementing the current applications, affect the implementation of future applications, which may have different properties than the current ones.

One solution to measure flexibility is to use synthetic circuits [11], [12]. It is based on techniques that examine a set of real circuits and generate new ones with similar properties [13]–[16]. They profile the initial circuits for basic properties such as type of logic, fanout, logic depth, number and type of interconnections, etc., and use graph construction techniques to create new circuits with similar characteristics. Then, the generated (synthetic) circuits can be used as a large set of example circuits to evaluate the flexibility of the architecture. This is accomplished by mapping the synthetic circuits onto the system and evaluating its efficiency to implement those circuits. However, the use of synthetic circuits as testing circuits may be dangerous. Since the synthetic circuits mimic some properties of the real circuits, it is possible some unmeasured but critical feature(s) of the real circuits may be lost. The correct approach is to generate the architecture using synthetic circuits and to measure the flexibility and efficiency of the generated architecture with real designs taken from the targeted application domain [11]. These two approaches are shown in Fig. 2.3.

Moreover, the use of synthetic circuits for generating architectures and evaluating their flexibility, offers an additional opportunity. We can manipulate the settings of the synthetic circuits' generator to check the sensitivity of the architecture for a

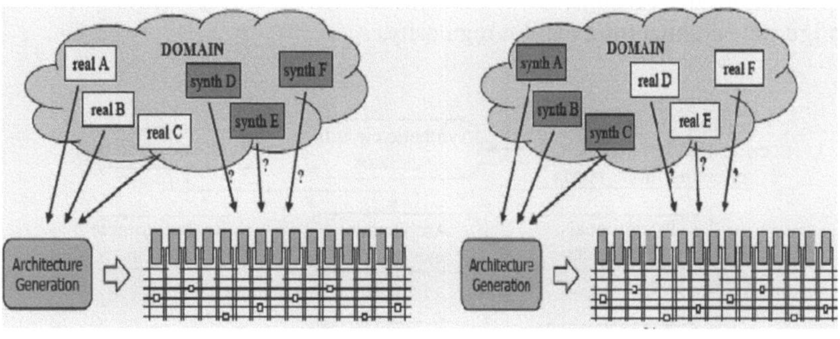

(a) (b)

Fig. 2.3 Flexibility measurement. (a) Use synthetic circuits for flexibility measurement. (b) Use real circuits for flexibility measurement [11]

number of design parameters [11], [12]. For instance, the designer may be concerned that future designs may have less locality and wants to examine whether a parameter of the architecture, for instance, the interconnection network is sensitive to this. To test this, the synthetic circuit generator can be fed benchmark statistics with artificially low values of locality, which reflects the needs of future circuits. If the generated architecture can support the current designs (with the current values of locality), this gives confidence that the architecture can also support future circuits with low locality. Figure 2.4 demonstrates how synthetic circuits can be used to evaluate the sensitivity of the architecture on critical design parameters.

2.4.1.2 Flexibility Enhancement

A major question that arises during the development of a coarse-grain reconfigurable system is how the system is designed to provide enough flexibility to implement new applications. The simplest and more efficient way in terms of area for implementing a set of multiple circuits is to generate an architecture that can be reconfigured realizing only these circuits. Such a system consists of processing units which perform only the required operations and are placed where-ever-needed, while special interconnections with limited programmability exist to interconnect the processing units. We call these systems application class-specific systems and discuss them in the following section. Unfortunately, such a so highly optimized, custom, and irregular architecture is able to implement only the set of applications for which it has been designed. Even slight modifications or bug fixes on the circuits used to generate the architecture are unlikely to fit.

To overcome the above limitations the architecture must characterized by *generality and regularity*. By *generality* it is meant that the architecture must not contain only the required number and types of processing units and interconnections for implementing a class of applications. It must also include additional resources that may be useful for future needs. Also, the architecture must exhibit *regularity* which means that the resources (reconfigurable units and interconnections) must be organized in regular structures. It must be stressed that the need of regular structures also stems from the fact that the dominant kernels, which are implemented by the reconfigurable architecture, exhibit regularity.

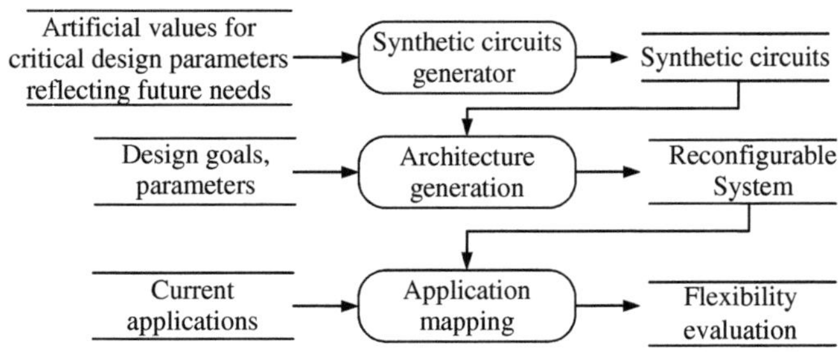

Fig. 2.4 Use of synthetic circuits and flexibility measurement to evaluate architecture's sensitivity on critical design parameters

Therefore, the enhancement of flexibility of the system can be achieved by developing the architecture using patterns of processing units and interconnections which are characterized by generality and regularity. Thus, instead of putting down individual units and wires, it is more preferable to select resources from a set of regular and flexible patterns and repeat them in the architecture. In that way, although extra resources and area are spent, due to regular and flexible structure of the patterns, the employed units and wires are more likely to be reused for new circuits and applications. Furthermore, the use of regular patterns makes the architecture scalable allowing adding extra resources easily.

For illustrations purposes Fig. 2.5 shows how a 1-D reconfigurable system is built using a single regular pattern. The pattern includes a set of basic processing units and a rich programmable interconnection network to enhance its generality. The resources are organized in a regular structure (1-D array) and the pattern is repeated building the reconfigurable system. In more complex cases different patterns may also be used. The number and types of the units and interconnections is critical design issues that affect the efficiency of the architecture. We discuss these issues in Section 2.4.2.

2.4.1.3 Flexibility-Based Classification of Coarse-Grain Reconfigurable Systems

According to flexibility coarse-grain reconfigurable systems can be classified in two categories, which are the application domain-specific and application class-specific systems.

An *Application Domain-Specific System* (ADSS) targets at implementing the applications of a certain application domain. It consists of proper CGRUs and reconfigurable interconnections, which are based on domain's needs, properly organized

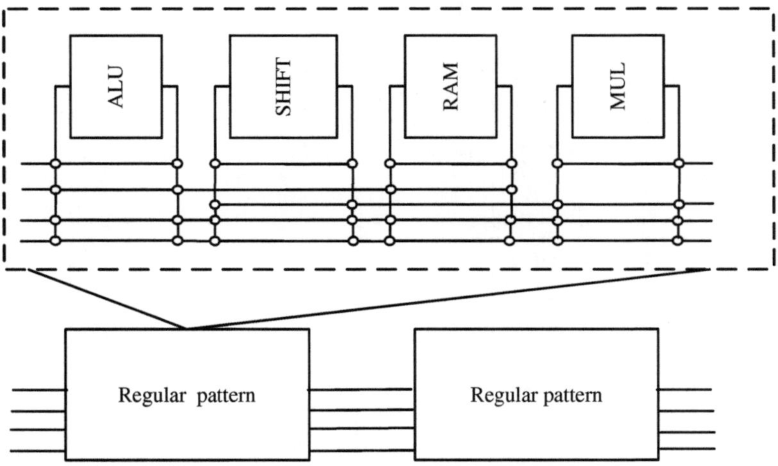

Fig. 2.5 Use of regular patterns of resources to enhance flexibility. Circles denote programmable interconnections

to retain flexibility for implementing efficiently the required circuits. The benefit of such system is its generality as it can be used to implement any circuit and application of the domain. However, due to the offered high flexibility, the complexity of designing such architecture increases. A lot of issues such as the type and amount of employed CGRUs and interconnections, the occupied area, the achieved performance, and power consumption must be concerned and balanced. The vast majority of the existing coarse-grain reconfigurable systems belong to this category. For illustration purposes the architecture of Montium [17], which targets at DSP applications, is shown in Fig. 2.6. It consists of a Tile Processor (TP) that includes five ALUs, memories, register files and crossbar interconnections organized in a regular structure to enhance its flexibility. Based on the demands of the applications and the targeted goals (e.g. performance) a number of TPs can be used.

On the other hand, *Application Class-Specific Systems* (ACSSs) are flexible ASIC-like architectures that have been developed to *support only a predefined set of applications* having limited reconfigurability. In fact the can be configured to implement only the considered set of applications and not all the applications of the domain. They consist of specific types and number of processing units and particular direct point-to-point interconnections with limited programmability. The reconfiguration is achieved by applying different configuration signals to the processing and programmable interconnections at each cycle according to the CDFG of the implemented kernels. An example of such architecture is shown in Fig. 2.7. A certain amount of CGRUs are used, while point-to-point and few programmable interconnections exist.

Although, ACSSs do not meet fully one of the fundamental properties of reconfigurable systems, namely the capability to support functionality upgrades and future applications, they offer many advantages. As they are been designed to im-

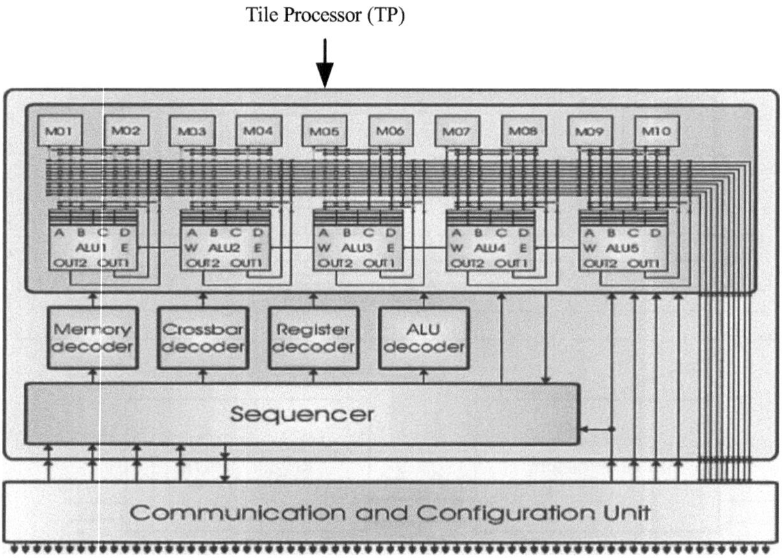

Fig. 2.6 A domain-specific system (the Montium Processing Tile [17])

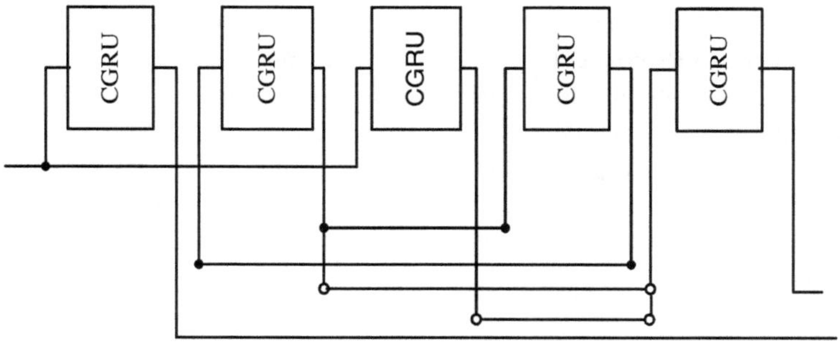

Fig. 2.7 An example of application class-specific system. White circles denote programmable interconnections, while black circles denote fixed connections

plement optimally a predefined set of circuits, this type of systems can be useful for cases where the exact algorithms and circuits are known in advanced, it is critical to meet strict design constraints, and no additional flexibility is required. Among others, examples of such architectures are the Pleiades architecture developed at Berkeley [18], [19], the cASICs developed by the Totem project [20], and the approach for designing reconfigurable datapaths proposed at Princeton [21]–[23].

As shown in Fig. 2.8, comparing ACSSs and ADSSs the former ones exhibit reduced flexibility and better performance. This stems form the fact that the class-specific systems are developed to implement a only a predefined class of applications, while the domain-specific ones are designed targeting at implementing the applications of certain application domain

2.4.2 Design Issues

The development of a coarse-grain domain-specific reconfigurable system involves a number of design issues that must be addressed. As CGRUs are more "expensive" than the logic blocks of an FPGA, the number of CGRUs, their organization, and the implemented operations are critical design parameters. Furthermore, the

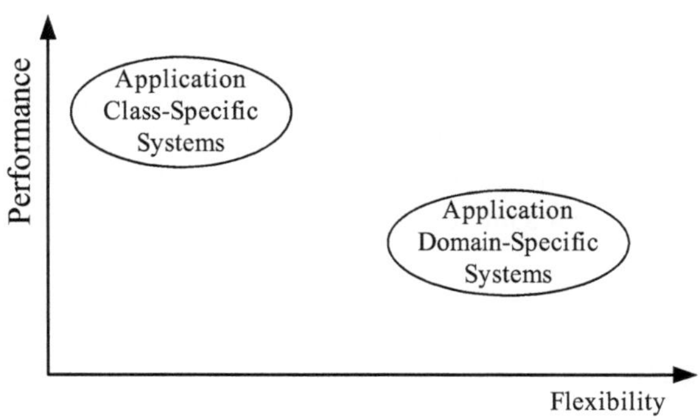

Fig. 2.8 Flexibility vs. performance for application class-specific and application domain-specific coarse-grain reconfigurable systems

structure of the interconnection network, the length of each routing channel, the number of nearest interconnections for each CGRU, as well as the reconfiguration mechanism, the coupling with the μP, and the communication with memory are also important issues that must be taken into account. In the following sections we study these issues and discuss the alternative decisions, which can be followed for each of them. Due to the different characteristics between class-specific and domain-specific coarse-grain reconfigurable systems, we divide the study in two sub-sections.

2.4.2.1 Application Class-Specific Systems

As it has been mentioned application class-specific coarse-grain reconfigurable systems are custom architectures target at implementing optimally only a predefined set (class) of applications. They consist of a fixed number and type of programmable interconnections and CGRUs usually organized in not so regular structures. Since these systems are used to realize a given set of applications with known requirements in terms of processing units and interconnections, the major issues concerning their development are: (a) the construction of the interconnection network, (b) the placement of the processing units, and (c) the reuse of the resources (processing units and interconnections). CGRUs must be placed optimally resulting in reduced routing demands, while the interconnection network must be developed properly offering the requiring flexibility so that the CGRUs to communicate each other according to the needs of the application of the target domain. Finally, reuse of resources is needed to reduce the area demands of the architecture. These can be achieved by developing separately optimal architectures for each application and merging them in one design which is able to implement the demanded circuits meeting the specifications in terms of performance, area, reconfiguration overhead, and power consumption. We discuss in details the development of class-specific architectures in Section 2.5.2.1.

2.4.2.2 Application Domain-Specific Systems

In contrast to ACSSs, ADSSs aim at implementing the applications of the whole domain. This imposes the development of a generic and flexible architecture which enforce to address a number of design issues. These are: (a) the organization of the CGRUs, (b) the number CGRUs, (c) the operations that are supported by each CGRU, and (d) the employed interconnections. We study these issues in the sections bellow.

Organization of CGRUs

According to the organization of CGRUs, the ADSSs are classified in two categories, namely the *mesh-based* and *linear array* architectures.

Mesh-Based Architectures

In mesh-based architectures the CGRUs are arranged in a rectangular 2-D array with horizontal and vertical connections that encourage Nearest Neighbor (NN) connections between adjacent CGRUs. These architectures are used to exploit the parallelism of data-intensive applications. The main parameters of the architecture are: (a) the number and type of CGRUs, (b) the supported operations of each CGRU, (c) the placement of CGRUs in the array, and (d) the development of the interconnection network. The majority of the proposed coarse-grain reconfigurable architectures such as Montium [17], ADRES [24], and REMARC [25], fall into this category. A simple mesh-based coarse-grain reconfigurable architecture is shown in Fig. 2.9 (a).

As these architectures aim at exploiting the inherent parallelism of data-intensive applications, a rich interconnection network that does not degrade performance is required. For that purpose a number of different interconnection structures have to be considering during the architecture's development. Besides the above simple structure where each CGRU communicates with the four NN units, additional schemes may be used. These include the use of horizontal and vertical segmented busses that can be configured to construct longer interconnections channels allowing the communication between distant units of a row or column. The number and length of the segmented buses per row and column, their direction (unidirectional or bi-directional) and the number of the attached CGRUs are parameters that must be determined considering the applications' needs of the targeted domain. An array that supports NN connections and 1-hop NN connections is shown in Fig. 2.9 (b).

Linear Array–Based Architectures

In linear-based architectures the CGRUs are organized in a 1-D array structure, while segmented routing channels of different lengths traverse the array. Typical examples of such coarse-grain reconfigurable architectures are RaPiD [26]–[30], PipeRench [31], and Totem [20]. For illustration purposes the RaPiD datapath

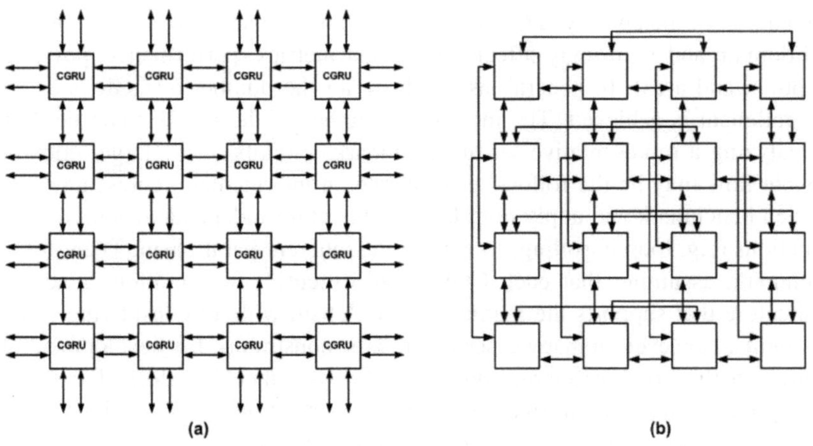

Fig. 2.9 (a) A single mesh-based (2-D) architecture, (b) 1-hop mesh architecture

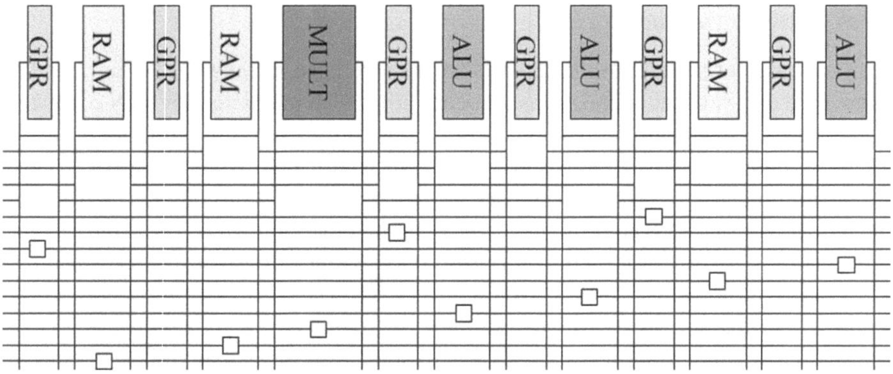

Fig. 2.10 A linear array architecture (RaPiD cell [26])

is shown Fig. 2.10. It contains coarse-grain units such as ALUs, memories, and multipliers arranged in a linear structure, while wires of different lengths traverse the array. Some of the used wires are segmented and can be programmed to create long wires for interconnecting distant processing units.

The parameters of such architecture are the number of the used processing units, the operations supported by each unit, the placement of the units in the array, as well as the number of programmable busses, their segmentation and the length of the segments. If the Control Data Flow Graph (CDFG) of the application have forks, which otherwise would require a 2-D realization, additional routing resources are needed like longer lines spanning the whole or a part of the array. These architectures are used for implementing streaming applications and easy mapping pipelines on these.

CGRUs Design Issues

Number of CGRUS

The number of the employed CGRUs depends on the characteristics of the considered domain and it strongly affects the design metrics (performance, power consumption, and area). In general, as much as are the number of CGRUs as much as parallelism is achieved. The maximum number of the CGRUs can be derived by analyzing a representative set of benchmarks circuits of the target domain. A possible flow may be the following. Generate an intermediate representation (IR) for each benchmark and apply high-level architecture-independent compiler transformations (e.g. loop unrolling) to expose the inherent parallelism. Then, for each benchmark, assuming that each CGRU can execute any operation, generate an architecture that supports the maximum parallelism without considering resource constraints. However, in many cases due to area constraints, the development of an architecture that contains a large number of CGRUs can not be afforded. In that case the mapping of applications onto the architecture must by performed by a methodology that ensure extensive reuse of the hardware in time to achieve the desired performance.

Operations Supported by a CGRU and Strength of CGRU

The arithmetic or logical operations that each CGRU executes is another design issue that has to be considered. Each CGRU may support any operation of the target domain offering high flexibility with the cost of possible wasted hardware if some operations are not frequently-appeared or they are characterized by reduced need for concurrent execution. For that reason, the majority of the employed CGRUs support basic and frequently-appeared operations, while complex and rarely-appeared operations are implemented by few units. Specifically, the majority of the existing systems, CGRUs are mainly ALUs that implement basic arithmetic (addition/subtraction) and logical operations and special-purpose shifting, while in many cases multiplication with a constant is also supported. More complex units such as multiplication, multiple-and-accumulate, are implemented by few units, which are placed in specific positions in the architecture. Also, memories and registers files may be included in the architecture to implement data-intensive applications.

The determination of the operations supported by CGRUs is a design aspect that should be carefully addressed since it strongly affects the performance, power consumption, and area from the implementation point of view and the complexity of the applied mapping methodology. This can be achieved by profiling extensively representative benchmarks of the considered domain and using a mapping methodology, measuring the impact of different decisions on the quality of the architecture, and determining the number of the units, the supported operations and their strengths.

Another issue that has to be considered is the strength of CGRU. It is referred to the number of the functional units included in each CGRU. Due to the routing latencies, it might be preferable to include a number of functional units in each CGRU rather than having them as separate ones. For that reason apart from ALUs, a number of architectures include additional units in the PEs. For instance, the reconfigurable processing units of ADRES and Montium include register files, while the cells of REMARC and PipeRench contain multipliers for performing multiplication with constant and barrel shifters.

Studies on CGRUs-Related Design Issues

Regarding with the organization of CGRUs, the interconnection topologies, and the design issues related to CGRUs, a lot of studies regarding have been performed.

In [32], a general 2-D mesh architectures was consider and a set of experiments on a number of representative DSP benchmarks were performed varying the number of functional units within PEs, the functionality of the units, the number of CGRUs in the architecture, and the delays of the interconnections. To perform the experiments, a mapping methodology based on a list-based scheduling heuristic, which takes into account the interconnection delays, was developed. Similar exploration was performed in [33] for the ADRES architecture using the DRESC framework for mapping applications on the ADRES architecture. The results of these experiments are discussed bellow.

Maximum Number of CGRUs and Achieved Parallelism

As reconfigurable systems are used to exploit the inherent parallelism of the application, a major question is how much is the inherent instruction level parallelism of the applications. For that reason loop unrolling was performed in representative loops used in DSP applications [32]. The results demonstrate that the performance improves rapidly as the unrolling factor is increased from 0 to 10. However, increasing further the unrolling factor the performance is not improved significantly due to dependency of some operations from previous loop iterations [32]. This is a useful result that can be used to determine the maximum number of the CGRUs that must be used to exploit parallelism and improve performance. In other words, to determine the required number of CGRUs required to achieve the maximum parallelism we have to perform loop unrolling up to 10 times. Comparisons between a 4×4 and 8×8 arrays, which include 16 and 64 ALUs respectively, show that due to inter-iteration dependencies the amount of concurrent operations is limited and the use of more units is aimless.

Strength of CGRUs

As mentioned, due to the interconnection delay it might be preferable to include more functional units in the employed coarse-grain PEs rather than using them separately. To study this issue, two configurations of 2-D mesh architecture were examined [32]. The first configuration is an 8×8 array with one ALU in each PE, while the second is a 4×4 array with 4 ALUs within a PE. In both cases 64 ALUs were used, the ALUs can perform every arithmetic (including multiplication) and logical operation, while the zero communication delay was considered for the units within the PE. The experimental results proved that the second configuration achieves better performance as the communication between the ALUs inside PEs does not suffer by interconnection delay. This indicates that as the technology improves and the speed of CGRUs outpaces that of interconnections, putting more functional units within each CGRU results in improved performance.

Interconnection Topologies

Instead of increasing the number of units we can increase the number of connections among CGRUs to improve performance. This issue was studied in [32], [33]. Three different interconnection topologies were examined, which are shown in Fig. 2.11: (a) the simple-mesh topology where the CGRUs are connected to their immediate neighbors in the same row and column, (b) the meshplus or 1-hop interconnection topology where the CGRUs are connected to their immediate neighbors and the next neighbor, and the (c) the Morphosys-like where each CGRU is connected to 3 other CGRUs in the same row and column.

The experiments on DSP benchmarks demonstrated a better performance of mesplus topology over the simple mesh due to the rich interconnection network of the second one. However, there is no significant improvement in performance when the meshplus and Morphosys-like topologies are compared, while the Morphosys-like topology requires more silicon area and configuration bits

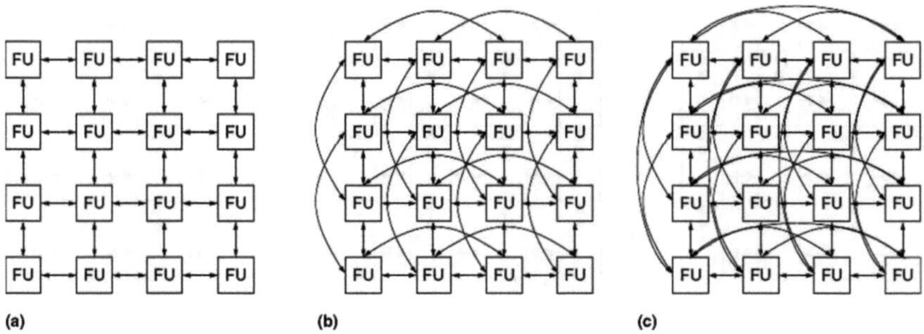

Fig. 2.11 Different interconnection topologies: (a) simple mesh, (b) meshplus, and (c) Morphosys-like

Concerning other interconnection topologies, the interested reader is referred to [34]–[36] where the crossbar, multistage interconnection networks, multiple-bus, and hierarchical mesh-based and other interconnection topologies are study in terms of performance and power consumption.

Interconnection Network Traversal

Also, the way the network topology is traversed while mapping operations to the CGRUs is a critical aspect. Mapping applications to such architectures is a complex task that is a combination of the operation scheduling, operation binding, and routing problems. Especially, the interconnections and their associated delays are critical concerns for an efficient mapping on these architectures. In [37], a study regarding with the aspects of three network topology on the performance was performed. Specifically, they studied: (a) the interconnection between CGRUs, (b) the way the array is traversed while mapping operations to the CGRUs, and (c) the communication delays on the interconnects between CGRUs. Concerning the interconnections, three different topologies were considered: (a) the CGRUs are connected to their immediate neighbours (NN) in the same row and column, (b) all the CGRUs are connected to their immediate and 1-hop NN connections, and (c) CGRUs are connected to all other CGRUs in the same row and same column. Regarding with the traversal of the array while mapping operations to the CGRUs three different strategies, namely the Zigzag, Reverse-S, and Spiral traversal, shown in Fig. 2.12 (a), (b), and (c), respectively, were studied.

Using an interconnect aware list-based scheduling heuristic to perform the network topology exploration, the experiments on a set of designs derived from DSP applications show that a spiral traversal strategy, which exploits better spatial and temporal locality, coupled with 1-hop NN connections leads to the best performance

2.4.3 Memory Accesses and Data Management

Although coarse-grain reconfigurable architectures offer very high degree of parallelism to improve performance in data-intensive applications, a major bottleneck

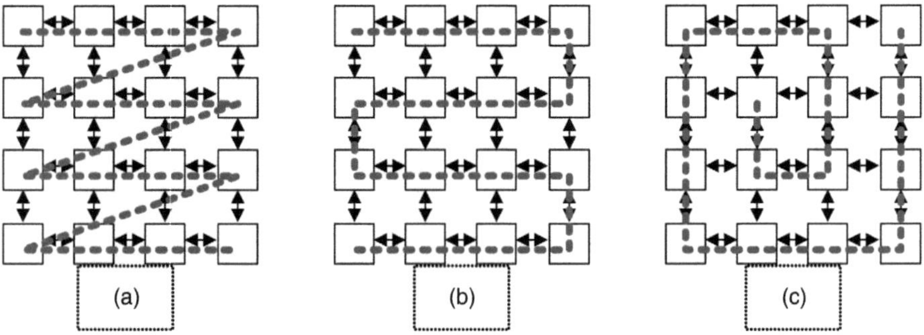

Fig. 2.12 Different traversal strategies: (a) Zigzag, (b) Reverse-S, (c) Spiral

arises because a large memory bandwidth is required to feed data concurrently to the underlying processing units. Also, the increase of memory ports results into power consumption increase. In [21] it was proved that the performance decreases as the number of the available memory ports is reduced. Therefore, proper techniques are required to alleviate the need for high memory bandwidth. Although, a lot of work has been performed in the field of compilers to address this issue, the compiler tools can not handle efficiently the idiosyncrasies of reconfigurable architectures, especially the employed interconnections and the associated delays.

In [38], [39] a technique has been proposed aiming at exploiting the opportunity the memory interface being shared by memory operations appearing in different iterations of a loop. The technique is based on the observation that if a data array is used in a loop, it is often the case that successive iterations of the loop refer to overlapping segment of the array. Thus, parts of data being read in an iteration of the loop have already been read in previous iterations. This redundant memory accesses can be eliminated if the iterations are executed in a pipeline fashion, by organizing the pipeline in such a way the related pipeline stages share the memory operations and save the memory interface resource. Proper conditions have been developed for sharing memory operations using generic 2-D reconfigurable mesh architecture. Also, a proper heuristic was developed to generate the pipelines with memory by properly assigning operations to processing units that use data which have already read for a memory in previous loop iterations. Experimental results show improvements of up to 3 times in throughput.

A similar approach that aims to exploit data reuse opportunities was proposed in [40]. The idea is to identify and exploit data reuse during the execution of the loops and to store the reused in scratch pad memory (local SRAM), which is equipped with a number of memory ports. As the size of the scratch pad memory is lower than that of main memory, the performance and energy cost of a memory access decreases. For that purpose a proper technique was developed. Specifically, by performing front-end compiler transformations the Data Dependency Reuse Graph (DDRG) is derived that handles the data dependencies and data reuse opportunities. Considering general 2-D mesh architecture (4×4 array) and the generated DDRG a list-based scheduling technique is used for mapping operations without performing pipeline taking into account the available resources and interconnections and the delays of interconnections. The experimental results show an improvement of 30 % in performance and memory accesses compared with the case where data reuse is not exploited.

2.5 Design Methodology for Coarse-Grain Reconfigurable Systems

In this section a design methodology for developing coarse-grain reconfigurable systems is proposed. The methodology targets at developing Application Domain-Specific Systems (ADSSs) or Application Class-Specific Systems (ACSSs). It consists of two stages that are the preprocessing stage and the architecture generation and mapping methodology development stage as shown in Fig. 2.13. Each stage includes a number of steps where critical issues are addressed. It must b stressed that the introduced methodology is a general one and some step may be removed or modified according to the targeted design goals.

The input to the methodology can be either a set of representative benchmarks of the targeted application domain, which are used for developing an ADSS, or the class of the applications, which are used for developing an ACSS described in a

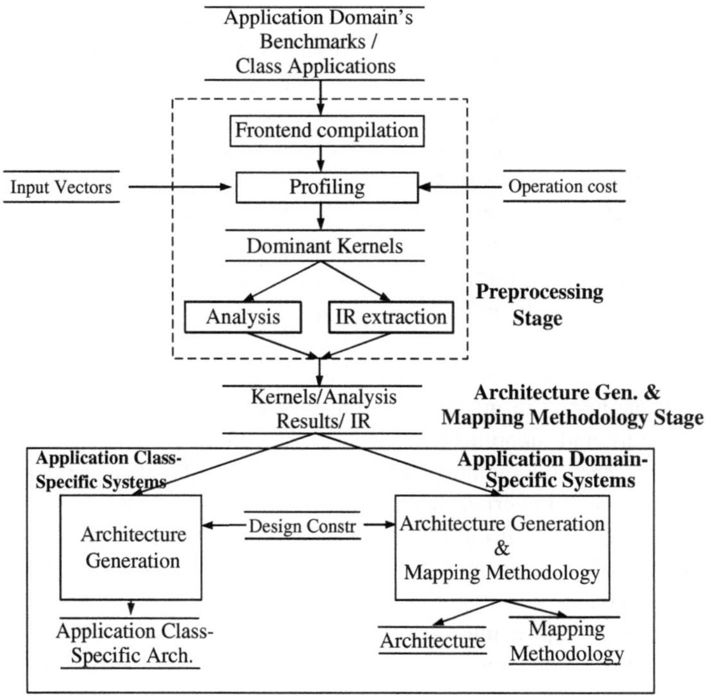

Fig. 2.13 Design methodology for developing coarse-grain reconfigurable systems

high-level language (e.g. C/C++). The goal of preprocessing stage is twofold. The first goal is to identify the computationally-intensive kernels that will be mapped onto the reconfigurable hardware. The second goal is to analyze the dominant kernels gathering useful information that is be exploited to develop the architecture and mapping methodology. Based on the results of preprocessing stage, the generation of the architecture and the development of the mapping methodology follow.

2.5.1 Preprocessing Stage

The preprocessing stage consists of three steps, which are: (a) the front-end compilation, (b) the profiling of the input descriptions to identify the computationally-intensive kernels, and (c) the analysis of the dominant kernels to gather useful information for developing the architecture and mapping methodology, and the extraction of an Internal Representation (IR) for each kernel. Initially, architecture-independent compiler transformations (e.g. loop unrolling) are applied to refine the initial description and to enhance parallelism.

Then, profiling in performed to identify the dominant kernels that will be implemented by the reconfigurable hardware. The inherent computational complexity (number of basic operations and memory accesses) is a meaningful measure for that purpose. To accomplish this, the refined description is simulated with appropriate input vectors, which represent standard operation, and profiling information is gathered at basic block level. The profiling information is obtained through a combination of dynamic and static analysis. The goal of dynamic analysis is to calculate the execution frequency of each loop and each conditional branch. Static analysis is performed at basic block level evaluating a base cost of the complexity for each basic block in terms of the performed operations and memory accesses. Since no implementation information is available, a generic cost is assigned to each basic operation and memory access. After performing simulation, the execution frequency of each loop and conditional branch, which are the outcomes of the dynamic analysis, is multiplied with the base cost of the corresponding basic block(s) and the cost of each loop/branch is obtained.

After the profiling step, the dominant kernels are analyzed to identify special properties and gather extra information that will be used during the development of the architecture and mapping methodology. The number of live-in and live-out signals of each kernel, the memory bandwidth needs, the locality of references, the data dependencies within kernels and the inter-kernel dependencies are included in the information obtained during the analysis step. The live-in/live-out signals are used during the switching from one configuration to another one and for the communication between the master processor and reconfigurable hardware, the memory bandwidth needs are taken into account to perform data management, while the intra- and inter-kernel dependencies are exploited for designing the datapaths, interconnections, and control units. Finally, an intermediate representation (IR), for instance, Control Data Flow Graphs (CDFGs) is extracted for each kernel.

2.5.2 Architecture Generation and Mapping Methodology Development

After the preprocessing stage, the stage of generating the reconfigurable architecture and mapping methodology follows. Since the methodology targets at developing either ADSSs or ACSSs, two separate paths can be followed, which are discussed bellow.

2.5.2.1 Application Class-Specific Architectures

As mentioned in Section 2.4.2.1 the design issues that should be addressed for developing ACSSs are: (a) the construction of the interconnection network (b) the placement of the processing units, and (c) the extensively reuse of the resources (processing units and interconnections) to reduce hardware cost. The steps for deriving an ACSS are shown in Fig. 2.14 [23]. Based on the results of preprocessing, an optimum datapath is extracted for each kernel. Then, the generated datapaths are combined into a single reconfigurable datapth. The goal is to derive a datapath with the minimum number of programmable interconnections, hardware units, and routing needs. Resource sharing is also performed so the hardware units to be reused by the considered kernels.

In [22], [23] a method for designing pipelined ACSSs was proposed. Based on the results of analysis a pipelined datapath is derived for each kernel. The datapath is generated with no resource constraints by direct mapping operations (i.e. software instructions) to hardware units and connecting all units according to data flow of the kernel. However, such a datapath may be not affordable due to design constraints (e.g. area, memory bandwidth). For instance, if the number of the available memory ports is lower than that generated datapath demands, then one memory port needs to be shared by different memory operations at different clock cycles. The same also holds for processing units which may need to be shared in time to perform different operations. The problem that must be solved is to schedule the operations under resource and memory constraints. An integer linear programming formulation

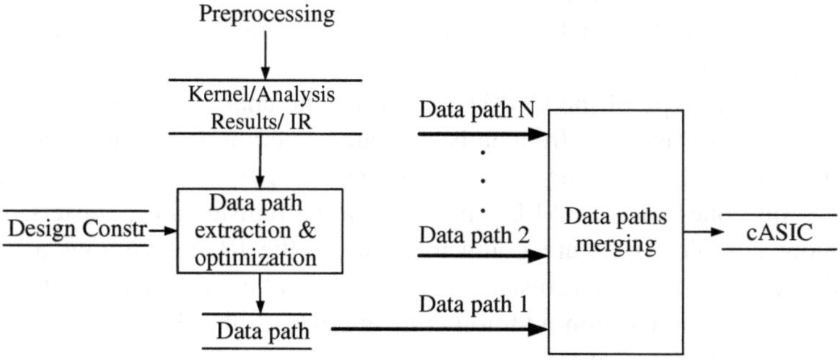

Fig. 2.14 Architecture generation of ACSSs [23]

was developed with three objective functions. The first one minimizes the iteration interval, the second minimizes the total number of pipeline stages, while the third objective function minimizes the total hardware cost (processing units and interconnections).

Regarding with the merging of datapaths and the construction of the final datapath, each datapath is modeled as a directed graph $Gi = (Vi, Ei)$, where a vertex, *Vi*, represents the hardware units in the datapath, while an arc, *Ei*, denotes an interconnection between two units. Afterwards, all graphs are merged in a single graph, G, and a compatibility graph, H, is constructed. Each node in H means a pair of possible vertex mappings, which share the same arc (interconnection) in G. To minimize the arcs in G, it is necessary to find the maximum number of arc mappings that are compatible with each other. This is actually the problem of finding the maximum clique of the compatibility graph H. An algorithm for finding the maximum clique between two graphs is proposed and the algorithm is applied iteratively to merge more graphs (datapaths).

Similar approaches proposed in [11], [41], [42] where bipartite matching and clique partitioning algorithms are proposed for constructing the graph G. Concerning the placement of the units and the generation of routing in each datapath, a simulated annealing algorithm was used targeting at minimizing the communication needs among the processing units.

2.5.2.2 Application Domain-Specific Architectures

The development of ADSS is accomplished in four steps as shown in Fig. 2.15. Each step includes a number of inter-dependent sub-steps.

Architecture Generation

The objective of the fist step is the generation of the coarse-grain reconfigurable architecture on which the dominant kernels of the considered application domain are implemented. The following issues must be addressed: (a) the determination of type and number of the employed CGRUs, (b) the organization of the CGRUs, (c) the selection of the interconnection topology, and (d) the addressing of data-management. The output of the architecture generation step is the model of the application domain-specific architecture.

Concerning the type of the CGRUs, based on the analysis results performed at the preprocessing stage, the frequently appeared operations are detected and the appropriate units implementing these operations are specified. The employed units may be simple ones such as ALUs, memory units, register files, shifters. In case where more complex units are going to be used, the IR descriptions are examined and frequently-appeared clusters of operations, called templates, such as MAC, multiple-multiple, or addition-addition units are extracted [43], [44]. The template generation is a challenging task involving a number of complex graph problems (template generation, checking graph isomorphism among the generated templates,

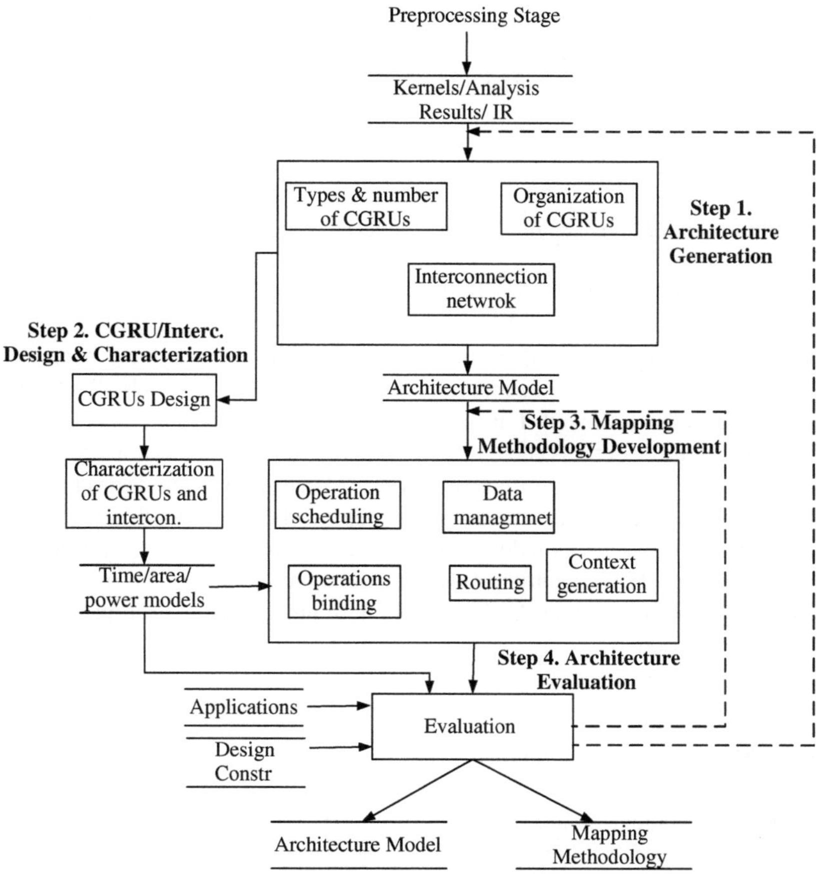

Fig. 2.15 Architecture generation and mapping methodology development for application domain-specific systems

and template selection). Regarding with the template generation task, the interested reader is referred to [43]–[47] for further reading.

As ADSSs are used to implement the dominant kernels of the whole application domain and high flexibility is required, the CGRUs should be organized in a proper manner resulting in regular and flexible organizations. When the system is going to be used to implement streaming applications, a 1-D organization should be adopted, while when data-intensive applications are targeted a 2-D organization may be selected. Based on the profiling/analysis (locality of references, operation dependencies within the kernels, and inter-kernel dependencies) and considering area and performance constraints, the number of the used CGRUs and their placement in the array are decided.

In addition, the type of employed interconnections (e.g. the number of NN connections, the length and number of the probably-used segmented busses, and the number of row/column busses) as well as the construction of the interconnection network (e.g. simple mesh, modified mesh, crossbar) are determined. Finally, decisions regarding the data fed to architecture are taken. For instance, if a lot of data needed to be read/written from/to the memory load/store units are placed in the first row a the 2-D array. Also, the number and type memory elements and their distribution into the array are determined.

CGRUs/Interconnections Design and Characterization

As mentioned CGRUs are optimally-designed hardwired units to improve performance, power consumption, and reduce area. So, the objective of the second step is the optimal design of CGRUs and interconnections, which have been determined in the previous step. To accomplish this, full-custom or standard-cell design approaches may be followed. Furthermore, the characterization of the employed CGRUs and interconnections and the development of performance, power consumption, and area models are performed at this step. According to the desired accuracy and complexity of the models several approaches may be followed. When high accuracy is demanded analytical models should be developed, while when reduced complexity is demanded low accuracy macro-models may be used. The output of this step it the optimally-designed CGRUs and interconnections and the performance, power, and area models.

Mapping Methodology Development

After the development of the architecture model and the characterization of the CGRUs and interconnections, the methodology for mapping kernels onto the architecture follows. The mapping methodology requires the development of proper algorithms and techniques addressing the following issues: (a) operations scheduling and binding to CGRUs, (b) data-management manipulation, (c) routing, and (d) context generation.

The scheduling of operations and their mapping onto the array is more complex task than the conventional high-level synthesis problem because the structure of the array has already determined, while the delays of the underlying interconnections must be taken into account. Several approaches have been proposed in the literature for mapping applications to coarse-grain reconfigurable. In [48], [49] a modulo scheduling algorithm that considers the structure of the array and the available CGRUs and interconnections proposed for mapping loops onto the ADRES reconfigurable architecture [24]. In [50], a technique for mapping DFGs on the Montium architecture is presented. In [37], concerning different interconnection delays, a list-based scheduling algorithm and traversal of the array was proposed for mapping DSP loops onto a 2-D coarse-grain reconfigurable architecture. In [51], [51] a compiler framework for mapping loops written in SA-C language to the Morphosys [52], [51] architecture was introduced. Also, as ADSSs are based on the systolic arrays, there is a lot of prior work on mapping applications to systolic arrays [53].

Architecture Evaluation

After the development of the architecture model and mapping methodology the evaluation phase follows. Mapping kernels taken from the considered application domain and taken into account performance, area, power constraints the architecture and design methodology are evaluated. If they do not meet the desired goals then a new mapping methodology must be developed or a new architecture must be derived. It is preferable to try first the development of a more efficient mapping methodology.

2.6 Coarse-Grain Reconfigurable Systems

In this section we present representative coarse-grain reconfigurable systems that have introduced in the literature. For each of them we discuss the target application domain, its architecture, the micro-architecture of the employed CGRUs, the compilation/ application mapping methodology, and the reconfiguration procedure.

2.6.1 REMARC

REMARC [25], which was designed to accelerate mainly multimedia applications, is a reconfigurable coarse-grain coprocessor coupled to a main RISC processor. Experiments performed on MPEG-2 decoding and encoding saw speedups ranging from a factor of 2.3 to 21 for the computational intensive kernels that are mapped and executed on REMARC coprocessor.

2.6.1.1 Architecture

REMARC consists of a global control unit and an 8×8 array of 16-bit identical programmable units called *nano processors (NP)*. The block diagram of REMARC and the organization of the nano processor are shown in Fig. 2.16. Each NP communicates directly to the four adjacent ones via dedicated connections. Also, 32-bit Horizontal (HBUS) and Vertical Buses (VBUS) exist to provide communication between the NPs of the same row or column. In addition, eight VBUSs are used to provide communication between the global control unit and the NPs.

The global control unit controls the nano processors and the data transfer between the main processor and them. It includes a 1024-entry global instruction RAM, data and control registers which can be accessed directly by the main processor. According to a global instruction, the control unit set values on the VBUSs, which are read by the NPs. When the NPs complete their execution, the control unit reads data from the VBUSs and stores them into the data registers.

The NP does not contain Program Counter (PC). Every cycle, according to the instruction stored in the global instruction RAM, the control unit generates a PC

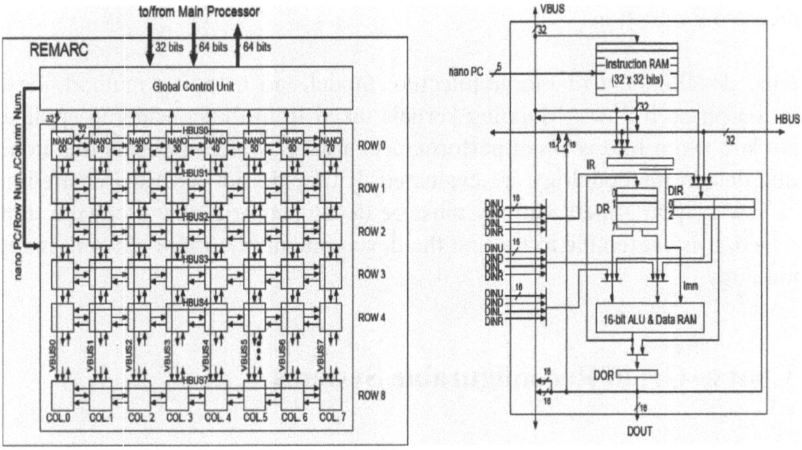

(a) (b)

Fig. 2.16 Block diagram of REMARC (a) and nano processor microarchitecture (b)

value which is received by all the nano processors. All NPs use the same nano PC value and execute the instructions indexed by the nano PC. However, each NP has its own instruction RAM, so different instructions can be stored at the same address of each nano instruction RAM. Thus, each NP can operate differently based on the stored nano instructions.

In that way, REMARC operates as a VLIW processor in which each instruction consists of 64 operations, which is much simpler than distributing execution control across the 64 nano processors. Also, by programming a row or a column with the same instruction, Single Input Multi Data (SIMD) operations are executed. To realize SIMD operations, two instruction types called HSIMD (Horizontal SIMD) and VSIMD (Vertical SIMD) are employed. In addition to the PC field, an HSIMD/VSIMD instruction has a column/row number filed that indicates which column/row is used to execute the particular instruction in SIMD fashion.

The instruction set of the coupled RISC main processor is extended by nine new instructions. These are: two instructions for downloading programs form main memory and storing them to the global and nano instruction RAMs, two instructions (load and store) for transfer data between the main memory and REMARC data registers, two instructions (load and store) for transfer data between the main processor and REMARC data registers, two instructions for transfer data between the data and control registers, and one instructions to start the execution of a REMARC program.

2.6.1.2 Nano Processor Microarchitecture

Each NP includes a 16-bit ALU, a 16-entry data RAM, a 32-entry instruction RAM (nano instruction RAM), an instruction register (IR), eight data registers (DR), four data input registers (DIR), and one data output register (DOR). The length of the data registers and IR is 16 and 32, respectively. The ALU executes 30 instructions including common arithmetic, logical and shift instructions, as well as special instructions for multimedia such as Minimum, Maximum Average with Rounding, Shift Right Arithmetic and Add, and Absolute and Add. It should be mentioned, that the ALU does not include a hardware multiplier. The Shift Right Arithmetic and Add instruction provides a primitive operation for constant multiplications instead.

Each NP communicates to the four adjacent ones through dedicated connections. Specifically, each nano processor can get data from the DOR register of the four adjacent nano processors via dedicated connections (DINU, DIND, SINL, and DINR) as shown in Fig. 2.16. Also, the NPs in the same row and the same column communicate via a 32-bit Horizontal Bus (HBUS) and 32-bit Vertical Bus (VBUS), respectively, allowing data broadcasting between non-adjacent nano processors.

2.6.1.3 Compilation and Programming

To program REMARC an assembly-based programming environment, along with a simulator developed. It contains a global instruction and a nano instruction assembler. The global instruction assembler starts with a global assembly code, which describes the nano instructions that will be executed by the nano processors, and generates configuration data and label information, while the nano assembler starts with nano assembly code and generates the corresponding configuration data. The global assembler also produces a file named *remarc.h* that defines labels for the global assembly code. Using "asm" compiler directive, assembly instructions are manually written to initial C code. Then the GCC compiler is used to generate intermediate code that includes instructions which are executed by the RISC core and the new instructions that are executed by REMARC. A special assembler is employed to generate the binary code for the new instructions. Finally, the GCC is used to generate executable code that includes the instructions of the main processor and the REMARC ones. It must be stressed that the global and nano assembly code is provided manually by the user, which means that the assignment and scheduling of operations are performed by the user. Also, the C code rewriting to include the "asm" directives is performed manually by the programmer.

2.6.2 RaPiD

RaPiD (Reconfigurable Pipelined Datapath) [26]–[29] is a reconfigurable coarse-grain architecture optimized to implement deep linear pipelines, much like those are appeared in DSP algorithms. This is achieved by mapping the computation into a pipeline structure using a 1-D linear array of coarse-grained units like ALUs, registers, and RAMs, which are communicate in nearest-neighbor fashion through a programmable interconnection network.

Compared to a general purpose processor, RaPiD can be treated as a superscalar architecture with a lot of functional units but with no cache, register file, or crossbar interconnections. Instead of a data cache, data are streamed in directly from an external memory. Programmable controllers are employed to generate a small instruction stream, which is decoded at run-time as it flows in parallel with the data path. Instead of a global register file, data and intermediate results are stored locally in registers and small RAMs, close to the functional units. Finally, instead of a crossbar, a programmable interconnect network, which consists of segmented buses, is used to transfer data between the functional units.

A key feature of RaPiD is the combination of static and dynamic control. While the main part of the architecture is configured statically, a limited amount of dynamic control is provided which greatly increases the range and capability of applications that can be mapped.

2.6.2.1 Architecture

As shown in Fig. 2.17, which illustrates a single RaPiD cell, the cell is composed of: (a) a set of application-specific function units, such as ALUs, multipliers, and shifters, (b) a set of memory units (registers and small data memories), (c) input and output ports for interfacing with the external environment, (d) a programmable interconnection network that transfers data among the units of the data path using a combination of configurable and dynamically controlled multiplexers, (e) an instruction generator that issues "*instructions*" to control the data path, and (f) a control path that decode the instruction and generates the required control signals for the data path. The number of cells and the granularity of ALUs are design parameters. A typical single-chip contains 8–32 of these cells, while the granularity of processing units is 16 bits.

The functional units are connected using segmented buses that run the length of the data path. Each functional unit output includes registers, which can be

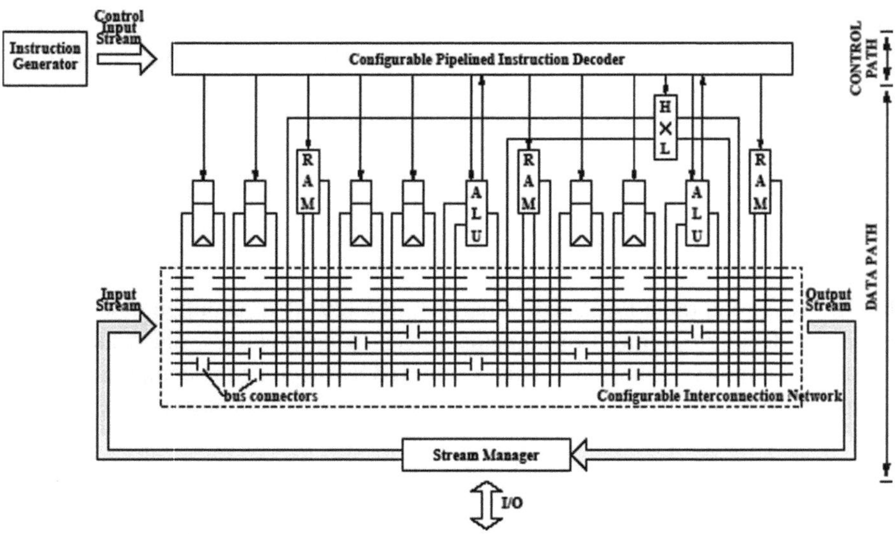

Fig. 2.17 The architecture of a RaPiD cell

programmed to accommodate pipeline delays, and tri-state drivers to feed their output onto one or more bus segments. The ALUs perform common word-level logical and arithmetic operations and they can also be chained to implement wide-integer computations. The multiplier produces a double-word result, which can be shifted to accomplish a given fixed-point representation. The registers are used to store constants and temporary values, as well. They are also used as multiplexers to simplify control and to connect bus segments in different tracks and/or for additional pipeline delays. Concerning buses, they are segmented into different lengths to achieve efficient use of the connection resources. Also, adjacent bus segments can be connected together via a bus connector. This connection can be programmed in either direction via a unidirectional buffer or can be pipelined with up to three register delays allowing data pipelines to be built in the bus itself.

In many applications, the data are grouped into blocks which are loaded once, saved locally, reused, and then discarded. The local memories in the data path serve this purpose. Each memory has a specialized data path register used as an address register. More complex addressing patterns can be generated using registers and ALUs in the data path. Input and output data enter and exit via I/O streams at each end of the data path. Each stream contains a FIFO filled with the required data or with the produced results. External memory operations are accomplished by placing FIFOs between the array and a memory controller, which generates sequences of addresses for each stream.

2.6.2.2 Configuration

During the configuration the operations of the functional units and bus connections are determined. Due to the similarity appeared in loop iterations, the larger part of the structure is statically configured. However, there is also a need for dynamic control signals to implement the differences among loop iterations. For that purpose, the control signals are divided into static and dynamic ones.

The static control signals, which determine the structure of the pipeline, are stored into a configuration memory, loaded when the application starts and remain constant for the entire duration of the application. On the other hand, the dynamic control signals are used to schedule the operations on the data path over time [27]. They are produced by a pipelined control path which stretches parallel with the data path as shown in Fig. 2.17.

Since applications usually need a few dynamic control signals and use similar pipeline stages, the number of the control signals in the control path is relatively small. Specifically, dynamic control is implemented by inserting a few context values in each cycle in the control path. The context values are inserted by an instruction generator at one end of the control path and are transmitted from stage to stage of the control path pipeline where they are fed to functional units. The control path contains 1-bit segmented buses, while the context values include all the information required to compute the required dynamic control signals.

2.6.2.3 Compilation and Programming

Programming is performed using RaPiD-C, a C-like language with extensions (e.g. synchronization mechanisms and conditionals to specify the first or last loop iteration) to explicitly specify parallelism, data movement and partitioning [28].

Usually, a high-level algorithm specification is not suitable to map directly to a pipelined linear array. The parallelism and the data I/O are not specified, while the algorithm must be partitioned to fit on the target architecture. Automating these processes is a difficult problem for an arbitrary specification. Instead, C-like language was proposed that requires the programmer to specify the parallelism, data movement, and partitioning. To the end, the programmer uses well known techniques of loop trans-formation and space/time mapping.

The resulting specification is a nested loop where outer loops specify time, while the innermost loop specifies space. The space loop refers to a loop over the stages of the algorithm, where a stage corresponds to one iteration of the innermost loop. The compiler maps the entire stage loop to the target architecture by unrolling the loop to form a flat netlist. Thus, the programmer has to permute and tile the loop-nest so that the computation required after unrolling the innermost loop will fit onto the target architecture. The remainder of the loop nest determines the number of times the stage loop is executed.

A RaPiD-C program as briefly described above clearly specifies the hardware requirements. Therefore, the union of all stage loops is very close to the required structural description. One difference from a true structural description is that stage loop statements are specified sequentially but execute in parallel. A netlist must be generated to maintain these sequential semantics in a parallel environment. Also, the control is not explicit but instead it is embedded in a nested-loop structure. So, it must be extracted into multiplex select lines and functional unit control. Then, an instruction stream must be generated which can be decoded to form this control. Finally, address generators must be derived to get the data to and from memory at the appropriate time. Hence, compiling RaPiD-C into a structural description consists of four components: netlist generation, dynamic control extraction, instruction stream/decoder generation, and I/O address generation. The compilation process produces a structural specification consisting of components on the underlying architecture. The netlist is then mapped to the architecture via standard FPGA mapping techniques including pipelining, retiming, place and route. Placement is done by simulated annealing, while routing is accomplished by Pathfinder [30].

2.6.3 PipeRench

PipeRench [31], [54], [55] is a coarse-grain reconfigurable system consisting of stages organized in a pipeline structure. Using a technique called *pipeline reconfiguration*, PipeRench provides fast partial and dynamic reconfiguration, as well as run-time scheduling of configuration and data streams, which improve the compilation and reconfiguration time and maximize hardware utilization. PipeRench is used as a coprocessor for data-stream applications. Comparisons with general purpose processor have shown significant performance improvement up to 190 × versus a RISC processor for the dominant kernels.

2.6.3.1 Architecture

PipeRench, the architecture of which is shown in Fig. 2.18, is composed of identical stages called *stripes* organized in a pipeline structure. Each stripe contains a number of Processing Elements (PE), an interconnection network, and pass registers. Each PE contains an ALU, barrel shifters, extra circuitry to implement carry-chains and zero-detection, registers, and the required steering logic for feeding data into the ALU. The ALU, which is implemented by LUTs, is 8-bit although the architecture does not impose any restriction. Each stripe contains 16 PEs with 8 registers each, while the whole fabric has sixteen stripes.

The interconnection network in each stripe, which is a cross-bar network, is used to transmit data to the PEs. Each PE can access data from the registered outputs of the previous stripe as well as the registered or unregistered outputs of the other PEs of the same stripe. Interconnect that directly skips over one or more stages is not allowed, nor are interconnections from one stage to a previous one. To overcome this limitation pass registers are included in the PE that create virtual connections between distant stages. Finally, global buses are used for transferring data and configuration streams. The architecture also includes on-chip configuration memory, state memory (to save the register contents of a stripe), data and memory bus controllers, and a configuration controller. The data transfer in and out of the array is accomplished using FIFOs.

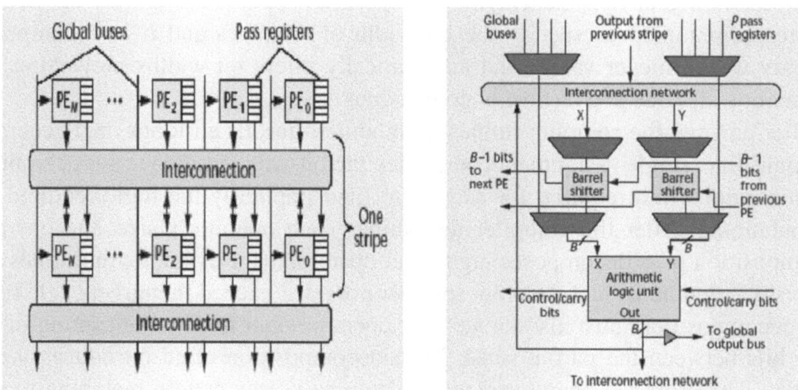

(a) (b)

Fig. 2.18 PipeRench Architecture: (a) Block diagram of a stripe, (b) Microarchitecture of a PE

2.6.3.2 Configuration

Configuration is done by a technique called *pipelined reconfiguration*, which allows performing large pieces of computations on a small piece of hardware through rapid reconfiguration. Pipelined reconfiguration involves virtualizing pipelined computations by breaking a single static configuration into pieces that correspond to pipeline stages of the application. Each pipeline stage is loaded every cycle making the computation possible, even if the whole configuration is never present in the fabric at one time. Since, some stages are configured while others are executed, reconfiguration does not affect performance. As the pipeline fills with data, the system configures stages for the needs of computations before the arrival of the data. So, even if there is no virtualization, configuration time is equivalent to the time of the pipeline and does not reduce throughput. A successful pipelined reconfiguration should configure a physical pipe stage in one cycle. To achieve this, a configuration buffer was included. A controller manages the configuration process.

Virtualization through pipelined reconfiguration imposes some constraints on the kinds of computations that can be accomplished. The most restrictive is that cyclic dependencies must fit within one pipeline stage. Therefore, allow direct connections are allowed only between consecutive stages. However, virtual connections are allowed between distant stages.

2.6.3.3 Compilation and Programming

To map applications onto PipeRench, a compiler that trades off configuration size for compilation speed was developed. The compiler starts by reading a description of the architecture. This description includes the number of PEs per stripe, the bit width of each PE, the number of pass registers per PE, the interconnection topology, the delay of PEs etc.

The source language is a *dataflow intermediate language* (DIL), which is a single-assignment language with C operators. DIL hides all notions of hardware resources, timing, and physical layout from programmers. It also allows, but doesn't require programmers to specify the bit width of variables and it can manipulate arbitrary width integer values and automatically infers bit widths preventing any information loss due to overflow or conversions.

After parsing, the compiler inlines all modules, unrolls all loops, and generates a straight-line, single-assignment code. Then the bit-value inference pass computes the minimum width required for each wire (and implicitly the logic required for computations). After the compiler determines each operator's size, the operator decomposition pass decomposes high-level operators (for example, multiplies become shifts and adds) and decomposes operators that exceed the target cycle time. This decomposition must also create new operators that handle the routing of the carry bits between the partial sums. Such decomposition often introduces inefficiencies. Therefore, an operator recomposition pass uses pattern matching to find subgraphs that it can map to parameterized modules. These modules take advantage of architecture-specific routing and PE capabilities to produce a more efficient set of operators.

The place-and-route algorithm is a deterministic, linear-time, greedy algorithm, which runs between two and three orders of magnitude faster than commercial tools and yields configurations with a comparable number of bit operations.

2.6.4 ADRES

ADRES (*Architecture for Dynamically Reconfigurable Embedded Systems*) is a reconfigurable template that consists of a VLIW processor and a coarse-grained reconfigurable matrix [24]. The reconfigurable matrix has direct access to the register files, caches, and memories of the system. This type of integration offers a lot of benefits including improved performance, simplified programming model, reduced communication cost, and substantial resource sharing. Also, a methodology for mapping applications described in C onto the ADRES template has been developed [48], [49]. The major characteristic of the mapping methodology is a novel modulo scheduling algorithm to exploit loop-level parallelism [56]. The target domain of the ADRES is multimedia and loop-based applications.

2.6.4.1 Architecture

The organization of the ADRES core and *Reconfigurable Cell* (RC) are shown in Fig. 2.19.

The ADRES core is composed by many basic components, including mainly Functional Units (FUs) and Register Files (RF). The FUs are capable of executing word-level operations. ADRES has two functional views, the VLIW processor and the reconfigurable matrix. The VLIW processor is used to execute the control parts of the application, while the reconfigurable matrix is used to accelerate data-flow kernels exploiting their inherent parallelism.

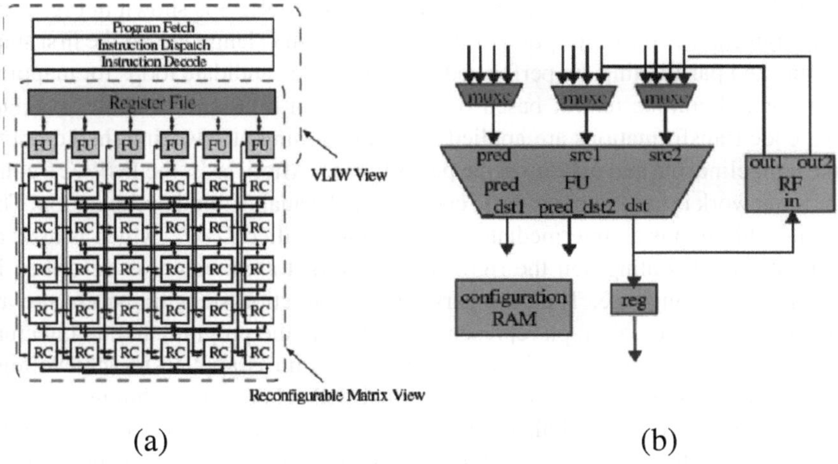

Fig. 2.19 The ADRES core (a) and the reconfigurable cell (b)

Regarding with the VLIW processor, several FUs are allocated and connected together through one multi-port register file. Compared with the counterparts of the reconfigurable matrix, these FUs are more powerful in terms of functionality and speed. Also, some of these FUs access the memory hierarchy, depending on available ports.

Concerning the reconfigurable matrix, besides the FUs and RF shared with the VLIW processor, there are a number of *reconfigurable cells* (RC) which basically consist of FUs and RFs (Fig. 2.19b). The FUs can be heterogeneous supporting different operations. To remove the control flow inside loops, the FUs support predicated operations. The configuration RAM stores a few configurations locally, which can be loaded on cycle-by- cycle basis. If the local configuration RAM is not big enough, the configurations are loaded from the memory hierarchy at the cost of extra delay. The behavior of a RC is determined by the stored reconfigurations whose bits control the multiplexers and FUs.

Local and global communication lines are employed for data transferring between the RCs, while the communication between the VLIW and the reconfigurable matrix takes place through the shared RF (i.e. the VLIW's RF) and the shared access to the memory.

Due to the above tight integration, ADRES has many advantages. First, the use of the VLIW processor instead of a RISC one as in other coarse-grain systems allows accelerating more efficiently the non-kernel code, which is often a bottleneck in many applications. Second, it greatly reduces both communication overhead and programming complexity through the shared RF and memory access between the VLIW and reconfigurable matrix. Finally, since the VLIW's FUs and RF can be also used by the reconfigurable matrix these shared resources reduce costs considerably.

2.6.4.2 Compilation

The methodology for mapping an application on the ADRES is shown in Fig. 2.20. The design entry is the description of the application in C language. In the first step, profiling and partitioning are performed to identify the candidate loops for mapping on the reconfigurable matrix based on the execution time and possible speedup. Next, code transformations are applied manually aiming at rewriting the kernel to make it pipelineable and maximize the performance. Afterwards, the IMPACT compiler framework is used to parse the C code and make analysis and optimization. The output of this step is an intermediate representation, called *Lcode*, which is used as the input for scheduling. On the right side, the target architecture is described in an XML-based language. Then the parser and abstraction steps transform the architecture to an internal graph representation. Taking the program and architecture representations as input, modulo scheduling algorithm is applied to achieve high parallelism for the kernels, whereas traditional ILP scheduling techniques are applied to gain moderate parallelism for the non-kernel code. Next, the tools generate scheduled code for both the reconfigurable matrix and the VLIW, which can be simulated by a co-simulator.

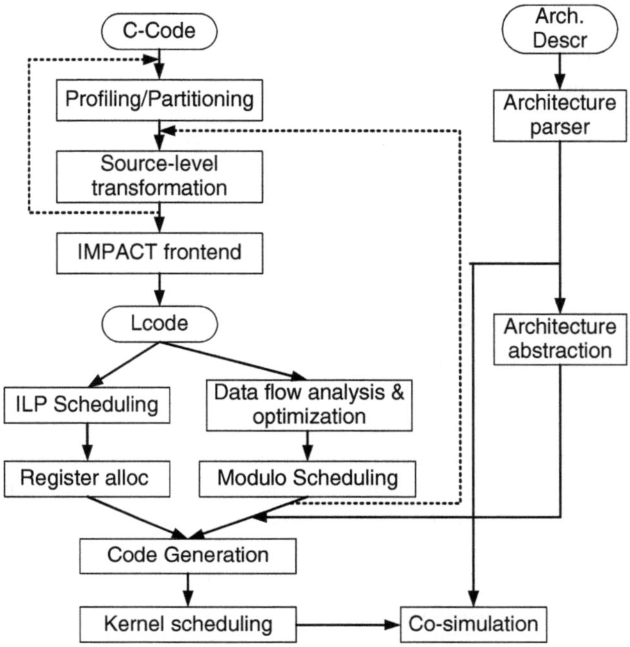

Fig. 2.20 Mapping methodology for ADRES

Due to the tight integration of the ADRES architecture, communication between the kernels and the remaining code can be handled by the compiler automatically with low communication overhead. The compiler only needs to identify live-in and live-out variables of the loop and assign them to the shared RF (VLIW RF). For communication through the memory space, we needn't do anything because the matrix and the VLIW share the memory access, which also eliminates the need for data copying.

Regarding modulo scheduling the adopted algorithm is an enhanced version of the original due to the constraints and features imposed by the coarse-grain reconfigurable matrix. Modulo scheduling is a pipeline technique that targets to improve parallelism by executing different loop iterations in parallel [57]. Applied to coarse-grained architectures, modulo scheduling becomes more complex, being a combination of placement and routing (P&R) in a modulo-constrained 3D space. An abstract architecture representation, modulo routing resource graph (MRRG) is used to enforce modulo constraints and describe the architecture. The algorithm combines ideas from FPGA placement and routing, and modulo scheduling from VLIW compilation.

2.6.5 Pleiades

Pleiades is a reusable coarse-grain reconfigurable template that can be used to implement domain-specific programmable processors for DSP algorithms [18], [19].

The architecture relies on an array of heterogeneous processing elements, optimized for a given domain of algorithms, which can be configured at run time to execute the dominant kernels of the considered domain.

2.6.5.1 Architecture

The Pleiades architecture is based on the template shown in Fig. 2.21. It is a template that can be used to create an instance of a domain-specific processor, which can be configured to implement a variety of algorithms of this domain. All instances of the template share a fixed set of control and communication primitives. However, the type and number of processing elements of an instance can vary and depend on the properties of the particular domain.

The template consists of a *control processor* (a general-purpose microprocessor core) surrounded by a heterogeneous array of autonomous, special-purpose processors called *satellites*, which communicate through a reconfigurable communication network. To achieve high performance and energy efficiency, the dominant kernels are executed on the satellites as a set of independent and concurrent threads of computation. The satellites have designed to implement the kernels with high performance and low energy consumption. As the satellites and communication network

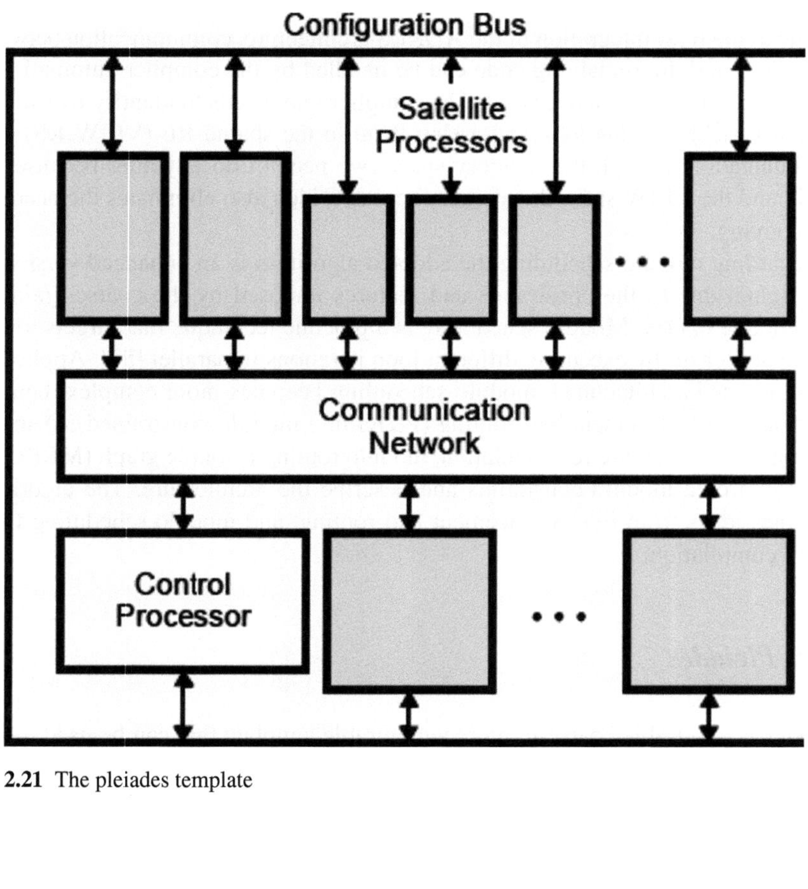

Fig. 2.21 The pleiades template

are configured at run-time, different kernels are executed at different times on the architecture.

The functionality of each hardware resource (a satellite or a switch of the communication network) is specified by its *configuration state*, which is a collection of bits that instruct the hardware resource what to do. The configuration state is stored locally in a storage element (register, register file or memory), which are distributed throughout the system. These storage elements belong to the memory map of the control processor and are accessed through the reconfiguration bus, which is an extension of the address/data/control bus of the control processor. Finally, all computation and communication activities are coordinated via a distributed data-driven control mechanism.

The Control Processor

The main tasks of the control processor are to configure the satellites and the communication network, to execute the control (non-intensive) parts of the algorithm, and to manage the overall control flow. The processor spawns the dominant kernels as independent threads of computation on the satellites and configures them and the communication network to realize the dataflow graph of the kernel(s) directly to the hardware. After the configuration of the hardware, the processor initiates the execution of the kernel by generating trigger signals to the satellites. Then, the processor can halt and wait for the kernel's completion or it can start executing another task.

The Satellite Processors

The computational core of Pleiades consists of a heterogeneous array of autonomous, special-purpose satellite processors that have designed to execute specific tasks with high performance and low energy. Examples of satellites are: (a) data memories that size and number depends on the domain, (b) address generators, (c) reconfigurable datapaths to implement the arithmetic operations required, (d) programmable gate array modules to implement various logic functions, (e) Multiply-Accumulate (MAC) units etc. A cluster of interconnected satellites, which implements a kernel, processes data tokens in a pipelined manner, as each satellite forms a pipeline stage. Also, multiple pipelines corresponding to multiple independent kernels can be executed in parallel. These capabilities allow efficient processing at very low supply voltages. For applications with dynamically varying throughput requirements, dynamic scaling of the supply voltage is used to meet throughput at the minimum supply voltage.

The Interconnection Network

The interconnection network is a generalization of the mesh structure. For a given placement of satellites, wiring channels are created along their sides. Switch-boxes are placed at the junctions between the wiring channels, and the required communication patterns are created by configuring these switch-boxes. The parameters of this mesh structure are the number of the employed buses in a channel and the functionality of the switch-boxes. These parameters depend on the placement of the satellite processors and the required communication patterns among the satellite processors.

Also, hierarchy is employed by creating clusters of tightly-connected satellites, which internally use a generalized-mesh structure. Communication among clusters is done by introducing inter-cluster switchboxes that allow inter-cluster communication. In addition, Pleiades uses reduced swing bus driver and receiver circuits to reduce the energy. A benefit of this approach is that the electrical interface through the communication network becomes independent of the supply voltages of the communicating satellites. This allows the use of dynamic scaling of the supply voltage, as satellites at the two ends of a channel can operate at independent supply voltages.

2.6.5.2 Configuration

Regarding configuration the goal is to minimize the reconfiguration time. This is accomplished with a combination of several strategies. The first strategy is to reduce the amount of configuration information. The word-level granularity of the satellites and the communication network is one contributing factor. Another factor is that the behavior of most satellite processors is specified by simple coarse-grain instructions choosing one of a few different possible operations supported by a satellite and a few basic parameters. In addition, in the Pleiades architecture a wide configuration bus is used to load the configuration bits. Finally, overlapping of the configuration and execution is employed. While some satellites execute a kernel some others can be configured by the control processor for the next kernel. This can be accomplished by allowing multiple configuration *contexts* (i.e. multiple sets of configuration store registers).

2.6.5.3 Mapping Methodology

The design methodology has two separate, but related, aspects that address different tasks. One aspect addresses the problem of deriving a template instance, while the other one addresses the problem of mapping an algorithm onto a processor instance.

The design entry is a description of the algorithm in C or C++. Initially, the algorithm is executed onto the control processor. The power and performance of this execution are used as reference values during the subsequent optimizations. A critical task is to identify the dominant kernels in terms of energy and performance. This is done by performing dynamic profiling in which the execution time and energy consumption of each function are evaluated. For that reason appropriate power models for processor's instructions are used. Also, the algorithm is refined by applying architecture-independent optimizations and code rewriting. Once dominant kernels are identified, they are ranked in the order of importance and addressed one at a time until satisfactory results are obtained. One important step at this point is to rewrite the initial algorithm description, so that kernels that are candidates for being mapped onto satellite processors are distinct function calls.

Next follows the implementation of a kernel on the array by directly mapping the kernel's DFG onto a set of satellite processors. In the created hardware structure, each satellite corresponds to a node(s) of the dataflow graph (DFG) and the links correspond to the arcs of the DFG. Each arc is assigned to a dedicated link via the communication network ensuring that temporal correlations of the data are preserved. Mapped kernels are represented using an intermediate form as C++ functions that replace the original functions allowing their simulation and evaluation with the rest of the algorithm within a uniform environment. Finally, routing is performed with advanced routing algorithms, while automate configuration code generation is supported.

2.6.6 Montium

Montium [17] is a reconfigurable coarse-grain architecture that targets the 16-bit digital signal processing domain.

2.6.6.1 Architecture

Figure 2.22 shows a single Montium processing tile that consists of a reconfigurable Tile Processor (TP), and a Communication and Configuration unit (CCU).

The five identical ALUs (ALU1-ALU5) can exploit spatial concurrency and locality of reference. Since, a high memory bandwidth is needed, 10 local memories (M01-M10) exist in the tile. A vertical segment that contains one ALU, its input register files, a part of the interconnections and two local memories is called Processing Part (PP), while the five processing parts together are called Processing Part Array (PPA). The PPA is controlled by a sequencer.

The Montium has a datapath width of 16-bits and supports both signed integer and signed fixed-point arithmetic. The ALU, which is an entirely combinational circuit, has four 16-bit inputs. Each input has a private input register file that can store up to four operands. Input registers can be written by various sources via a flexible crossbar interconnection network. An ALU has two 16-bit outputs, which are connected to the interconnection network. Also, each ALU has a configurable instruction set of up to four instructions.

The ALU is organized in two levels. The upper level contains four function units and implements general arithmetic and logic operations, while the lower level contains a MAC unit. Neighboring ALUs can communicate directly on level 2. The West-output of an ALU connects to the East-input of the ALU neighboring on the left. An ALU has a single status output bit, which can be tested by the sequencer.

Each local SRAM is16-bit wide and has 512 entries. An Address Generation Unit (AGU) accompanies each memory. The AGU contains an address register that can

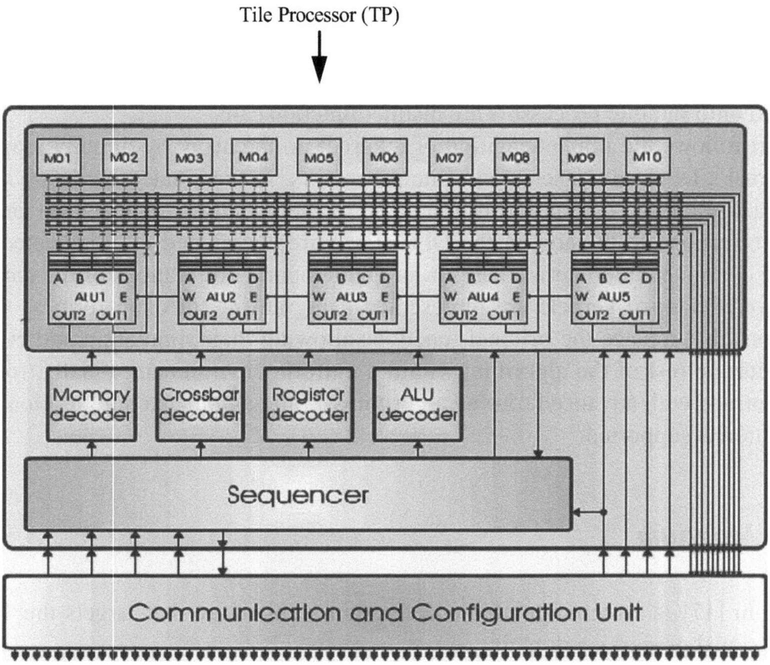

Fig. 2.22 The Montium processing tile

be modified using base and modify registers. It is also possible to use the memory as a LUT for complicated functions that cannot be calculated using an ALU (e.g. sine or division). At any time the CCU can take control of the memories via a direct memory access interface. The configuration of the interconnection network can change at every clock cycle. There are ten busses that are used for inter-processing part communication. The CCU is also connected to the busses to access the local memories and to handle data in streaming algorithms.

The flexibility of the above datapath results in a vast amount of control signals. To reduce the control overhead a hierarchy of small decoders is used. Also, the ALU in a PP has an associated configuration register. This configuration register contains up to four local instructions that the ALU can execute. The other units in a PP, (i.e. the input registers, interconnect and memories) have a similar configuration register for their local instructions. Moreover, a second level of instruction decoders is used to further reduce the amount of control signals. These decoders contain PPA instructions. There are four decoders: a memory decoder, an interconnect decoder, a register decoder and an ALU decoder. The sequencer has a small instruction set of only eight instructions, which are used to implement a state machine. It supports conditional execution and can test the ALU status outputs, handshake signals from the CCU and internal flags. Other sequencer features include support for up to two nested manifest loops at the time and not nested conditional subroutine calls. The sequencer instruction memory can store up to 256 instructions.

2.6.6.2 Compilation

Figure 2.23 shows the entire C to MONTIUM design flow [50]. First the system checks whether a kernel (C code) is already in the library, if so the MONTIUM configurations can be generated directly. Otherwise, a high-level C program is translated into an intermediate CDFG Template language and a hierarchical CDFG is obtained. Next, this graph is cleaned by applying architecture independent transformations (e.g. dead code elimination and common sub-expression elimination transformations).

The next steps are architecture dependent. First the CDFG is clustered. These clusters constitute the 'instructions' of the reconfigurable processor. Examples of clusters are: a butterfly operation for a FFT and a MAC operation for a FIR filter. Clustering is critical step as these clusters (='instructions') are application dependent and should match the capabilities of the processor as close as possible. More information on our clustering algorithm can be found in [58]. Next the clustered graph is scheduled taking the number of ALUs into account. Finally, the resources such as registers, memories and crossbar are allocated. In this phase also some Montium specific transformations are applied, for example, conversion from array index calculations to Montium AGU (Address Generation Unit) instructions, transformation of the control part of the CDFG to sequencer instructions.

Once the graph has been clustered, scheduled, allocated and converted to the Montium architecture, the result is outputted to MontiumC, a cycle true 'human readable' description of the configurations. This description, in an ANSI C++ compatible format, can be compiled with a standard C++ compiler Montium processor.

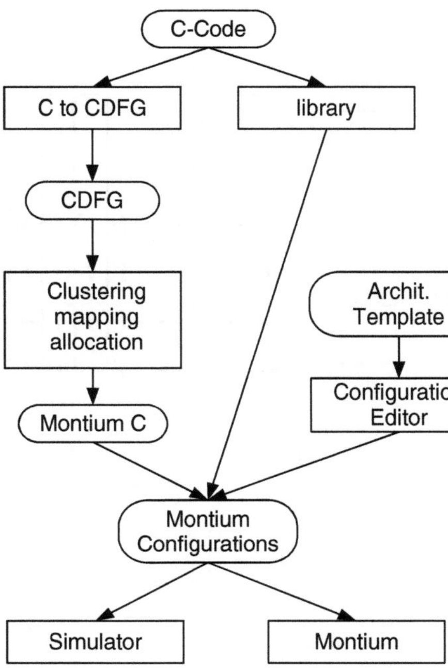

Fig. 2.23 Compilation flow for Montium

2.6.7 PACT XPP

The eXtreme Processing Platform (XPP) [59]–[61] architecture is a runtime reconfigurable data processing technology that consists of a hierarchical array of coarse-grain adaptive computing elements and a packet oriented communication network.

The strength of the XPP architecture comes from the combination of massive array (parallel) processing with efficient run-time reconfiguration mechanisms. Parts of the array can be configured rapidly in parallel while neighboring computing elements are processing data. Reconfiguration is triggered externally or by special event signals originating within the array enabling self-reconfiguration. It also incorporates user transparent and automatic resource management strategies to support application development via high-level programming languages like C.

The XPP architecture is designed to realize different types of parallelism: pipelining, instruction level, data flow, and task level parallelism. Thus, XPP technology is well suited for multimedia, telecommunications, digital signal processing (DSP), and similar stream-based applications.

2.6.7.1 Architecture

The architecture of an XPP device, which is shown in Fig. 2.24, is composed by: an array of 32-bit coarse-grain functional units called Processing Array Elements (PAEs), which are organized as Processing Arrays (PAs), a packet-oriented communication network, a hierarchical Configuration Manager (CM) and high-speed I/O modules.

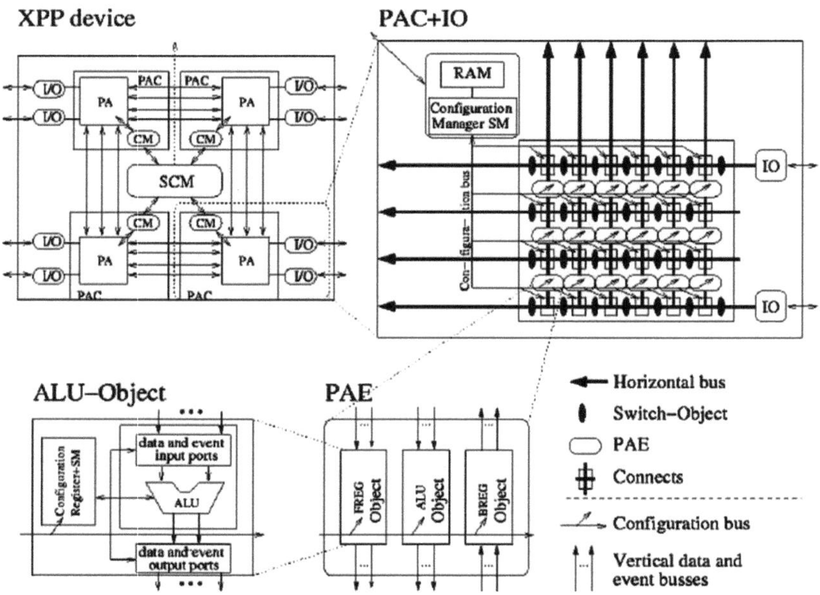

Fig. 2.24 XPP architecture with four Processing Array Clusters (PACs)

An XPP device contains one or several PAs. Each PA is attached to a CM which is responsible to write configuration data into the configurable objects of the PA. The combination of a PA with CM is called Processing Array Cluster (PAC). Multi-PAC devices contain additional CMs for concurrent configuration data handling, forming a hierarchical tree of CMs. The root CM is called the Supervising CM (SCM) and it is equipped with an interface to connect with an external configuration memory.

The PAC itself contains a configurable bus which connects the CM with PAEs and other configurable objects. Horizontal busses are used to connect the objects within a PAE in a row using switches for segmenting the horizontal communication lines. Vertically, each object can connect itself to the horizontal busses using Register-Objects integrated into the PAE.

2.6.7.2 PAE Microarchitecture

A PAE is a collection of configurable objects. The typical PAE contains a back (BREG) and a forward (FREG) register, which are used for vertical routing, and an ALU-object.

The ALU-object contains a state machine (SM), CM interfacing and connection control, the ALU itself and the input and output ports. The ALU performs 32-bit fixed-point arithmetical and logical operations and special three-input operations such as multiply-add, sort, and counters. The input and output ports are able to receive and transmit data and event packets. Data packets are processed by the ALU, while event packets are processed by the state machine. This state machine also receives status information from the ALU, which is used to generate new event packets.

The BREG and FREG objects are not only used for vertical routing. The BREG is equipped with an ALU for arithmetical operations such as add and subtract, and support for normalization, while the FREG has functions which support counters and control the flow of data based on events.

Two types of packets flow through the XPP array: data packets and event packets. Data packets have a uniform bit width specific to the processor type, while event packets use one bit. The event packets are used to transmit state information to control execution and data packet generation. Hardware protocols are used to avoid loss of packets, even during pipelining stalls or configuration cycles.

2.6.7.3 Configuration

As it has been mentioned the strength of the XPP architecture comes from the supported configuration mechanisms, which are presented bellow.

Parallel and User-Transparent Configuration: For rapid reconfiguration, the CMs operate independently and they are able to configure their respective parts of the array in parallel. To relieve the user of synchronizing the configurations, the leaf CM locally synchronizes with the PAEs in the PAC it configures. Once a PAE is configured, it changes its state to "configured" preventing the CM to reconfigure it.

The CM caches the configuration data in its internal RAM until the required PAEs become available. Thus, no global synchronization in needed.

Computation and configuration: While loading a configuration, all PAEs start the computations as soon as they are in "configured" state. This concurrency of configuration and computation hides configuration latency. Additionally, a pre-fetching mechanism is used. After a configuration is loaded onto the array, the next configuration may already be requested and cached in the low-level CMs' internal RAM and in the PAEs.

Self-reconfiguration: Reconfiguration and pre-fetching requests can be issued also by event signals generated in the array itself. These signals are wired to the corresponding leaf CM. Thus, it is possible to execute an application consisting of several phases without any external control. By selecting the next configuration depending on the result of the current one, it is possible to implement conditional execution of configurations and even arrange configurations in loops.

Partial reconfiguration: Finally, XXP also supports partial reconfiguration. This is appropriate for applications in which the configurations do not differ largely. For such cases, partial configurations are much more effective then the complete one. As opposed to complete configurations, partial configurations only describe changes with respect to a given complete configuration.

2.6.7.4 Compilation and Programming

To exploit the capabilities of the XXP architecture an efficient mapping framework is necessary. For that purpose the Native Mapping Language (NML), a PACT proprietary structural language with reconfiguration primitives, was developed [61]. It gives the programmer direct access to all hardware features. Additionally, a complete XPU Development Suite (XDS) has been implemented for NML programming. The tools include a compiler and mapper for the NML, a simulator for the XPP processor models and an interactive visualization and debugging tool. Additionally, a Vectorizing C compiler (XXP-VC) was developed. This translates the C to NML modules and uses vectorization techniques to execute loops in a pipeline fashion. Furthermore, an efficient temporal partitioning technique is also included for executing large programs. This technique spit the original program in several consecutive temporal partitions which executed consecutive by the XXP.

2.6.8 XiRisc

XiRisc (*eXtended Instruction Set RISC*) [62], [63] is a reconfigurable processor that consists of a VLIW processor and a gate array, which is tightly integrated within the CPU instruction set architecture, behaving as part of the control unit and the datapath. The main goal is the exploitation of the instruction level parallelism targeting at a wide range of algorithms including DSP functions, telecommunication, data encryption and multimedia.

2.6.8.1 Architecture

XiRisc, the architecture of which is shown in Fig. 2.25, is a VLIW processor based on the classic RISC five-stage pipeline. It includes hardwired units for DSP calculations and a pipelined run-time configurable datapath (called PiCo gate array or PiCoGA), acting as a repository of application-specific functional units. XiRisc is a load/store architecture, where all data loaded from memory are stored in the register file before they are used by the functional units. The processor fetches two 32-bit instructions each clock cycle, which are executed concurrently on the available functional units, determining two symmetrical separate execution flows called data channels.

General-purpose functional units perform typical DSP calculations such as 32-bit multiply–accumulation, SIMD ALU operations, and saturation arithmetic. On the other hand, the PiCoGA unit offers the capability of dynamically extending the processor instruction set with application-specific instructions achieving run-time configurability. The architecture is fully bypassed, to achieve high throughput.

The PiCoGA is tightly integrated in the processor core, just like any other functional unit, receiving inputs from the register file and writing back results to the register file. In order to exploit instruction-level parallelism, the PiCoGA unit supports up to four source and two destination registers for each instruction issued. Moreover, PiCoGA can hold an internal state across several computations, thus reducing the pressure on connection from/to the register file. Elaboration on the two

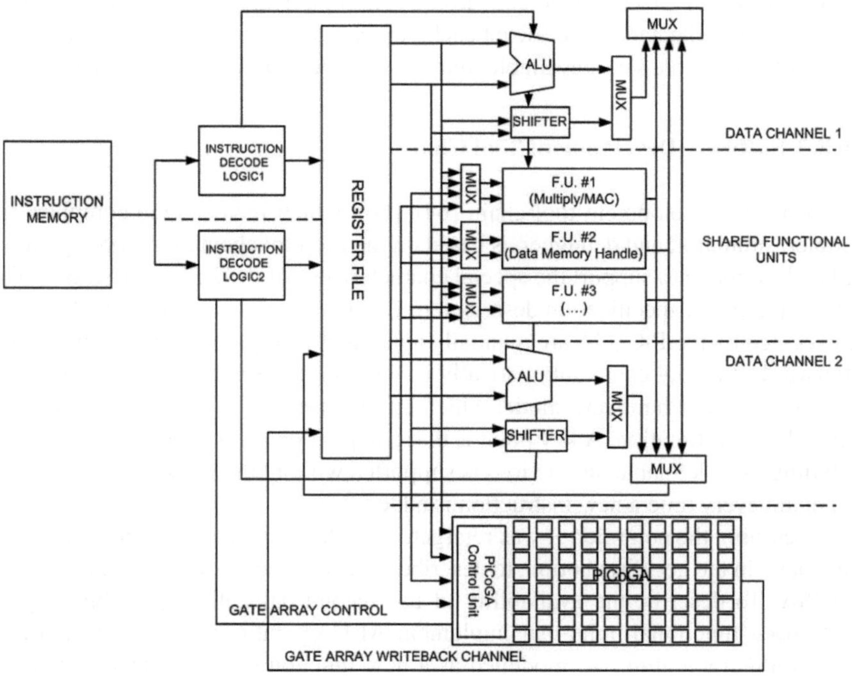

Fig. 2.25 The architecture of XiRisc

hardwired data channels and the reconfigurable data path is concurrent, improving parallel computations. Synchronization and consistency between program flow and PiCoGA elaboration is granted by hardware stall logic based on a register locking mechanism, which handles read-after-write hazards.

Dynamic reconfiguration is handled by a special assembly instruction, which loads a configuration inside the array reading from an on-chip dedicated memory called configuration cache. In order to avoid stalls due to reconfiguration when different PiCoGA functions are needed in a short time span, data of several configurations may be stored inside the array, and are immediately available.

2.6.8.2 Configuration

As the employed PiCoGA is a fine grain reconfigurable, to overcome the associated reconfiguration cost three different approaches have been adopted First, the PiCoGA is provided with a first-level cache, storing four configurations for each reconfigurable logic cell (RLC). Context switch is done in a single clock cycle, providing four immediately available PiCoGA instructions. Moreover, increases in the number of functions simultaneously supported by the array can be obtained exploiting partial run-time reconfiguration, which gives the opportunity for reprogramming only the portion of the PiCoGA needed by the configuration. Second, the PiCoGA may concurrently execute one computation and one reconfiguration instruction which configures the next instruction to be performed. Finally, reconfiguration time can be reduced exploiting a wide configuration bus to the PiCoGA. The RLCs of the array in a row are programmed concurrently through dedicated wires, taking up to 16 cycles. A dedicated second-level cache on chip is used to provide such a wide bus, while the whole set of available functions can be stored in an off-chip memory.

2.6.8.3 Software Development Tool Chain

The software development tool chain [64]–[66], which includes the compiler, assembler, simulator, and debugger, is based on the gcc tool chain that properly modified and extended to support the special characteristics of the XiRisc processor. The input is the initial specification described in C, where sections of the code that must be executed by the PiCoGA are manually annotated with proper *pragma* directives. Afterwards, the tool chain automatically generates the assembler code, the simulation model, and a hardware model which can be used for instruction latency and datapath cost estimation. A key point is that compilation and simulation of software including user-definable instructions is supported without the need to recompile the tool chain every time a new instruction is added.

Concerning the compiler, it was retargeted by changing the machine description files found in the gcc distribution, to describe the extensions to the DLX architecture and ISA. To describe the availability of the second datapath, the multiplicity of all existing functional units that implement ALU operations was doubled, while the reconfigurable unit was modelled as a new function unit. To support different user-defined instruction on the FPGA unit, the FPGA instructions was classified according to their latency. Thus the FPGA function unit was defined as a pipelined resource with a set of possible latencies.

The gcc assembler, is responsible for three main tasks: i) expansion of macro instructions into sequences of machine instructions, ii) scheduling of machine instructions to satisfy constraints, and iii) generation of binary object code. The scheduler was properly modified to handle the second data-path. This contains only an integer ALU, and hence it is able to perform only arithmetic and logical operations. Loads, stores, multiply, jumps, and branches are performed on the main data-path, and hence such 16-bit instructions must be placed at addresses that are multiple of 4. For that reason, nop instructions are inserted whenever an illegal instruction would be emitted at an address that is not a multiple of 4. Also, nop instructions are inserted to avoid scheduling on the second data path an instruction that reads an operand written by the instruction scheduled on the first data path. Also, the file that contains the assembler instruction mnemonics and their binary encodings was modified. This is required to add three classes of instructions: i) the DSP instructions that are treated just as new MIPS instructions and assigned some of the unused op-codes, ii) the FPGA instructions, that have a 6-bit fixed opcode identifying the FPGA instruction class, and an immediate field that defines the specific instruction, and iii) two instructions, called *tofpga* and *fmfpga*, that are used with the simulator to emulate the FPGA instructions with a software model.

Regarding with simulator to avoid recompilation of the simulator every time a new instruction is added to the FPGA, new instructions are modelled as a software function to be compiled and linked with the rest of the application, and interpreted by the simulator. The simulator can be run stand-alone to generate traces, or it can be attached to *gdb* with all standard debugging features, such as breakpoints, step by step execution, source level listing, inspection and update of variables and so on.

2.6.9 ReRisc

Reconfigurable RISC (ReRisc) [67], [68] is an embedded processor extended with a tightly-coupled coarse-grain reconfigurable functional unit (RFU) aiming mainly at DSP and multimedia applications. The efficient integration of the RFU with the control unit and the datapath of the processor eliminate the communication overhead. To improve performance, the RFU exploits Instruction Level Parallelism (ILP) and spatial computation. Also, the integration of the RFU efficiently exploits the pipeline structure of the processor, leading to further performance improvements. The processor is supported by a development framework which is fully automated, hiding all reconfigurable-hardware related issues from the user.

2.6.9.1 Architecture

The processor is based on standard 32-bit, single-issue, five-stage pipeline RISC architecture that has been extended with the following features: a) Extended ISA to support three types of operations performed by the RFU, which are complex computations, complex addressing modes, and complex control transfer operations, b) an interface supporting the tightly coupling of an RFU to the processor pipeline, and c) an RFU array of Processing Elements (PEs).

The RFU is capable to execute complex instructions which are Multiple Input single Output (MISO) clusters of the processor instructions. Exploiting the clock slack and instruction parallelism, the execution of the MISO clusters by the RFU leads in a reduced latency, compared to the latency when these instructions are sequentially executed by the processor core. Also, both the execution (EX) and memory (MEM) stages of the processor's pipeline are used to process a reconfigurable instruction.

On each execution cycle an instruction is fetched from the Instruction Memory. If the instruction is identified (based on special bit of the instruction word) as reconfigurable its opcode and instruction operands from the register file are forwarded to the RFU. In addition, the opcode is decoded and produces the necessary control signals to drive the Core/RFU interface and pipeline. At the same time the RFU is appropriately configured by downloading the necessary configuration bits from a local configuration memory with no extra cycle penalty.

The processing of the reconfigurable instruction is initiated in the execution pipeline stage. If the instruction has been identified as addressing mode or control transfer then its result is delivered back to the execution pipeline stage to access the data memory or the branch unit, respectively. Otherwise, the next pipeline stage is also used in order to execute longer chains of operations and improve performance. In the final stage results are delivered back to the register file. Since instructions are issued and completed in-order, while all data hazards are resolved in hardware, the architecture does not require any special attention by the compiler.

2.6.9.2 RFU Organization and PE Microarchitecture

The Processing & Interconnect Layers of the RFU consist of a 1-Dimension array of PEs (Fig. 2.27a). The array features an interconnection network that allows

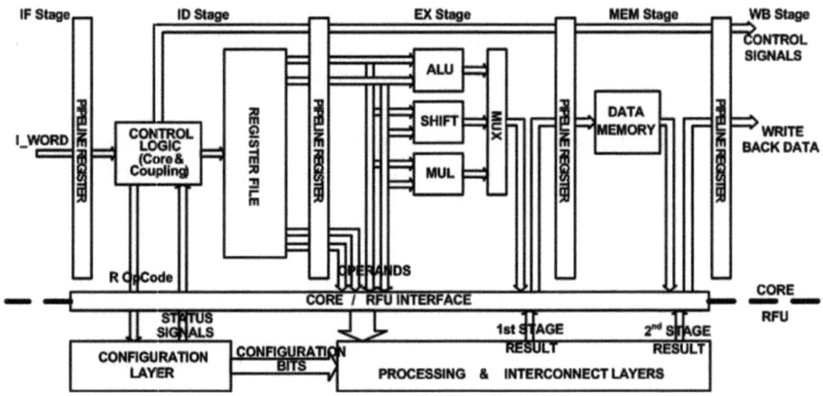

Fig. 2.26 The architecture of ReRisc

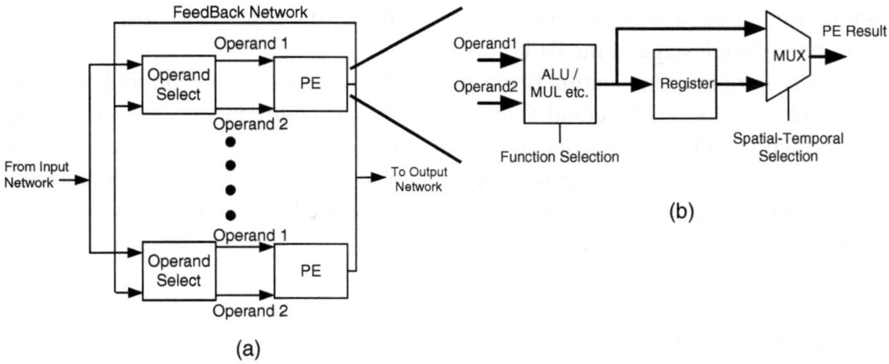

Fig. 2.27 (a): The organization of the RFU and (b): the microarchitecture of the PE

connection of all PEs to each other. The granularity of PEs is 32-bit allowing the execution of the same word-level operations as the processor's datapath. Furthermore, each PE can be configured to provide its un-register or register result (Fig. 2.27b). In the first case, spatial computation is exploited (in addition to parallel execution) by executing chains of operations in the same clock cycle. When the delay of a chain exceeds the clock cycle, the register output is used to exploit temporal computation by providing the value to the next pipeline stage.

2.6.9.3 Interconnection Layer

The interconnection layer (Fig. 2.28) features two global blocks for the intercommunication of the RFU: Input Network and Output Network. The former is responsible to receive the operands from the register file and the local memory and delivers to the following blocks their registered and unregistered values. In this way, operands for both execution stages of the RFU are constructed. The Output Network

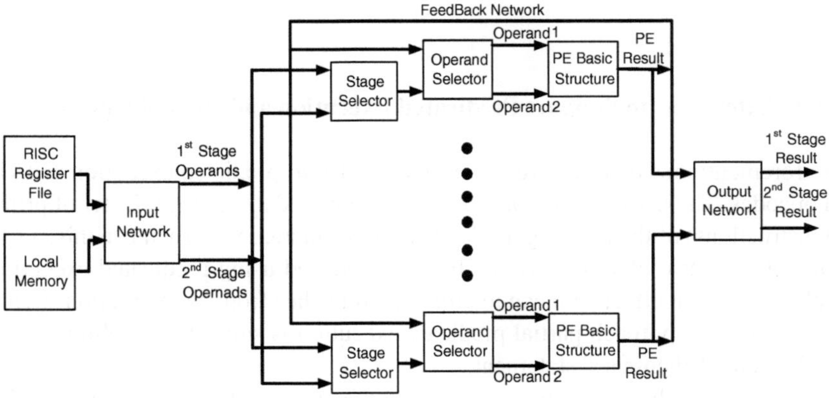

Fig. 2.28 The interconnection layer

can be configured to select the appropriate PE result that is going to be delivered to the output of each stage of the RFU.

For the intra-communication between the PEs, two blocks are offered for each PE: Stage Selector and Operand Selector. The first is configured to select the stage from which the PE receives operands. Thus, this block is the one that configures the stage that each PE will operate. Operand Selector receives the final operands, in addition with feedbacks from each PE and is configured to forward the appropriate values.

2.6.9.4 Configuration Layer

The components of the Configuration layer are shown in Fig. 2.29. On each execution cycle the opcode of the reconfigurable instruction is delivered from the core processor's Instruction Decode stage to the RFU. The opcode is forwarded to a local structure that stores the configuration bits of the locally available instructions. If the required instruction is available the configuration bits for the processing and interconnection layers are retrieved. In a different case, a control signal indicates that new configuration bits must be downloaded from an external configuration memory to the local storage structure and the processor execution stalls.

In addition, as part of the configuration bit stream of each instruction, the storage structure delivers two words, that each one indicates the Resources Occupation required for the execution of the instruction on the corresponding stage. These words are forwarded to the Resource Availability Control Logic, that stores for one cycle the 2nd Stage Resource Occupation Word. On each cycle the logic compares the 1st Stage Resource Occupation of the current instruction with the 2nd of the previously instruction. If a resource conflict is produced, a control signal indicates to the processor core to stall the pipeline execution for one cycle. Finally, the retrieved configuration bits moves through pipeline registers to the first and second execution stage of the RFU. A multiplexer controlled by the Resource Configuration bits, selects the correct configuration bits for each PE and its corresponding interconnection network.

2.6.9.5 Extensions to Support Predicated Execution and Virtual Opcode

The aforementioned architecture has been extended to predicted execution and virtual opcodes. The performance can be further improved if the size (cluster of primitive instructions) of the reconfigurable instructions increases. To achieve this, a way is size of the basic blocks. This can be accomplished using predicated execution, which provides an effective mean to eliminate branches from an instruction stream. In the proposed approach partial predicate execution is supported to eliminate the branch in an "if-then-else" statement.

As mentioned, the explicitly communication between the processor and the RFU involves the direct encoding of reconfigurable instructions to the opcode of the

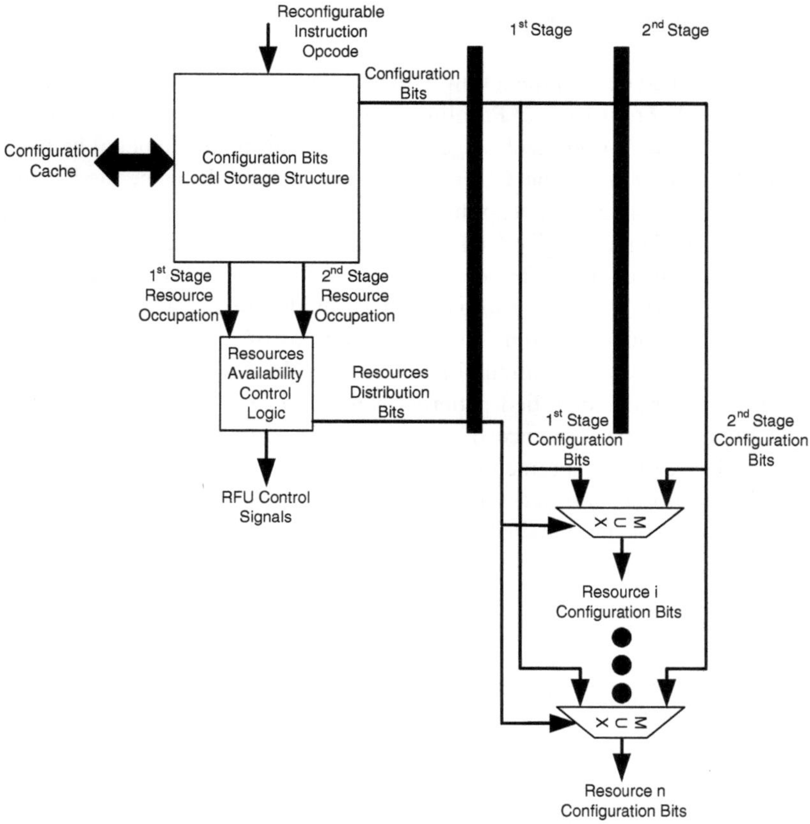

Fig. 2.29 The configuration layer

instruction word. This fact limits the number of reconfigurable instructions that can be supported, leaving unutilized available performance improvements. On the other hand, the decision to increase the opcode space requires hardware and software modifications. Such modifications may be in general unacceptable. To address this problem an enhancement at the architecture called "*virtual opcode*" is employed. Virtual opcode aims at increasing the available opcodes without increasing the size of the opcode bits or modify the instruction's word format. Each virtual opcode consists of two parts. The first is the native opcode contained in the instruction word that has been fetched for execution in the RFU. The second is a value indicating the region of the application in which this instruction word has been fetched. This value is stored in the configuration layer of the RFU for the whole time the application execution trace is in this specific region. Combining the two parts, different instructions can be assigned to the same native opcode across different regions of the application featuring a virtually "*unlimited*" number of reconfigurable instructions.

2.6.9.6 Compilation and Development Flow

The compilation and development flow, which shown in Fig. 2.30, is divided in five stages, namely: 1) Front-End, 2) Profiling, 3) Instruction Generation, 4) Instruction Selection, and 5) Back-End. Each stage of the flow is presented in detail below.

At the *Front-End stage* the CDFG of the application IIR generated, while a number of machine-independent optimizations (e.g. dead code elimination, strength reduction) are performed on the CDFG. At the *Profiling stage* using proper SUIF passes profiling information for the execution frequency of the basic blocks is collected. The *Instruction Generation* stage is divided in two steps. The goal of the first step is the identification of complex patterns of primitive operations that can be merged into one reconfigurable instruction. In the second step, the mapping of the previously identified patterns in the RFU is performed and to evaluate the impact on performance of each possible reconfigurable instruction as well as to derive its requirements in terms of hardware and configuration resources. At the *Instruction Selection*: stage the new instructions are selected. To bound the number of the new instructions graph isomorphism techniques are employed.

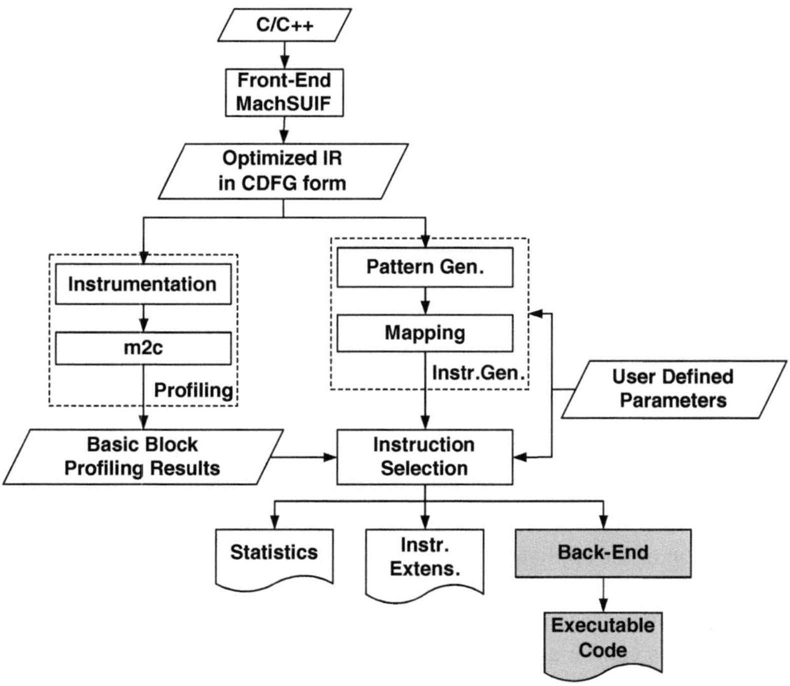

Fig. 2.30 Compilation flow

2.6.10 Morphosys

Morphosys [52] is a coarse-grain reconfigurable systems targeting mainly at DSP and multimedia applications. Because it is presented in details in a separate chapter in this book, we discuss briefly only its architecture.

2.6.10.1 Architecture

MorphoSys consists of a core RISC processor, an 8 × 8 reconfigurable array of identical PEs, and a memory interface as shown in Fig. 2.1. At intra-cell level, each PE is similar to a simple microprocessor except that an instruction is replaced with a context word and there is no instruction decoder or program counter. The PE comprised of an ALU-multiplier and a shifter connected in series. The output of the shifter is temporarily stored in an output register and then goes back to the ALU/multiplier, to a register file, or to other cells. Finally, for the inputs of the ALU/multiplier muxes are used, which select the input from several possible sources (e.g. register file, neighboring cells). The bitwidth of the functional or storage units is at least 16 bits except the multiplier, which supports multiplication of 16 × 12 bits. The function of the PEs is configured by a context word, which defines the opcode and an optional constant and the control signals.

At inter-cell level, there are two major components: the interconnection network and the memory interface. Interconnection exists between the cells of either the same row or the same column. Since the interconnection network is symmetrical and every row (column) has the same interconnection with other rows (columns), it is enough to define only interconnections between the cells of one row. For a row, there are two kinds of connections. One is dedicated interconnection between two cells of the row. This is defined between neighboring cells and between cells of every 4-cell group. The other kind of connection is called *express lane* and provides a direct path from any one of each group to any one in the other group. The memory interface consists of Frame Buffer and memory buses. To support a high

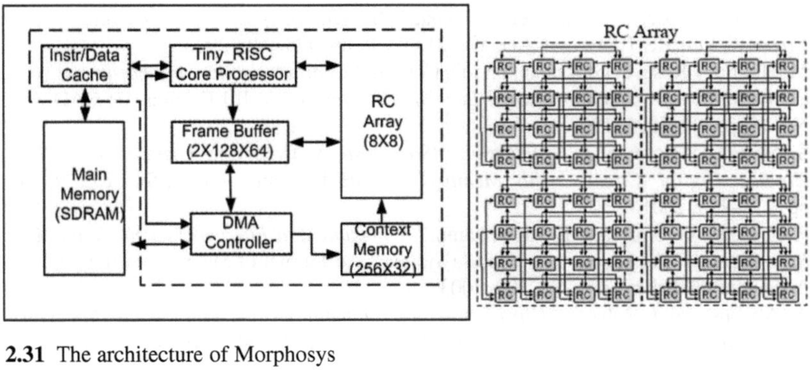

Fig. 2.31 The architecture of Morphosys

bandwidth, the architecture uses a DMA unit, while overlapping of the data transfer with computation is also supported.

The context memory has 32 *context planes*, with a context plane being a set of context words to program the entire array for one cycle. The dynamic reloading of any of the context planes can be done concurrently with the RC Array execution.

References

1. K. Compton and S. Hauck, "*Reconfigurable Computing a Survey of Systems and Software*", in ACM Computing Surveys, Vol. 34, No. 2, pp.171–210, June 2002.
2. A. De Hon and J. Wawrzyenk, "*Reconfigurable Computing" What, Why and Implications of Design Automation*", in Proc. of DAC, pp. 610–615, 1999.
3. R. Hartenstein, "*A Decade of Reconfigurable Computing: a Visionary Perspective*", in Proc. of DATE, pp. 642–649, 2001.
4. A. Shoa and S. Shirani, "*Run-Time Reconfigurable Systems for Digital Signal Processing Applications: A Survey*", in Journal of VLSI Signal Processing, Vol. 39, pp. 213–235, 2005, Springer Science.
5. P. Schaumont, I.Verbauwhede, K. Keutzer, and Majid Sarrafzadeh, "*A Quick Safari Through the Reconfigurable Jungle*", in Proc. of DAC, pp. 172–177, 2001.
6. R. Hartenstein, "*Coarse Grain Reconfigurable Architectures*", in. Proc. of ASP-DAC, pp. 564–570, 2001.
7. F. Barat, R.Lauwereins, and G. Deconick, "*Reconfigurable Instruction Set Processors from a Hardware/Software Perspective*", in IEEE Trans. on Software Engineering, Vol. 28, No.9, pp. 847–862, Sept. 2002.
8. M. Sima, S. Vassiliadis, S. Cotofana, J. Eijndhoven, and K. VIssers, "*Field-Programmable Custom Computing Machines–A Taxonomy-*", in Proc. of Int. Conf. on Field Programmable Logic and Applications (FLP), pp. 77–88, Springer-Verlag, 2002.
9. I. Kuon, and J. Rose, "*Measuring the Gap Between FPGAs and ASICs*", in IEEE Trans. on CAD, vol 26., No 2., pp. 203–215, Feb 07.
10. A. De Hon, "*Reconfigurable Accelerators*", Technical Report 1586, MIT Artificial Intelligence Laboratory, 1996.
11. K. Compton, "*Architecture Generation of Customized Reconfigurable Hardware*", Ph.D Thesis, Northwestern Univ, Dept. of ECE, 2003.
12. K. Compton and S. Hauck, "*Flexibility Measurement of Domain-Specific Reconfigurable Hardware*", in Proc. of Int. Symp. on FPGAs, pp. 155–161, 2004.
13. J. Darnauer and W.W.-M. Dai, "*A Method for Generating Random Circuits and its Application to Routability Measurement*", in Proc. of Int. Symp. on FPGAs, 1996.
14. M. Hutton, J Rose, and D. Corneli, "*Automatic Generation of Synthetic Sequential Benchmark Circuits*", in IEEE Trans. on CAD, Vol. 21, No. 8, pp. 928–940, 2002.
15. M. Hutton, J Rose, J. Grossman, and D. Corneli, "*Characterization and Parameterized Generation of Synthetic Combinational Benchmark Circuits*:" in IEEE Trans. on CAD, Vol. 17, No. 10, pp. 985–996, 1998.
16. S. Wilton, J Rose, and Z. Vranesic, "*Structural Analysis and Generation of Synthetic Circuits Digital Circuits with Memory*", in IEEE Trans. on VLSI, Vol. 9, No. 1, pp. 223–226, 2001.
17. P. Heysters, G. Smit, and E. Molenkamp, "*A Flexible and Energy-Efficient Coarse-Grained Reconfigurable Architecture for Mobile Systems*", in Journal of Supercomputing, 26, Kluwer Academic Publishers, pp. 283–308, 2003.
18. A. Abnous and J. Rabaey, "*Ultra-Low-Power Domain-Specific Multimedia Processors*", in proc. of IEEE Workshop on VLSI Signal Processing, pp. 461–470, 1996.

19. M. Wan, H. Zhang, V. George, M. Benes, A. Arnous, V. Prabhu, and J. Rabaey, "*Design Methodology of a Low-Energy Reconfigurable Single-Chip DSP System*", in Journal of VLSI Signal Processing, vol. 28, no. 1–2, pp. 47–61, May-June 2001.
20. K. Compton, and S. Hauck, "*Totem: Custom Reconfigurable Array Generation*": in IEEE Symposium on FPGAs for Custom Machines, pp. 111–119, 2001.
21. Z. Huang and S. Malik, "*Exploiting Operational Level Parallelism through Dynamically Reconfigurable Datapaths*", in Proc. of DAC, pp. 337–342, 2002.
22. Z. Huang and S. Malik, "*Managing Dynamic Reconfiguration Overhead in Systems –on-a-Chip Design Using Reconfigurable Datapaths and Optimized Interconnection Networks*", in Proc. of DATE, pp. 735–740, 2001.
23. Z. Huang, S. Malik, N. Moreano, and G. Araujo, "*The Design of Dynamically Reconfigurable Datapath Processors*", in ACM Trans. on Embedded Computing Systems, Vol. 3, No. 2, pp. 361–384, 2004.
24. B. Mei, S. Vernadle, D. Verkest, H. De Man, and R. Lauwereins, "*ADRES: An Architecture with Tightly Coupled VLIW Reconfigurable Processor and Coarse-Grained Reconfigurable Matrix*", in Proc. of Int. Conf. on Field Programmable Logic and Applications (FLP), pp. 61–70, 2003.
25. T. Miyamori and K. Olukotun, "*REMARC: Reconfigurable Multimedia Array Coprocessor*", in Proc. of Int. Symp. on Field Programmable Gate Arrays (FPGA), pp. 261, 1998.
26. D. Gronquist, P. Franklin, C. Fisher, M. Figeoroa, and C. Ebeling, "*Architecture Design of Reconfiguable Pipeline Datapaths*", in Proc. of Int. Conf. on Advanced VLSI, pp. 23–40, 1999.
27. C. Ebeling, D. Gronquist, P. Franklin, J. Secosky and, S. Berg, "*Mapping Applications to the RaPiD configurable Architecture*", in Proc. of Int. Symp. on Field-Programmable Custom Computing Machines (FCCM), pp. 106–115, 1997.
28. D. Gronquist, P. Franklin, S. Berg and, C. Ebeling, "*Specifying and Compiling Applications on RaPiD*", in Proc. of Int. Symp. on Field-Programmable Custom Computing Machines (FCCM), pp. 116, 1998.
29. C. Ebeling, C. Fisher, C. Xing, M. Shen, and H. Liu, "Implementing an OFDM Receiver on the Rapid Reconfigurable Architecture", in IEEE Trans. on Cmputes, Vol. 53, No. 11., pp. 1436–1448, Nov. 2004.
30. C. Ebeling, L. Mc Murchie, S. Hauck, and S. Burns, "*Placement and Routing Tools for the Triptych FPGA*", in IEEE Trans. on VLSI Systems, Vol. 3, No. 4, pp. 473–482, Dec. 1995.
31. S. Goldstein, H. Schmit, M. Moe, M.Budiu, S. Cadambi, R. Taylor, and R. LEfer, "*PipeRench: A Coprocessor for Streaming Multimedia Acceleration*", in Proc. of International Symposium on Computer Architecture (ISCA), pp. 28–39, 1999.
32. N. Bansal, S. Goupta, N. Dutt, and A. Nicolaou, "*Analysis of the Performance of Coarse-Grain Reconfigurable Architectures with Different Processing Elements Configurations*", in Proc. of Workshop on Application Specific Processors (WASP), 2003.
33. B. Mei, A. Lambrechts, J-Y. Mignolet, D. Verkest, and R. Lauwereins, "*Architecture Exploration for a Reconfigurable Architecture Template*", in IEEE Design and Test, Vol. 2, pp. 90–101, 2005.
34. H. Zang, M. Wan, V. George, and J. Rabaey, "*Interconnect Architecture Exploration for Low-Energy Reconfigurable Single-Chips DSPs*", in proc. of Annual Workshop on VLSI, pp. 2–8, 1999.
35. K. Bondalapati and V. K. Prasanna, "*Reconfigurable Meshes: Theory and Practice*", in proc. of Reconf. Architectures Workshop, International Parallel Processing Symposium, 1997.
36. N. Kavalgjiev and G. Smit, "*A survey for efficient on-chip communication for SoC*", in Proc. PROGRESS 2003 Embedded Systems Symposium, October 2003.
37. N. Bansal, S. Goupta, N. Dutt, A. Nicolaou and R. Goupta, "*Network Topology Exploration for Mesh-Based Coarse-Grain Reconfigurable Architectures*", in Proc. of DATE, pp. 474–479, 2004.
38. J. Lee, K. Choi, and N. Dutt, "*Compilation Approach for Coarse-Grained Reconfigurable Architectures*", in IEEE Design & Test, pp. 26–33, Jan-Feb. 2003.
39. J. Lee, K. Choi, and N. Dutt, "*Mapping Loops on Coarse-Grain Reconfigurable Architectures Using Memory Operation Sharing*", Tech. Report, Univ. of California, Irvine, Sept. 2002.

40. G. Dimitroulakos, M.D. Galanis, and C.E. Goutis, "*A Compiler Method for Memory-Conscious Mapping of Applications on Coarse-Grain Reconfigurable Architectures*", in Proc. of IPDPS 05.
41. K. Compton and S. Hauck, "*Flexible Routing Architecture Generation for Domain-Specific Reconfigurable Subsystems*", in Proc. of Field-Programming Logic and Applications (FPL), pp. 56–68, 2002.
42. K. Compton and S. Hauck, "*Automatic Generation of Area-Efficient Configurable ASIC Cores*", submitted to IEEE Trans. on Computers.
43. R. Kastner et al., "*Instruction Generation for Hybrid Reconfigurable Systems*", in ACM Transactions on Design Automation of Embedded Systems (TODAES), vol 7., no.4, pp. 605–627, October, 2002.
44. J. Cong et al., "*Application-Specific Instruction Generation for Configurable Processor Architectures*", in Proc. of ACM International Symposium on Field-Programmable Gate Arrays (FPGA 2004), 2004.
45. R. Corazao et al., "*Performance Optimization Using Template Mapping for Data-path-Intensive High-Level Synthesis*", in IEEE Trans. on CAD, vol.15, no. 2, pp. 877–888, August 1996.
46. S. Cadambi and S. C. Goldstein, "*CPR: a configuration profiling tool*", in Symposium on Field-Programmable Custom Computing Machines (FCCM), 1999.
47. K. Atasu, et al., "*Automatic application-specific instruction-set extensions under microarchitectural constraints*", in Proc. of Design Automation Conference (DAC 2003), pp. 256–261, 2003.
48. B. Mei, S. Vernadle, D. Verkest, H. De Man., and R. Lauwereins, "*DRESC: A Retargatable Compiler for Coarse-Grained Reconfigurable Architectures*", in Proc. of Int. Conf. on Field Programmable Technology, pp. 166–173, 2002.
49. B. Mei, S. Vernadle, D. Verkest, and R. Lauwereins, "*Design Methodology for a Tightly Coupled VLIW/Reconfigurable Matrix Architecture: A Case Study*", in proc. of DATE, pp. 1224–1229, 2004.
50. P. Heysters, and G. Smit, "*Mapping of DSP Algorithms on the MONTIUM Architecture*", in Proc. of Engin. Reconfigurable Systems and Algorithms (ERSA), pp. 45–51, 2004.
51. G. Venkataramani, W. Najjar, F. Kurdahi, N. Bagherzadeh, W. Bohm, and J. Hammes, "*Automatic Compilation to a Coarse-Grained Reconfigurable System-on-Chip*", in ACM Trans. on Embedded Computing Systems, Vol. 2, No. 4, November 2003, Pages 560–589.
52. H. Singh, M-H Lee, G. Lu, F. Kurdahi, N. Begherzadeh, and E.M.C. Filho, "*MorphoSys: an Integrated Reconfigurable System for Data Parallel and Computation-Intensive Applications*", in IEEE Trans. on Computers, 2000.
53. Quinton and Y. Robert, "*Systolic Algorithms and Architectures*", Prentice Hall, 1991.
54. H. Schmit *et al.*, "*PipeRech: A Virtualized Programmable Datapath in 0.18 Micron Technology*", in Proc. of Custom Integrated Circuits, pp. 201–205, 2002.
55. S. Goldstein *et al.*, "*PipeRench: A Reconfigurable Architecture and Compiler*", in IEEE Computers, pp. 70–77, April 2000.
56. B. Mei, S. Vernadle, D. Verkest, H. De Man., and R. Lauwereins, "*Exploiting loop-Level Parallelism on Coarse-Grain Reconfigurable Architectures Using Modulo Scheduling*", in proc. of DATE, pp. 296–301, 2003.
57. B.R. Rao, "*Iterative Modulo Scheduling*", Technical Report, Hewlett-Packard Lab:HPL-94–115, 1995.
58. Y. Guo, G. Smit, P. Heysters, and H. Broersma "*A Graph Covering Algorithm for a Coarse Grain Reconfigurable System*", in Proc. of LCTES 2003, pp. 199–208, 2003
59. V. Baumgarte, G. Ehlers, F. May, A. Nuckel, M. Vorbach, and W. Weinhardt, "*PACT XPP-A Self-Reconfigurable Data Processing Architecture*", in Journal of Supercomputing, Vol. 26, pp. 167–184, 2003, Kluwer Academic Publishers.
60. [60]"*The XPP White Paper*", available at http://www.pactcorp.com.
61. J. Cardoso and M. Weinhardt, "*XPP-VC: A C Compiler with Temporal Partitioning for the PACT-XPP Architecture*", in Proc. of Field-Programming Logic and Applications (FPL), pp. 864–874, Springer-Verlag, 2002.

62. A. Lodi, M. Toma, F. Campi, A. Cappelli, R. Canegallo, and R. Guerrieri, "*A VLIW Processor With Reconfigurable Instruction Set for Embedded Applications*", in IEEE journal of solid-state circuits, vol. 38, no. 11, November 2003, pp. 1876–1886.
63. A. La Rosa, L. Lavagno, and C. Passerone, "Implementation of a UMTS Turbo Decoder on a Dynamically Reconfigurable Platform", in IEEE trans. on CAD, Vol. 24, No. 3, pp. 100–106, Jan. 2005.
64. A. La Rosa, L. Lavagno, and C. Passerone, "*Software Development Tool Chain for a Reconfigurable Processor*", in proc. of CASES, pp. 93–88, 2001.
65. A. La Rosa, L. Lavagno, and C. Passerone, "Hardware/*Software Design Space Exploration for a Reconfigurable Processor*", in proc. of DATE, 2003.
66. A. La Rosa, L. Lavagno, and C. Passerone, "*Software Development for High-Performance, Reconfigurable, Embedded Multimedia Systems*", in IEEE Design & Test of Computers, Jan-Feb 2005, pp. 28–38.
67. N. Vassiliadis, N. Kavvadias, G. Theodoridis, and S. Nikolaidis, "*A RISC Architecture Extended by an Efficient Tightly Coupled Reconfigurable Unit*", in International Journal of Electronics, Taylor & Francis, vol.93, No. 6., pp. 421–438, 2006 (Special Issue Paper of ARC05 conference).
68. N. Vassiliadis, G. Theodoridis, and S. Nikolaidis, "*Exploring Opportunities to Improve the Performance of a Reconfigurable Instruction Set Processor*", accepted for publication in *International Journal of Electronics, Taylor & Francis,* (Special Issue Paper of ARC06 conference).
69. S. Cadambi. J. Weener, S. Goldstein. H. Schmit, and D. Thomas, "*Managing Pipeline-Reconfigurable FPGAs*", in Proc. of Int. Symp. on Field Programmable Gate Arrays (FPGA), pp. 55–64, 1998.

Part II
Case Studies

Chapter 3
Amdrel

A Low-Energy FPGA Architecture and Supporting CAD Tool Design Flow*

D. Soudris, K. Tatas, K. Siozios, G. Koutroumpezis, S. Nikolaidis, S. Siskos, N. Vasiliadis, V. Kalenteridis, H. Pournara and I. Pappas

This chapter describes a complete system for the implementation of digital logic in a fine-grain reconfigurable platform (FPGA). The energy-efficient FPGA architecture is designed and simulated in STM 0.18μm CMOS technology. The detailed design and circuit characteristics of the Configurable Logic Block and the interconnection network are determined and evaluated in terms of energy, delay and area. A number of circuit-level low-power techniques are employed because power consumption is the primary concern.

Additionally, a complete tool framework for the implementation of digital logic circuits in FPGA platforms is introduced. The framework is composed of i) non-modified academic tools, ii) modified academic tools and iii) new tools. The developed tool framework supports a variety of FPGA architectures. Qualitative and quantitative comparisons with existing academic and commercial architectures and tools are provided, yielding promising results.

3.1 Introduction

FPGAs have recently benefited from technology process advances to become significant alternatives to Application Specific Integrated Circuits (ASICs). An important feature that has made FPGAs, particularly attractive is a logic mapping and implementation flow similar to the ASIC design flow (from VHDL or Verilog down to the configuration bitstream) provided by the industrial sector [1, 2]. However, in order to implement real-life applications on an FPGA platform, embedded or discrete, increasingly performance and power-efficient FPGA architectures are required. Furthermore, efficient architectures cannot be used effectively without a complete set

* This work was partially supported by the project IST-34793-AMDREL which is funded by the E.C.

S. Vassiliadis, D. Soudris (eds.), *Fine- and Coarse-Grain Reconfigurable Computing.*

of tools for implementing logic while utilizing the advantages and features of the target device.

Consequently, research has lately focused on the development of FPGA architectures [3, 4, 5, 6, 7, 8] as mentioned in Chapter I. Many solid efforts for the development of a complete tool design flow from the academic sector have also taken place ([6, 9, 10]). The above design groups have focused on the development of tools that can target a variety of FPGA architectures, while keeping the tools open-source.

Despite the above efforts, there is a gap in the complete design flow (from VHDL to configuration bit-stream) provided by existing academic tools. This is due to the lack of an open-source synthesizer and a FPGA configuration bit-stream generation tool. Therefore, there is no existing complete academic system capable of implementing, in a FPGA, logic specified in a hardware description language. Just an assortment of various fine-grain architectures and tools that cannot be easily integrated into a complete system.

In this chapter, such a complete system is presented. The hardware design of an efficient FPGA architecture is presented in detail. An exploration in terms of power, delay and area at both Configurable Logic Block (CLB) design and interconnection architecture has been applied in order to make appropriate architecture decisions. Particularly, Basic Logic Element (BLE) using gated clock approach is investigated, at CLB level, while at the interconnect network level, new research results about the type and sizing of routing switches are presented in 0.18μm process. This investigation is mostly focused on minimizing power dissipation, since this was our primary target in this FPGA implementation, without significantly degrading delay and area.

Additionally, a complete toolset for mapping logic on the FPGA mentioned above is presented, starting from a VHDL circuit description down to the FPGA configuration bitstream. The framework is composed of i) non-modified academic tools, ii) modified academic tools and iii) new tools. The developed tool framework supports a variety of FPGA architectures. The FPGA architecture and tools were developed as part of the AMDREL project [11] and the tools can be run on-line at the AMDREL website.

The rest of the chapter is organized as follows: Section 3.2 describes the FPGA hardware platform in detail, while Sect. 3.3 is a brief presentation of the tools. Sect. 3.4 provides a number of quantitative and qualitative comparisons with existing academic and commercial approaches in order to evaluate the entire system of tools and platform. Conclusions are further discussed in Sect. 3.5.

3.2 FPGA Architecture

The architecture that was designed is an island-style embedded FPGA [5] (Fig. 3.1). The main design consideration during the design of the FPGA platform was power minimization under the delay constraints, while maintaining a reasonable silicon area. The purpose of this chapter is to present the entire system of hardware architecture and software tools and not to focus on each design parameter detail. Therefore,

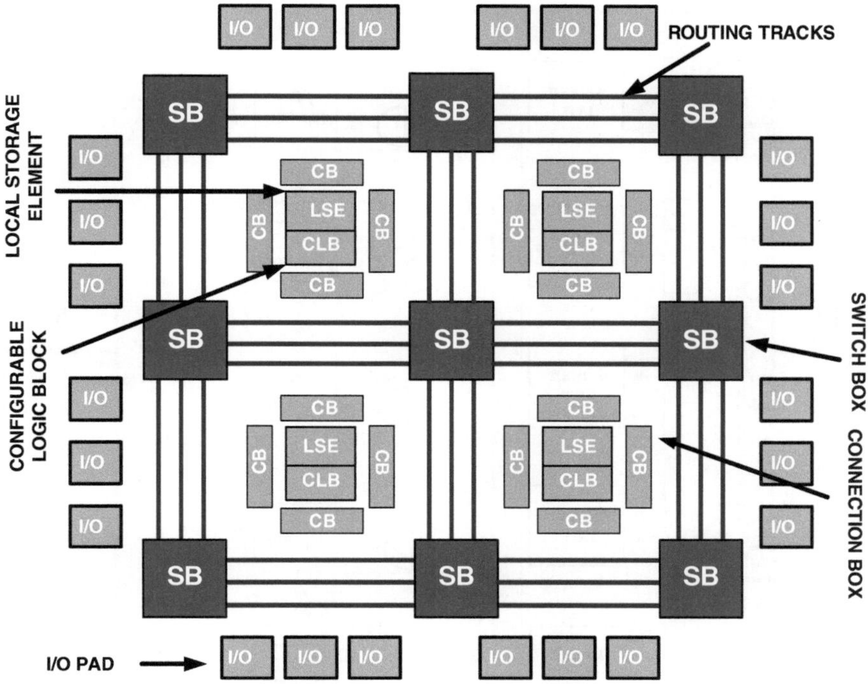

Fig. 3.1 AMDREL FPGA structure

the FPGA design parameters, which were selected through exploration in terms of power, delay and area in [12, 13, 14], are briefly described here.

3.2.1 Configurable Logic Block (CLB) Architecture

The design of the CLB architecture is crucial to the CLB granularity, performance, and power consumption. The developed CLB consists of a collection of Basic Logic Elements (BLEs), which are interconnected by a local network, as shown in Fig. 3.2. A number of parameters have to be determined: a) the number of the Look-Up Table (LUT) inputs, K, b) the number of BLEs per CLB (cluster size), N and c) the number of CLB inputs, I.

3.2.1.1 LUT Inputs (K)

The LUT is used for the implementation of logic functions. It has been demonstrated in [8] that 4-input LUTs lead to the lowest power consumption for the FPGA, providing an efficient area-delay product.

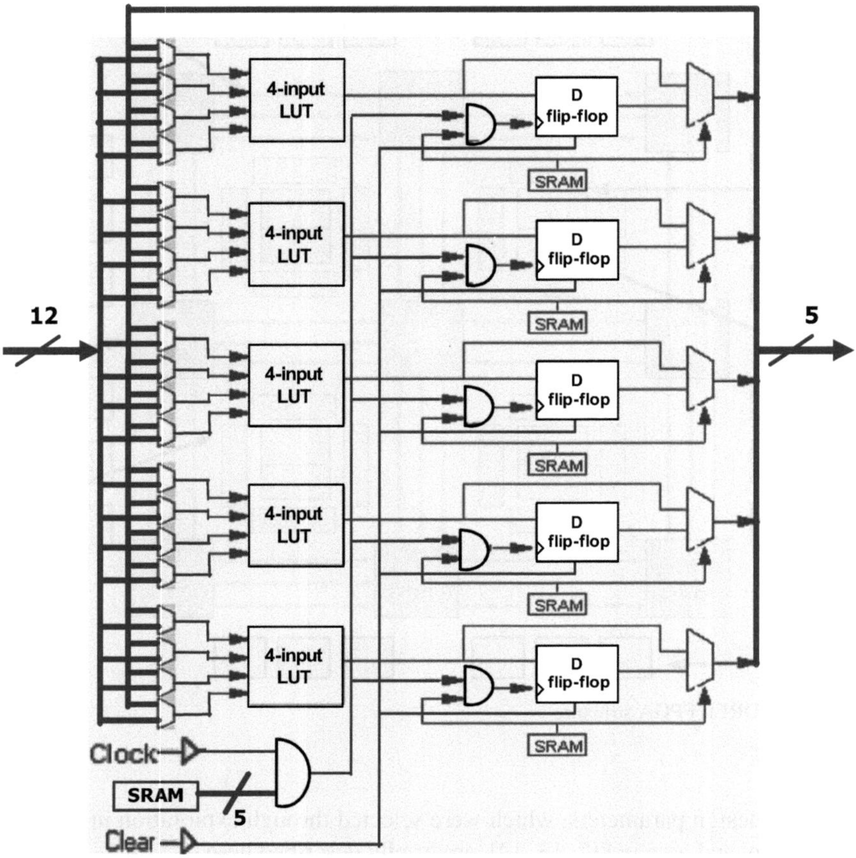

Fig. 3.2 CLB structure

3.2.1.2 CLB Inputs (*I*)

An exploration for finding the optimal number of CLB inputs, which provides 98 % utilization of all the BLEs [8], results in an almost linear dependency with the number of LUT inputs (*K*), and the cluster size (*N*), considering the formula: $\boldsymbol{I} = (K/2) \times (N+\boldsymbol{1})$

The above design parameter decisions affect the tools seen in the next section.

3.2.1.3 Cluster Size (*N*)

The Cluster Size corresponds to the number of BLEs within a CLB. Taking into account mostly the minimization of power consumption, our design exploration showed that a cluster size of 5 BLEs leads to the minimization of power consumption (Fig. 3.2) [12, 14].

3.2.2 CLB Circuit Design

The CLB was designed at transistor level in order to obtain the maximum power savings. It is well-known that the minimization of the effective circuit capacitance leads to low power consumption. This is achieved by using minimum-sized transistors, at the cost of some delay time. Power consumption minimization involves some techniques such as logic threshold adjustment in critical buffers and gated clock technique. Simulations were performed in Cadence framework using 0.18 STM technology.

3.2.2.1 LUT and Multiplexer Design

The 4-input LUT is implemented by using a multiplexer (MUX), as shown in Fig. 3.3. The main difference from a typical MUX is that the control signals are the inputs to the LUT and the inputs to the multiplexer are stored in memory cells ($S_0 - S_{15}$). The LUT and MUX structures with the minimum-sized transistors were

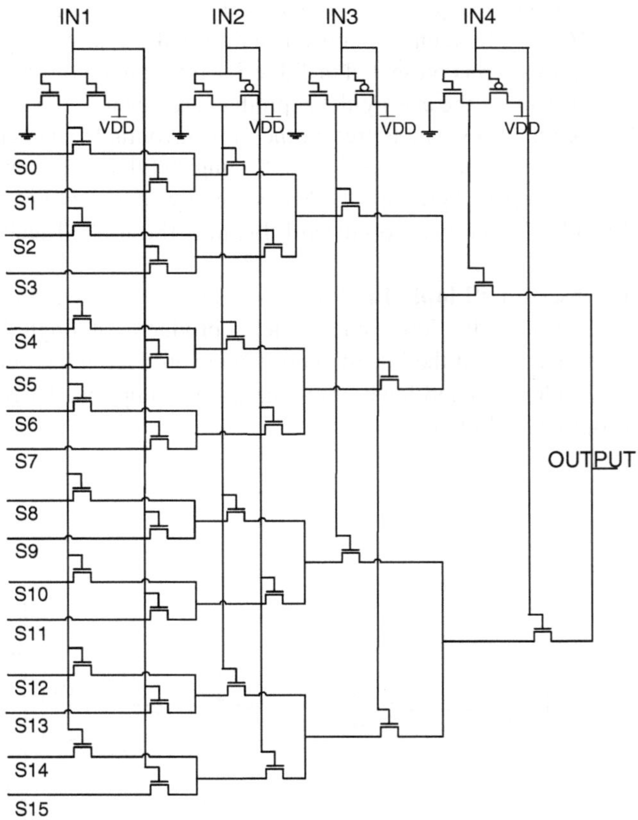

Fig. 3.3 Circuit design of the LUT

adopted, since they lead to the lowest power consumption without degradation in delay. Transistors of minimum size are also used for the 2-to-1 MUX at the output of the BLE.

3.2.2.2 D-Flip/Flop Design

A significant reduction in power consumption can be achieved by using Double Edge-Triggered Flip-Flop (DETFF), since it maintains the data throughput rate while working at half frequency, and the power dissipation on the clock network is halved. Five alternative implementations of the most popular DETFFs in literature were designed and simulated in STM 0.18μm process, in order to determine the optimal one. The one that was finally used is a modified version of the F/F developed in [15], using nMOS transistors instead of transmission gates, because it exhibits low power consumption.

Two versions of the Chung F/F proposed in [13] (Chung1 and Chung2) and of Llopis F/F [12, 14], (Llopis1 and Llopis2) were evaluated, depending on tri-state inverter type, as shown in Fig. 3.4. Another DETFF type has been proposed by Strollo et al. in [16]. The total energy consumed during the application of the input sequence shown in Fig. 3.5 is presented in Table 3.1 for all five F/F types. Also, the worst case delay as well as the energy-delay product are given.

As it is observed the F/Fs which present the most favourable characteristics are the "Llopis1" [12, 14] and the "Chung2" [13]. "Chung 2" F/F has the lowest energy-delay product and "Llopis 1" presents the lowest energy consumption. In Table 3.2 the simulated results for the two optimized F/Fs, in 0,18μm STM technology, are listed.

As it can be observed, **"Llopis-1a"** F/F presents the lower energy consumption but **"Chung-2a"** presents the lower energy-delay product. Although **"Llopis-1a"** F/F Fig. 3.6, does not present the lowest energy-delay product, it has simpler structure leading in smaller area and lower total energy consumption. Therefore it was selected as the optimal solution.

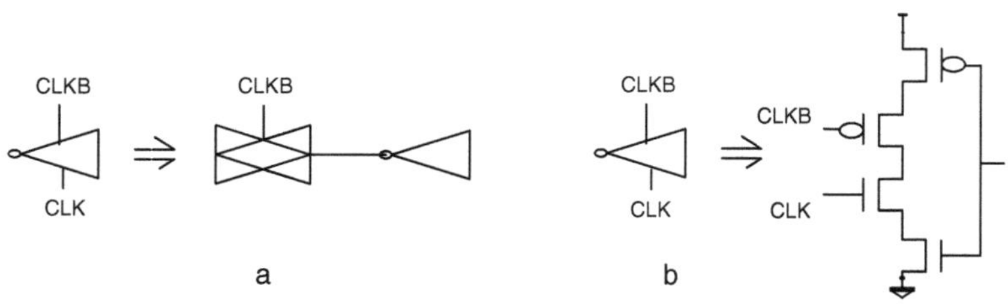

Fig. 3.4 Type of tri-state inverters

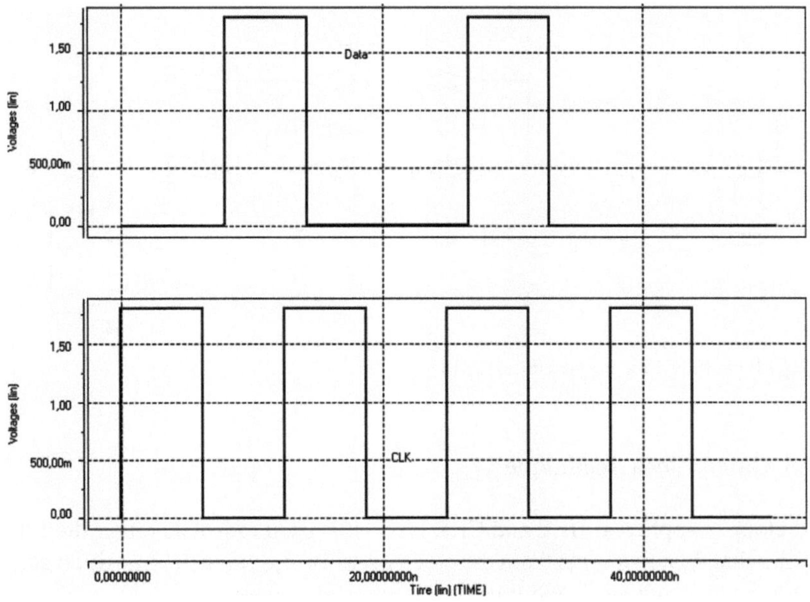

Fig. 3.5 Input pulses to the Flip/Flops for simulation

Table 3.1 Energy consumption, delay and energy-delay product

Cell	Total Energy (fJoules)	Delay (psec)	Energy-Delay Product
Chung 1	433.1	163.5	$70.8*10^{-24}$
Chung 2	457.2	135.3	$61.9*10^{-24}$
Llopis 1	409.0	217.2	$88.8*10^{-24}$
Llopis 2	429.7	241.8	$104*10^{-24}$
Strollo	413.5	270.0	$112*10^{-24}$

Table 3.2 Energy consumption, delay and energy-delay product for optimized F/Fs

Cell	Total Energy (fJoules)	Delay (psec)	Energy-Delay Product
Chung-2a	436.3	138.5	$60.4*10^{-24}$
Llopis-1a	387.7	194.8	$75.5*10^{-24}$

3.2.2.3 Logic Threshold Adjustment in Critical Buffers

The use of pass logic causes the problem of voltage drop across the switch, *Vdd-Vth* instead of *Vdd*, at the output node of a pass transistor. Consequently, the pmos transistor of the next inverter, which accepts as input the logic '1', is not completely "OFF". This may lead, depending on the relation of the threshold voltages of the nmos and pmos transistors, to a significant leakage current. To reduce this leakage current the width of the pmos transistor is minimized.

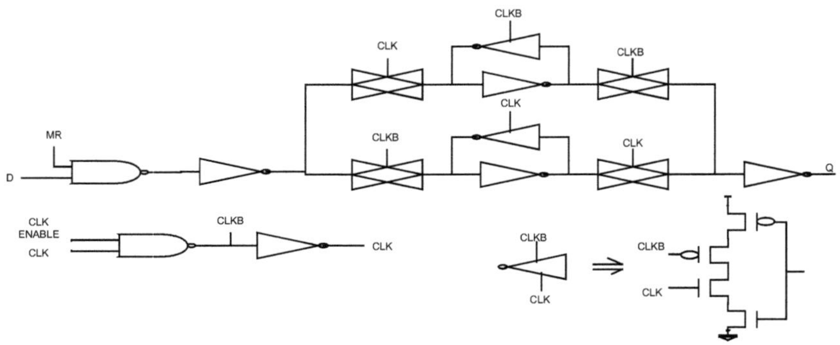

Fig. 3.6 Llopis Flip- Flop proposed in [12, 14]

3.2.2.4 Gated Clock Technique

Gated clock is applied at BLE and CLB level. It is used to isolate either the F/Fs or the BLEs that do not operate from the clock network, reducing the transition activity on the local clock network and thus the effective capacitance.

3.2.2.5 a) BLE Level

At BLE level when the clock enable, *CLK_ENABLE,* is '0', the F/F is "OFF" and is not triggered. The circuit structures that are used for simulation are given in Fig. 3.7,

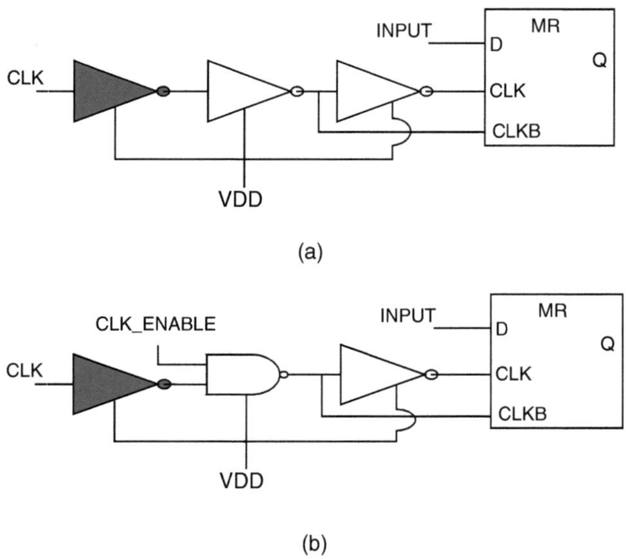

Fig. 3.7 a) Single clock signal b) Gated clock signal

where the shaded inverters in the chain are set for measuring the effect of the input capacitance of the NAND gate, on the energy consumption.

For the structure in Fig. 3.7a the average energy amount consumed for a positive and a negative output transition of the F/F is measured. In case of using gated clock (Fig. 3.7b) the same measurement is taken, considering both '0' and '1' for the*CLK_ ENABLE* signal. The results are given in Table 3.3.

As it can be observed, significant energy savings about 77 % can be achieved when *CLK_ ENABLE* is '0' (the D-FF is "OFF"). However, when *CLK_ ENABLE* is '1' there is a slight increase in energy consumption (6.2 %) caused by the larger capacitance of the NAND gate than the inverter's one.

3.2.2.6 b) Gated Clock at CLB Level.

A gated clock at CLB level can minimize the energy at the local clock network when all F/Fs of the CLB are idle. In this case, the gated clock inputs of the F/Fs and the local clock network of the CLB are constantly at '0' and no dynamic energy is consumed on this. The circuit structures that are used to measure the energy consumption for the single and the gated clock cases are shown in Fig. 3.8.

The energy consumption has been obtained by simulation for various conditions. The simulation results are given in Table 3.4.

As shown, the gated clock signal achieves a 83 % energy consumption reduction when all the flip-flops (F/Fs) are "OFF" and a quite smaller increase in energy when one or more F/Fs are "ON". The conclusion that the adoption of the gated clock at the CLB level is reasonable when the probability of all the flip/flops in the CLB to be OFF is higher than 1/3, is derived from these results.

3.2.2.7 Selected CLB Architecture

The architecture selection was based on the results mentioned in the previous sections and those reported in the literature. Consequently, the features of the CLB are:

a) Cluster of 5 BLEs,
b) 4-inputs LUT per BLE
c) One double edge-triggered Flip-Flop per BLE
d) One gated clock signal per BLE and CLB
e) 12 inputs and 5 outputs provided by each CLB
f) All 5 outputs can be registered

Table 3.3 Energy consumption for single and gated clock

Single clock	gated clock
E=40.76fJ	Clock_enable: "1", E=43.44 fJ
	Clock_enable : "0", E= 9.31 fJ

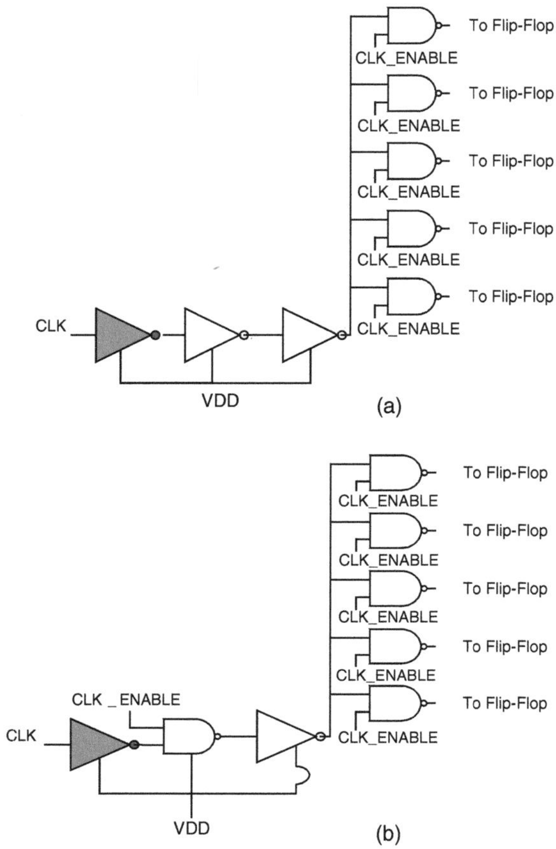

Fig. 3.8 a) Single clock circuit at CLB level b) Gated clock array at CLB level

g) A fully connected CLB resulting to 17-to-1 multiplexing in every input of a LUT
h) A single asynchronous clear signal for the whole CLB
i) A single clock signal for the whole CLB.

The placement and routing tool described in the next section is indifferent to the exact low-level implementations (transistor level), allowing us to employ several transistor-level low-power techniques.

Table 3.4 Energy consumption for single and gated clock at CLB level

Condition	Single Clock	Gated Clock (NAND)
all F/Fs "OFF"	E=23.1fJ	E=3.9fJ
one F/F "ON"	E= 24.1fJ	E= 32.1fJ
all F/Fs "ON"	E= 27.8fJ	E= 35.8fJ

3.2.3 Interconnect Network Architecture

A RAM-based, island-style interconnection architecture [5] was designed; this style of FPGA interconnect is employed by Xilinx [1], Lucent Technologies [17] and the Vantis VF1 [18]. In this interconnection style, the logic blocks are surrounded by vertical and horizontal metal routing tracks, which connect the logic blocks, via programmable routing switches. These switches contribute significant capacitance and combined with the metal wire capacitance are responsible for the greatest amount of dissipated power. Routing switches are either pass transistors or pairs of tri-state buffers (one in each direction) and allow wire segments to be joined in order to form longer connections [19]. The effect of the routing switches on power, performance and area was explored in [6].

Alternative configurations for different segment lengths and for three types of the Switch Box (SB), namely Disjoint, Wilton and Universal [6], were tested. A number of ITC benchmark circuits were mapped on these architectures and the power, delay and area requirements were measured. Another important parameter is the routing segment length. A number of general benchmarks were mapped on FPGA arrays of various sizes and segment lengths and the results were evaluated [12, 13, 14].

Figure 3.9 shows the energy delay products (EDPs) for the three types of SB and various segment lengths. For small segment lengths Disjoint and Universal SBs exhibit almost similar EDPs with the Disjoint topology being slightly better. Also, the lower EDP results correspond to the L1 segment length, meaning that the track has a span of one CLB.

Exploration results for energy consumption, performance and area for the Disjoint switch box topology for various FPGA array sizes and wire segments, are shown in Figs (3.10)–(3.12), respectively. Based on these exploration results among others provided in [6], an interconnect architecture with the following features was selected:

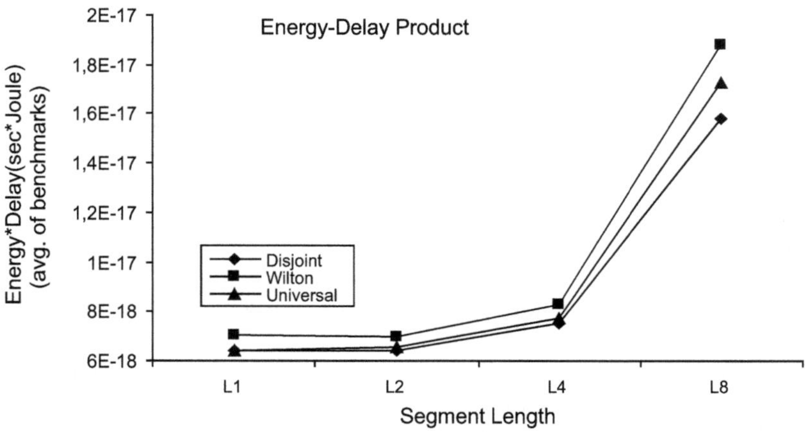

Fig. 3.9 Impact of the SB type and the segment length on energy-delay product

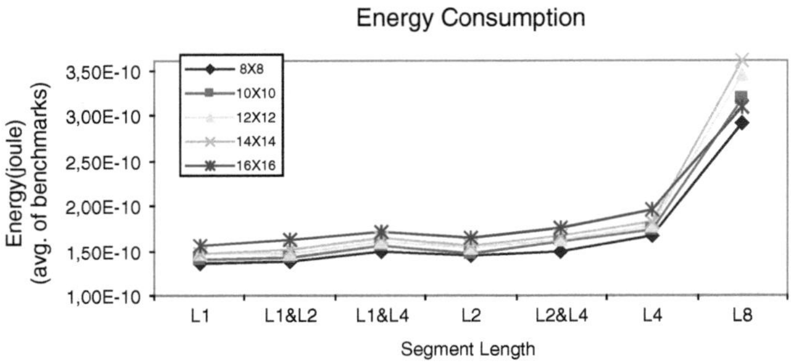

Fig. 3.10 Energy consumption exploration results

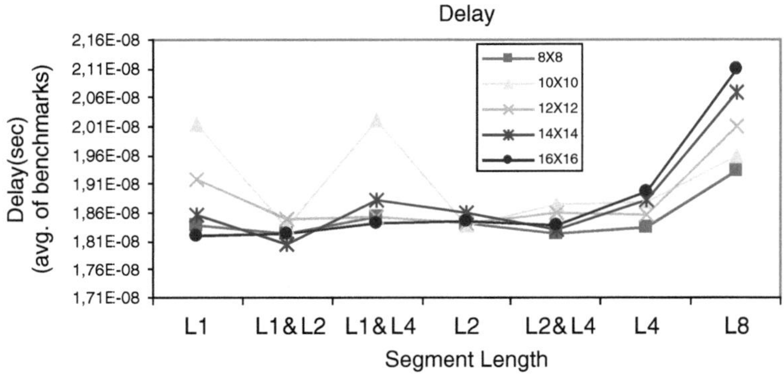

Fig. 3.11 Performance exploration results

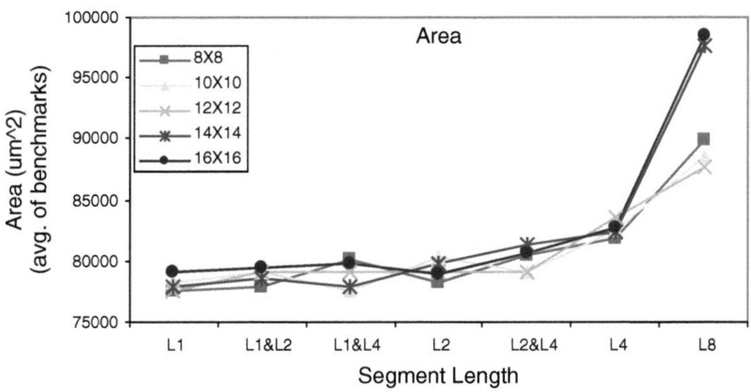

Fig. 3.12 Area exploration results

- Disjoint Switch-Box Topology with switch box connectivity $F_s = 3$ [12, 14].
- Segment Length L1 [13]
- Connection-Box (CB) connectivity equal to one ($F_c = 1$) for input and output Connection-Boxes [12, 13, 14]
- Full Population for Switch and Connection-Boxes
- The size of the CB output's and SB's transistors is $W_n/L_n = 10{*}0.28/0.18$ [13].

The clock network features H-tree topology and low-swing signaling [13].

The circuits of low-swing signaling driver and receiver are shown in Fig. 3.13.

3.2.4 Circuit-Level Low-Power Techniques

Since low-power consumption of the FPGA architecture was the dominant design consideration, a number of circuit-level low power techniques were employed, including the following:

- Double edge triggered F/Fs
- Gated clock at BLE level (up to 77 % savings)
- Gated clock at CLB level (up to 83 % savings)
- Adjustment of the logic threshold of the buffers
- Minimum transistor size for the multiplexers
- Appropriate transistor sizing for buffers
- Selection of the optimal F/F structure for performance and power consumption [15]
- Low-swing signaling (up to 33 % savings on the interconnect network, 47 % on the clock signal)
- Minimum width-double spacing in the metal routing tracks
- Interconnection network is realized using the lowest capacitance 3rd metal layer

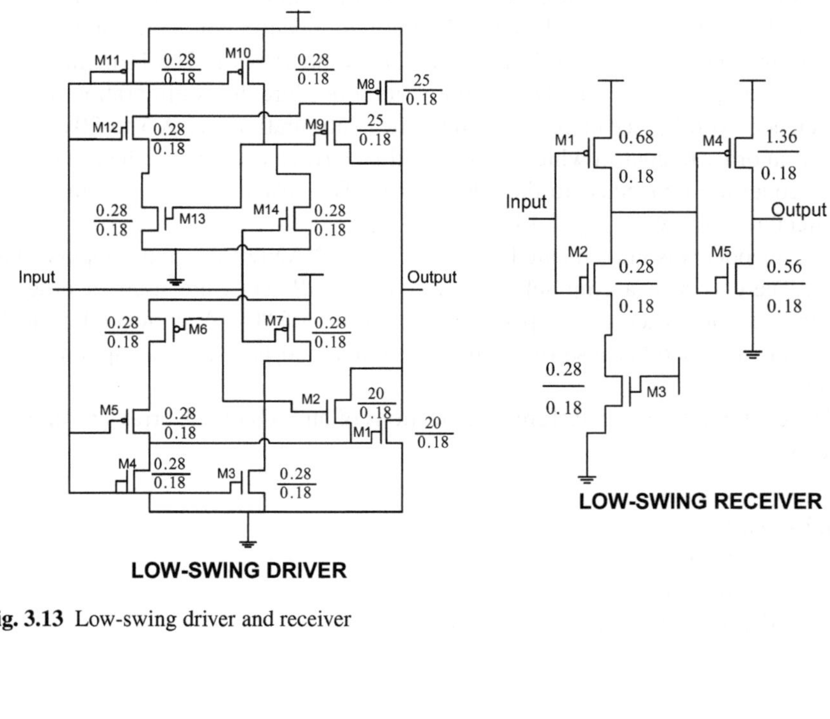

Fig. 3.13 Low-swing driver and receiver

3.2.5 Configuration Architecture

The developed configuration architecture consists of the following components: the memory cell, where the programming bits are stored, the local storage element for each tile (a tile consists of a CLB with its input and output connection boxes, a Switch Box plus the memory for its configuration) and the decoder which controls the configuration procedure of all the FPGA.

3.2.5.1 Memory Cell

The memory cell which is used in the configuration architecture is based on a typical 6T memory cell with all transistors having minimum size. The written data are stored in cross-coupled inverters. Transmission gates were used instead of pass transistors because of their stability. The memory cell is provided with a reset mechanism to disable the switch to which it is connected. This prevents the short-circuit currents that can occur in a FPGA if it operates with unknown configuration states at start-up. The memory cell can only be written into; the contents cannot be read back. That is why it is sufficient to have a simple latch to store the configuration.

3.2.5.2 Configuration Element and Configuration Procedure

Each tile includes a storage element in which the configuration information of the tile is stored. Assuming an 8×8 FPGA physical implementation, the configuration element has 480 memory cells since the tile requires 465 configuration bits. The array of the memory cells consists of 30 columns and 16 rows. The 16 memory bits of a row compose a "word". During the write procedure the configuration bits are written per "word" since there is a 16-bit write configuration bus. A 5-to-30 decoder is used in order to control which "word" will be written each time. The 5-inputs of the decoder are connected to the address bus. The structure of the configuration architecture is shown in Fig. 3.14.

The decoder was implemented by using 5-input NAND gates and 2-inputs NOR gates. There is also a chip select signal. The NOR gates are used in order to idle the decoder when the chip select has a value of "0". A pre-decoding technique was not used because of the increased area and power consumption that it produces.

The configuration architecture specifications of an 8×8 FPGA array are summarized as:

- 4.2Kb size
- 16-bits data bus
- 12-bits address bus
- 1.4ns delay for writing a row of 16 memory cells

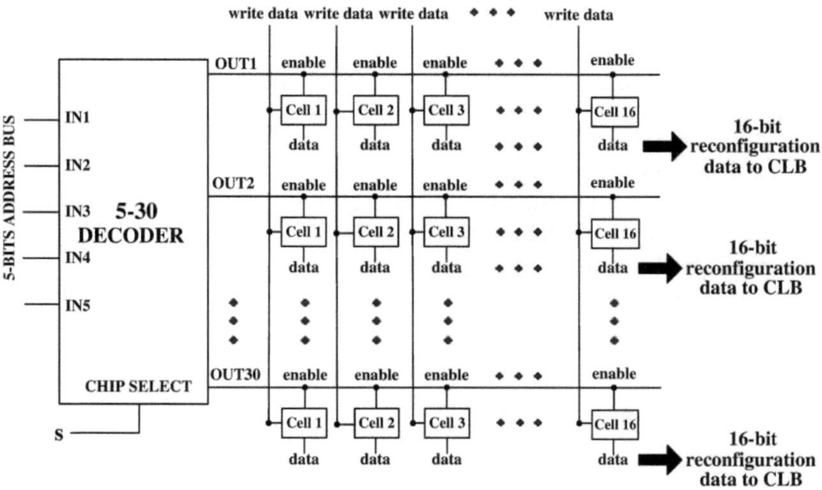

Fig. 3.14 The configuration architecture

- 2100 cycles for entire FPGA configuration
- Independent configuration of each tile, allowing partial and dynamic reconfiguration

The layout of a single tile can be seen in Fig. 3.15.

A prototype full-custom FPGA was designed in a 0.18um STM process technology. The prototype features:

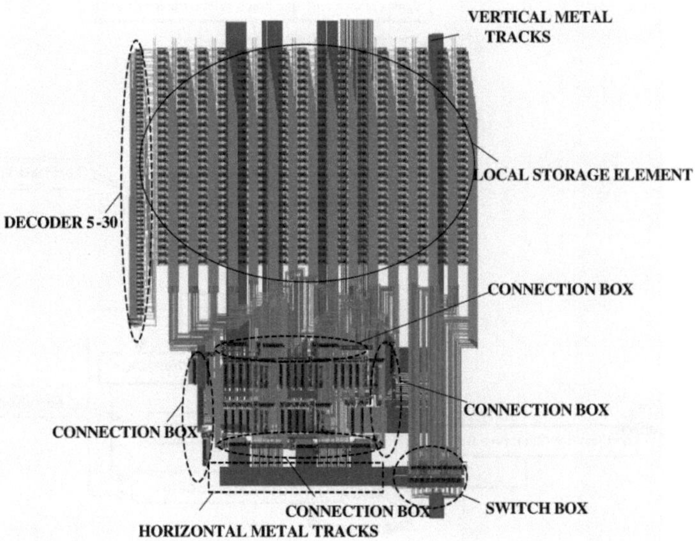

Fig. 3.15 Tile layout

- 8×8 array size (320 LUTs, 320 F/Fs, 96 I/Os)
- 1.8 V power supply
- $4.86 \times 5.28mm^2$ area
- 6 metal layers
  - metal1: Short Connections, Power supply
  - metal2: Short, Intra-cluster, Inter-cluster connections, buses, ground supply
  - metal3: Intra-cluster, Main interconnections
  - metal4: Clock signal, Configuration
  - metal5: Configuration
  - metal6: Configuration
- 2.94 μsec configuration time
- RAM configuration

3.3 Design Framework

Equally important to an FPGA platform is a tool set, which supports the implementation of digital logic on the developed FPGA. Therefore, such a design flow was realized. It comprises a sequenced set of steps employed in programming an FPGA chip, as shown in Fig. 3.16. The input is the RTL-VHDL circuit description, while

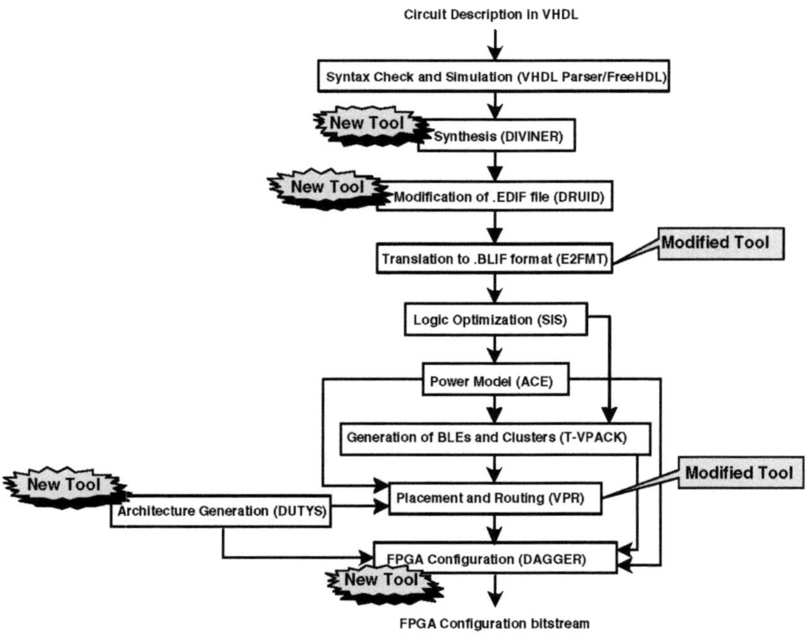

Fig. 3.16 The developed design framework

the output of the CAD flow is the bitstream file that can be used to configure the FPGA. Three different types of tools comprise the flow: i) non-modified existing tools, ii) modified existing tools, iii) and new tools. It is the first complete academic design flow beginning from an RTL description of the application and producing the actual configuration bitstream. Additionally, the developed tool framework can be used in architecture-level exploration, i.e. in finding the appropriate FPGA array size (number of CLBs) and routing track parameters (SB, CB, etc.) for the optimal implementation of a target application. The tools are available at the AMDREL website [11].

All tools can be executed both from the command line and the GUI (Graphical User Interface) presented in a following subsection. It should be noted, that the developed design framework possesses the following attractive features:

- Linux Operating System
- Source description in C/C++ language
- Input format: RTL VHDL, Structural VHDL, EDIF, BLIF
- Output: FPAG Configuration Bitstream
- Implementation Process Technology Independence
- Portability (e.g. i386, SPARC)
- Modularity: each tool can run as a standalone tool
- Graphical User Interface (GUI)
- Capability of running on a local machine or through the Internet/Intranet
- Power Consumption Estimation
- Minimum requirements: x486, 64 MB RAM, 30 MB HD

The following paragraphs provide a short description of each tool.

3.3.1 VHDL Parser

VHDL Parser [20] is a tool that performs syntax checking of VHDL input files.

Input: VHDL source.
Output: Syntax check message.
Usage: This tool is used to check the correctness of the VHDL file compared to the VHDL-93 standard [21].

3.3.2 DIVINER

Democritus Un**i**versity of Thrace RTL Sy**n**thesiz**er** (DIVINER) is a new software tool that performs the basic function of the RTL synthesis procedure. It converts a VHDL description to an EDIF format netlist, similar to one produced by commercial synthesis tools such as Leonardo [22] and Synplicity [23].

At present, DIVINER supports a subset of VHDL as all synthesis tools do. DIVINER supports virtually any combinational and sequential circuit, but the combinational part should be separated in the code from the sequential part. In other words, combinational logic should not be described in clocked processes. This imposes no limitations on the digital circuits that can be implemented, it simply may lead to slightly larger VHDL code. DIVINER does not presently support enumerated types in state machines.

DIVINER only performs a partial syntax check of input VHDL files, therefore the input files should be compiled first using any VHDL simulation tool, commercial (Modelsim) or open-source (FreeHDL). Additionally, at this stage, DIVINER does not perform Boolean optimization, therefore the designer should be careful in the VHDL coding of Boolean expressions.

DIVINER outputs a generic EDIF format netlist, which can then be used with technology mapping tools in order to implement the digital system in any ASIC or FPGA technology and not necessarily in the developed fine-grain reconfigurable hardware platform.

More info about the DIVINER, can be found in the tool manual [24].

Input: VHDL source.
Output: EDIF netlist (commercial tool format).
Usage: The DIVINER tool is used as a synthesizer of behavioral VHDL language.

3.3.3 DRUID

Democ**R**itus **U**niversity of Thrace ED**I**F to E**D**IF translator (**DRUID**) is a new tool that converts the EDIF format netlist produced by a commercial synthesis tool or DIVINER to an equivalent EDIF format netlist compatible with the next tool of the design flow.

DRUID [24] serves a threefold purpose: i) it modifies the names of the libraries, cells etc, found in the input EDIF file, ii) it simplifies the structure of the EDIF file in order to make it compatible to our tool framework and iii) it constructs, in the simplest way possible, the cells and generated modules that are included in the input EDIF file and are not found in the libraries of the following tools.

Without DRUID, the hardware architectures that could be processed by the developed framework would be the ones specified in structural level by using only basic components (inverter, AND, OR and XOR gates of 8 inputs maximum, a 2-input multiplexer, a latch and a D-type F/F without set and reset). Moreover, signal vectors are not supported. Obviously, DRUID is necessary in order to implement real-life applications on the developed FPGA.

Input: EDIF netlist (commercial tool format).
Output: EDIF netlist (T-VPack format).
Usage: The DRUID tool is used to modify the EDIF [25] output file that is produced during the synthesis step, so that is can be used by the following tools of the design flow.

3.3.4 E2FMT

Input: EDIF netlist.
Output: BLIF netlist.
Usage: Translation of the netlist from EDIF to BLIF [26] format.

3.3.5 SIS

Input: BLIF netlist (generic components).
Output: BLIF netlist (LUTs and FFs).
Usage: SIS [27] is used for mapping the logic described in generic components (such as gates and arithmetic units) into the elements of the developed FPGA.

3.3.6 T-VPack

Input: BLIF netlist (gate and F/Fs).
Output: T-VPack netlist (LUTs and F/Fs).
Usage: The T-VPack tool [10] is used to group a LUT and an F/F to form BLE or a cluster of BLEs

3.3.7 DUTYS

DUTYS (**D**emocritus **U**niversity of **T**hrace Architecture file generator-s**y**nthe**s**izer) is a new tool that creates the architecture file of the FPGA that is required by VPR [10]. The architecture file contains a description of various parameters of the FPGA architecture, including size (array of CLBs), number of pins and their positions, number of BLEs per CLB, plus interconnection layout details such as relative channel widths, switch box type, etc. It has a GUI that helps the designer select the FPGA architecture features and then automatically creates the architecture file in the required format. Each line in an architecture file consists of a keyword followed by one or more parameters. A comprehensive description for the DUTYS parameters, as well as the execution both from command line and through the GUI are stated to the tool's manual [24].

Input: FPGA features.
Output: FPGA architecture file.
Usage: Generates the architecture file description of the target FPGA.

3.3.8 PowerModel (ACE)

Input: BLIF netlist, Placement and routing file.
Output: Power estimation report.

Usage: The PowerModel tool [9] estimates the dynamic, static and short-circuit power consumption of an island-style FPGA. It was modified and extended in order to also calculate leakage current power consumption.

3.3.9 VPR

Input: T-VPack netlist (LUTs and F/Fs), FPGA architecture file.
Output: Placement and routing file.
Usage: placement and routing of the target circuit into the FPGA. VPR [10] was extended by adding a model that estimates the area of the device in mm^2 assuming STM 0.18 μm technology.

3.3.10 DAGGER

DAGGER (**D**EMOCRITUS UNIVERSITY OF THR**A**CE E-FP**G**A BITSTREAM **GE**NE**R**ATOR) is a *new* FPGA configuration bitstream generator. This tool has been designed and developed from scratch. To our knowledge there is no other available academic implementation of such a tool.

The main features of DAGGER are:

- Technology independence: DAGGER has no constraint about the device design technology.
- Partial and real-time reconfiguration: The DAGGER tool supports both run-time and partial reconfiguration, as long as the target device does also. In any case, reconfiguration must be done as efficiently and as quickly as possible. This is in order to ensure that the reconfiguration overhead does not offset the benefit gained by hardware acceleration. Using partial reconfiguration can greatly reduce the amount of configuration data that must be transferred to the FPGA device.
- Bitstream reallocation (defragment): Partially reconfigurable systems have advantages over the single context, but problems may occur if two partial configurations are supposed to be located at overlapping physical locations on the FPGA. DAGGER features the bitstream reallocation technique. This gives DAGGER the ability to defrag the reconfigurable device, preventing such potential problems.
- Bitstream compression: When multiple contexts or configurations have to be loaded in quick succession then the system's performance may not be satisfactory. In such a case, the delay incurred is minimized when the amount of data transferred from the processor to the reconfigurable hardware is minimized. A technique that could be used in order to compact this configuration information is the configuration compression. The developed compression algorithm that is used by DAGGER is based on Run Length Encoding (RLE), on the LZW compression algorithm while taking into consideration the regularity of the bitstream file contents.
- Error detection: Error detection is important whenever a non-zero chance of data is getting corrupted. DAGGER incorporates the Cyclic Redundancy Checking (CRC) algorithm. CRC is used to check the data written to any configuration register. It is calculated according to a specific algorithm, each time a configuration

register is written. When the write CRC command is executed, the value of this command is checked against the internal value. In case the value is incorrect, the device is put in an ERROR condition by cancelling the module execution, preventing in this way any damage to the device. DAGGER uses the 16-bit CRC algorithm.

- Readback: Read-back is the process of reading all the data from the FPGA device in the internal configuration memory. This feature also founds in commercial FPGAs, for instance Virtex-II Pro. This can be used both to verify that the current configuration data is valid, as well as to read the current state of all internal CLBs, connection and switch boxes.
- Bitstream encryption: The DAGGER output file could be encrypted both for security to the FPGA device, as well as for the program running on it. The encryption is responsible for the protection of configuration data from unauthorized examination and modification. Without protection, the hardware IP stored in the configuration memory can be copied and potentially reverse engineered. The build-in encryption algorithm of the DAGGER tool is the Triple Data Encryption Standard (TDES), which performs three successive encryption-decryption-encryption operations using three different (56-bit) key sets. The TDES is a symmetric encryption standard, which means that the same keys are used both for encryption and decryption. Because of the key strength ($3,7 \bullet 10^{50}$), this method is considered absolutely secure. Beside the TDES, the user could use any other encryption algorithm. For this, the designer has to compile the DAGGER tool with the appropriate source library (which includes the desired encryption).
- Low-power techniques: DAGGER employs certain low-power techniques both in the source code of the tool, as well as at the way that the FPGA is programmed. The source code is written in a way that minimizes the I/O requests from the memories. Similar, the bitstream file is composed by a series of bits that are used for representing connections inside the FPGA and null bits. These null bits have no meaning and could have value '0' or '1', without affecting the circuit functionality. Taking into consideration that any transition from '0' to '1' (and vice versa) accompanies switching activity, we have to minimize the unnecessary transitions. Based on this, the DAGGER tool sets the value for these bits (indifferent bits) to the last used value.

The basic operation of DAGGER can be seen in the following pseudo-code.

```
open FPGA architecture file;
while (not end of FPGA architecture parameters)
{read channel_x_width, channel_y_width, switch_topology,
pad_per_clb, pins_topology;}
close FPGA architecture file;
open placement file;
read FPGA required array size;
read the possition of each CLB to the FPGA array;
```

```
close placement file;
open routing file;
while (not end of file)
{read the routing for a net;
while (not end the routing for this net)
{
if (routing_line includes an I/O pad)
{
if (its position is x_array_max/x_array_min)
then modify routing channel at RIGHT/LEFT [current_y];
if (its position is y_array_max/y_array_min)
then modify routing channel at TOP/BOTTOM[current_x];
}

/* Routing of switch box */
if (routing with two successive CHAN_ROUTE_X (or
CHAN_ROUTE_Y))
{
if (its position is x_array_max/x_array_min)
then modify switch routing at RIGHT/LEFT [current_y];
if (its position is y_array_max/y_array_min)
then modify switch routing at TOP/BOTTOM[current_x];
if (its position is y_array_min<y<y_array_max)
{
modify switch routing at CENTER[current_x][current_y];
}}

/* Routing of connection box */
if (CHAN_ROUTE_X or CHAN_ROUTE_Y) and (OUTPUT_PIN or
INPUT_PIN))
{
if (its position is x_array_max/x_array_min)
then modify switch routing at RIGHT/LEFT [current_y];
if (its position is y_array_max/y_array_min)
then modify switch routing at TOP/BOTTOM[current_x];
   if (its position is (y_array_min<y<y_array_max) AND
(x_array_min<x<x_array_max))
{
modify connection box at CENTER[current_x][current_y];
}} } }
close routing file;
open netlist file;
read inputs/output of each BLE;
check if the F/F is used;
close netlist file;
```

```
open function file;
read the boolean function of each LUT;
close function file;
Terminate program;
```

Input: PowerModel output file, Placement and Routing file, FPGA architecture file, T-VPack netlist.
Output: FPGA configuration bit stream file.
Usage: The DAGGER tool is used to generate the bitstream file.

3.3.11 Graphical User Interface

The Graphical User Interface (GUI) provides the designer with the opportunities to easily use all (or some of the tools) that are included in the developed design flow. It consists of six independent stages: i) the File Upload, ii) the Synthesis, iii) the Format Translation, iv) the Power Estimation, v) the Placement and Routing and vi) the FPGA configuration stage. Until now, there is no other academic implementation of such a complete graphical design chain. It is possible to run it from a local PC or through the Internet/Intranet, and the source code can be easily modified in order to add more tools. The tools can also be executed on-line at http://vlsi.ee.duth.gr:8081.

3.4 Comparisons

A complete FPGA system (H/W, S/W) includes a plethora of interdependent parameters, e.g. #CLBs, LUT size, SB type, etc. On one hand, we tried to qualitatively evaluate the tool framework by comparing the features it provides with the corresponding features (or lack thereof) of other commercial and academic tool frameworks. On the other hand, quantitative experimental results on different circuit benchmarks were obtained for a FPGA employing similar resources with commercial ones.

3.4.1 Qualitative Comparisons

Qualitative comparisons in terms of provided features among the developed, XILINX [1], TORONTO [6] and ALLIANCE [28] tool frameworks are provided in Table 3.5. The (✓) symbol indicates that the corresponding feature is available in the design framework, while the (×) symbol indicates that the specific feature is not supported by the design framework. The (×) symbol indicates that the corresponding feature is not provided, but not necessary for the completeness of that framework either.

Table 3.5 Qualitative comparison among tool frameworks

FEATURE	AMDREL	[1]	[6]	[28]
Input Format	VHDL/VERILOG	VHDL/VERILOG	BLIF	VHDL
Synthesis	✓	✓	×	✓
Format Translation	✓	-	×	×
Power Estimation	✓	✓	×	×
Architecture description	✓	✓	×	×
Placement	✓	✓	✓	✓
Routing	✓	✓	✓	✓
Bitstream Generation	✓	✓	×	×
Back-annotation	×	✓	×	×
GUI	✓	✓	×	×
Remote Access to GUI	✓	×	×	×
User Manual	✓	✓	✓	✓
Operating System	LINUX	SOLARIS/WINDOWS	SOLARIS	LINUX

Table 3.5 shows that the developed design framework provides implementation from as high-level a description as possible (RTL) down to the FPGA configuration file, while it also provides power consumption estimation, and configuration bitstream generation which the other academic frameworks do not. It also features a GUI (while academic frameworks do not) and remote access to it (which no other framework, commercial or academic does). The only limitations of the developed framework are that it does not currently support back-annotation, but no other academic tool frameworks does either, and it is available for the Linux operating system.

It is evident that the developed tool framework is the most complete academic tool framework, and is at least in terms of provided features comparable with commercial tools. It contains the only known academic implementation of a configuration bitstream generation tool. Additionally, the remote access to GUI feature allows the user to run the framework without even having the tools installed in his/her own computer.

3.4.2 Quantitative Comparisons

Various benchmarks from ITC'99 [29] (part of the MCNC benchmarks) were implemented in the AMDREL FPGA array described previously, using the AMDREL design framework and in Xilinx devices of similar resources using Xilinx ISE tools. The benchmarks range from a few gates to tens of thousands and include combinational logic, sequential logic and Finite State Machines (FSMs). Benchmarks b01 to b11 were mapped to the developed 8×8 FPGA device, while benchmarks b12 to b21_1 were mapped to the smallest fitting array, namely from 18×18 to 48×48.

Figure 3.17 shows the number of 4-input LUTs used to implement the same benchmarks in the developed and Xilinx environments. It can be seen that the resulting number of LUTs in the developed framework is greater. This is mainly due to the fact that the E2FMT tool libraries do not support many basic modules that had to be added by DRUID described at gate level, which leads to larger netlists and

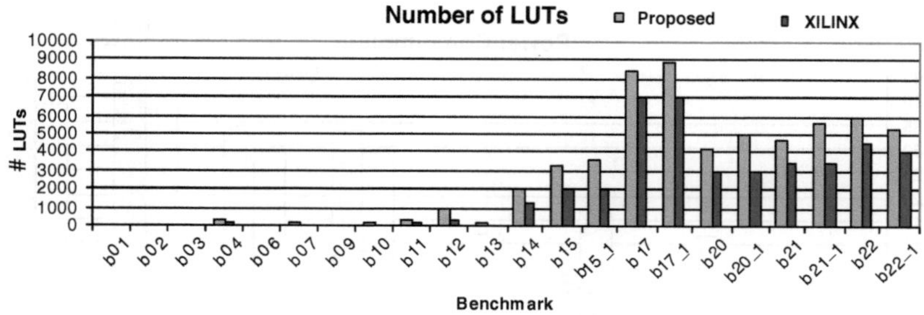

Fig. 3.17 LUT mapping comparison

therefore greater number of LUTs. This can only be efficiently remedied if E2FMT is drastically modified.

Figure 3.18 shows the maximum frequencies obtained by the two frameworks and devices. It can be seen that both frameworks perform similarly, with the developed one outperforming Xilinx in certain benchmarks Xilinx outperforming the developed one in others. More specifically, up to benchmark b11 which is in the order of tens of thousands of gates (the benchmarks get progressively larger in gate count), the developed framework outperforms Xilinx. For larger benchmarks (about a hundred thousand gates) Xilinx performs somewhat better. This is rather due to inherent limitations of the tools than lack of efficiency on the part of the FPGA architecture. More specifically, the main reason for the somewhat greater delay of the developed system is due to the greater number of LUTs required to implement the same benchmark in the developed flow, discussed above. Still, the frequencies achieved by the developed framework and device are of the same order as the ones reached by Xilinx Virtex devices.

Figure 3.19 provides power consumption figures for some of the benchmarks mentioned above. It can be seen that the power consumption of the developed architecture is somewhat greater than that of the Xilinx architecture. Once again, this is due to the tool limitations that lead to an increased number of LUTs. Still, it can be seen that the relative increase in power consumption per benchmark is smaller than the relative increase in number of LUTs (25 and 35 % respectively in the case

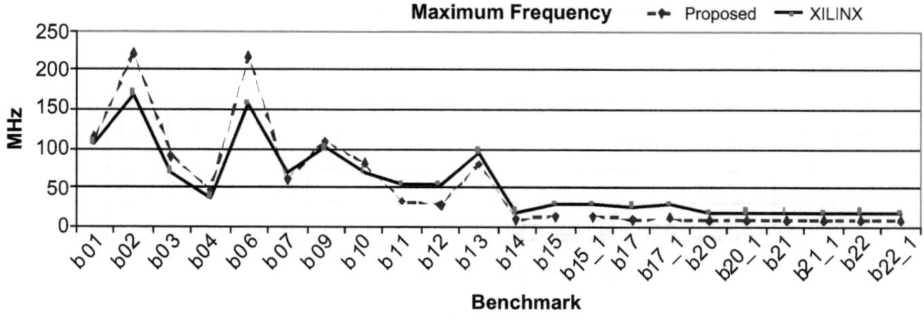

Fig. 3.18 Maximum frequency comparison

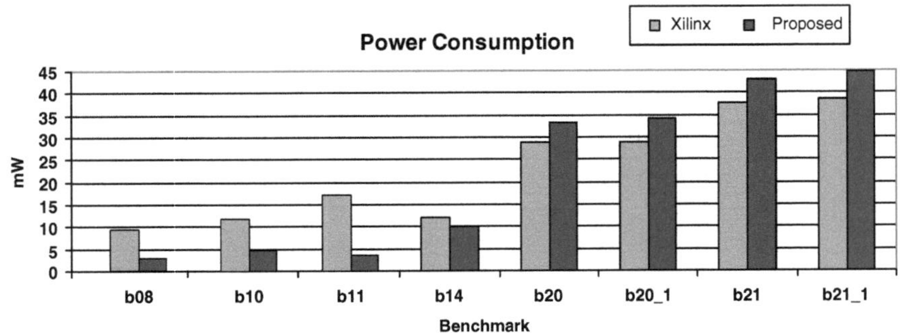

Fig. 3.19 Power consumption comparison

of benchmark b_20) which confirms the efficiency of the employed circuit-level techniques. In order to improve the power efficiency of the developed system, the LUT-mapping process of E2FMT and DRUID will have to be improved.

Figure 3.20 shows the power consumption for a number of benchmarks with and without the employed low-swing scheme, estimated using PowerModel [9]. It can be seen that the power saved by employing the low-swing technique is significant.

Table 3.6 shows the results from applying the DAGGER strategy for partial bitstream reconfiguration to the developed FPGA array for a number of benchmarks. The second column represents the smallest FPGA array required to implement the corresponding benchmark, the third column shows the number of CLBs required to implement each benchmark. The fourth column shows the required number of bits for programming the optimal array without employing the features of DAGGER, such as compression and partial reconfiguration while the fifth column gives the number of bits produced by DAGGER. Finally, the last column gives the percentage gain of the DAGGER bitstream file size, compared to the uncompressed bitstream required to configure the optimal array.

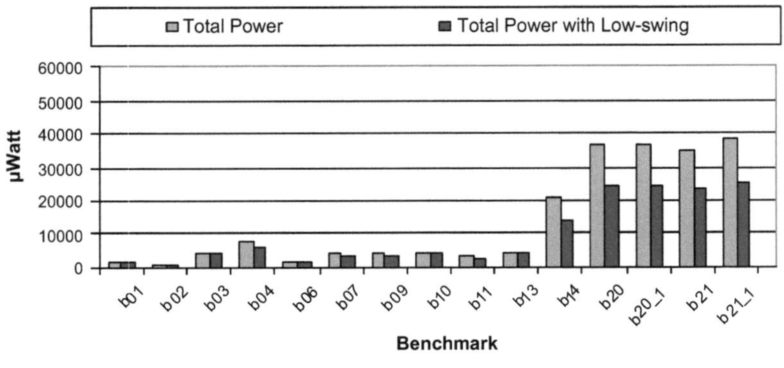

Fig. 3.20 Low-swing power savings

Table 3.6 DAGGER bitstream

Bench-mark	Optimal Array	#CLB	Bitstream Size for Optimal Array	DAGGER Bitstream file	% Gain
Add5and2	2×2	2	2640	1200	54
addsub_3	2×2	1	2640	600	77
decrem9	2×2	2	2640	1200	54
fft16pt	5×5	20	13800	10140	26
fft256pt	5×5	20	13800	10140	26
mul5and2	2×2	3	2640	1800	31
mux2_if	2×2	1	2640	600	77
mux4	2×2	1	2640	600	77
mux7	2×2	3	2640	1800	31
mux32	5×5	19	13800	9720	29
mux48	6×6	27	19440	13980	28
Subtract4	2×2	2	2640	1200	54
umin_8bit	2×2	3	2640	1740	34
b01	3×3	6	5400	3120	42
b02	2×2	2	2640	1200	54
b03	5×5	17	13800	8520	38
b04	8×8	58	33600	28560	15
b06	2×2	4	2640	2220	15
b07	6×6	32	19440	15720	19
b09	4×4	15	9120	7440	18
b10	5×5	25	13800	12660	8
b11	8×8	57	33600	27900	16
b13	6×6	29	19440	14580	25

3.5 Conclusions

A novel FPGA architecture (CLB, interconnect and configuration architecture) with low-power features was presented together with complete tool framework for implementing logic in this platform. The developed system of the FPGA (implemented in 0.18 μm STM technology) and tool framework showed promising results when compared with commercial products using a number of benchmarks.

References

1. http://direct.xilinx.com/bvdocs/publications/ds003.pdf
2. http://www.altera.com/products/devices/dev-index.jsp
3. http://www-cad.eecs.berkeley.edu/
4. http://ballade.cs.ucla.edu/
5. G. Varghese, J. M. Rabaey, "Low-Energy FPGAs –Architecture and Design", Kluwer Academic Publishers, 2001
6. V. Betz, J. Rose and A. Marquardt, "Architecture and CAD for Deep-Submicron FPGAs", Kluwer Academic Publishers, 1999
7. V. George, H. Zhang, and J. Rabaey, "The Design of a Low Energy FPGA", In Proc. Int. Symp. On Low Power Electronics and Design (ISLPED'99), pp. 188–193, San Diego, California, August 16–17, 1999.
8. V. Betz, and J. Rose, "FPGA Routing Architecture: Segmentation and Buffering to Optimize

Speed and Density", in ACM/SIGDA Int. Symp. on Field Programmable Gate Arrays, Monterey, CA (1999), pp. 59–68
9. K. Poon, A. Yan, and S. Wilton, "A Flexible Power Model for FPGAs", in Proc. Field-Programmable Logic and Applications (FPL) 2002, Montpellier, France, 2002, pp. 312–321.
10. http://www.eecg.toronto.edu/~vaughn/vpr/vpr.html
11. http://vlsi.ee.duth.gr/AMDREL
12. V. Kalenteridis et al. "An Integrated FPGA Design Framework: Custom Designed FPGA Platform and Application Mapping Toolset Development", in Proc. of the Reconfigurable Architectures Workshop (RAW 2004), April 26–27, 2004, Santa Fe, New Mexico, USA.
13. H. Pournara et al., "Energy Efficient Fine-Grain Reconfigurable Hardware", Proceedings of 12th IEEE Mediterranean Electrotechnical Conference (MELECON) 2004.
14. V. Kalenteridis, H. Pournara, K. Siozos, K. Tatas, G. Koutroumpezis, N. Vassiliadis, I.Pappas, S. Nikolaidis, S.Siskos, D. J. Soudris and A. Thanailakis, "A Complete Platform and Toolset for System Implementation on Fine-Grain Reconfigurable Hardware" Journal on Microprocessors and Microsystems, Vol 29/6 pp. 247–259, 2005.
15. R. Peset Llopis, and M. Sachdev, "Low Power, Testable Dual Edge Triggered Flip-Flops", Proceedings of IEEE International Symposium on Low Power Electronics and Design, August 1996, Monterey, USA
16. A.G.M. Strollo, E. Napoli, and C. Cimino, "Analysis of Power Dissipation in Double Edge Triggered Flip-Flops", IEEE Transaction on VLSI Systems, Vol. 8, No. 5, October 2000, pp. 624–628
17. http://www.lucent.com
18. http://www.vantis.com
19. V. Betz, and J. Rose, "Circuit Design, Transistor Sizing and Wire Layout of FPGA Interconnect", IEEE Custom Integrated Circuits Conference, (CICC), San Diego, California, 1999
20. http://search.cpan.org/author/GSLONDON/Hardware-Vhdl-Parser-0.12/
21. http://opensource.ethz.ch/emacs/vhdl93_syntax.html
22. http://www.mentor.com/leonardospectrum/datasheet.pdf
23. http://www.synplicity.com/products/synplifypro/ index.html
24. http://vlsi.ee.duth.gr:8081/help/{DIVINER, DRUID, DUTYS, DAGGER}_manual.pdf
25. http://www.edif.org
26. http://www.bdd-portal.org/docu/blif/blif.html
27. M. Sentovich, K. J. Singh, and L. Lavagno, et al.: "SIS: A System for Sequential Circuit Synthesis", UCB/ERL M92/41 (1992)
28. http://www-asim.lip6.fr/recherche/alliance/
29. Ken McElvain, "Benchmarks tests files", Proc. MCNC International Workshop on Logic Synthesis 1993, ftp://ftp.mcnc.org/pub/benchmark/Benchmark_dirs/LGSynth93/ LGSynth93.tar
30. K. Siozios et al., "A Novel FPGA Configuration Bitstream Generation Algorithm and Tool Development", in Proceedings of 13th International Conference on Field Programmable Logic and Applications (FPL), pp. 1116–1118, August 30 - September 1, 2004, Antwerp, Belgium, pp. 1116–1118.
31. K. Tatas et al. "FPGA Architecture Design and Toolset for Logic Implementation", in Proceedings of 13th International Workshop, pp. 607–616, PATMOS 2003, Turin, Italy, September 2003.
32. K. Siozios, G. Koutroumpezis, K. Tatas, D. Soudris and A. Thanailakis, "DAGGER: A Novel Generic Methodology for FPGA Bitstream Generation and its Software Tool Implementation", 12th Reconfigurable Architectures RAW 2005 Colorado, USA, April 4–5, 2005.
33. T. Lo, W. Man Chung, and M Sachdev, "A Comparative Analysis of Dual Edge Triggered Flip-flops", IEEE Transactions on VLSI Systems, Vol.10, No.6, December 2002, pp. 913–918

Chapter 4
A Coarse-Grain Dynamically Reconfigurable System and Compilation Framework

Marcos Sanchez-Elez, Milagros Fernandez, Nader Bagherzadeh, Roman Hermida, Fadi Kurdahi and Rafael Maestre

4.1 Introduction

Traditionally, there have been two classes of computing: general-purpose systems and application-specific embedded systems. The former is based on the usage of general-purpose processors wherein a single processor is used to execute a variety of applications. The latter category is based on the use of application specific integrated circuits (ICs) or ASICs, each of which is custom designed for a specific application. However, there is also a third class of computing, reconfigurable computing, which has recently come forth into prominence. Reconfigurable computing refers to computing systems that have reconfigurable architectures or are based on reconfigurable hardware components. The underlying principle of these systems is that the hardware organization, functionality or interconnections may be customized after fabrication of the system on silicon. This customization helps satisfy specific computational requirements of different applications.

A reconfigurable architecture or a system consisting of reconfigurable devices exhibits the following features:

- Configurable functionality: This implies that the processing unit can execute different functions at different times based on different configuration information.
- Configurable interconnect: This refers to the ability of the interconnections to be changed to meet different communication patterns or other application constraints.
- Configurable Storage: This indicates that the storage mechanisms (e.g. memories) inside the reconfigurable system can be changed in structure or functionality in order to best suit the data movements for a given application.
- Configurable I/O: This refers to the ability of the system to change the bitwidth, number and sequencing of input data depending on the application being processed.

The key point behind reconfigurable architectures is that in some special applications, one such system may replace several ASICs corresponding to different configurations of the former. This replacement clearly results in considerable silicon, power, and weight savings, when ASICs and reconfigurable architectures are contrasted with each other. Reconfigurable architectures, in return, suffer from some

S. Vassiliadis, D. Soudris (eds.), *Fine- and Coarse-grain Reconfigurable Computing.*

performance degradation. This, however, is easily tolerated in many cases. The so-called expensive (i.e. complicated) features of general purpose processors, on the other hand, are not usually exploited thoroughly in our intended applications. Therefore, the major asset of these processors is not utilized efficiently, and hence their application is nearly ruled out, as well. A reconfigurable system may be optimized for multiple applications. Different functions and variable interconnect patterns enable the execution of a range of applications. The combined effect is that applications execute faster (with performance levels close to that of ASICs) and more applications can be executed on the same system.

In summary, a reconfigurable architecture lends itself to on-board signal processing, offering small silicon size and physical weight, low power dissipation, and sufficient performance, for many applications. These upsides are useful in many real-world scenarios. Consider a system designed for a given class of applications (image processing, data encryption, etc.). Each application may have a complex and heterogeneous nature and comprises several subtasks with varying characteristics. For instance, a multimedia application may include a data-parallel task, a bit-level task, irregular computations, high-precision word operations, and a real-time component. For such complex applications with wide-ranging subtasks, the ASIC approach would lead to an uneconomical die size or a large number of separate chips. Also, most general-purpose processors would very likely not satisfy the performance or power constraints for the entire application. However, a reconfigurable system may be optimally reconfigured for each subtask, thus meeting the application constraints within the same chip. The same reconfigurable system can be reused to perform the other tasks in this application class while still satisfying performance and power constraints.

In this chapter, first we will discuss the MorphoSys architecture, followed by its first implementation, called the M1 chip. Later, we discuss the application execution model over MorphoSys as a result of the architecture constrains and application characteristics. Then, the compilation framework specifically created to efficient implement application in this type of architectures is described. Finally a brief example of implementing an application onto MorphoSys and its compilation framework is presented.

4.2 MorphoSys Architecture

The MorphoSys design model incorporates a reconfigurable component (to handle high-volume data-parallel operations) on the same die with a general-purpose reduced instruction set computer (RISC) processor (to perform sequential processing and control functions), and a high bandwidth memory interface.

The MorphoSys architecture [1] comprises five major components: the reconfigurable cell array (RC array), control processor (TinyRISC), context memory, frame buffer, and a direct memory access (DMA) controller. Fig. 4.1 shows the organization of the integrated MorphoSys reconfigurable computing system.

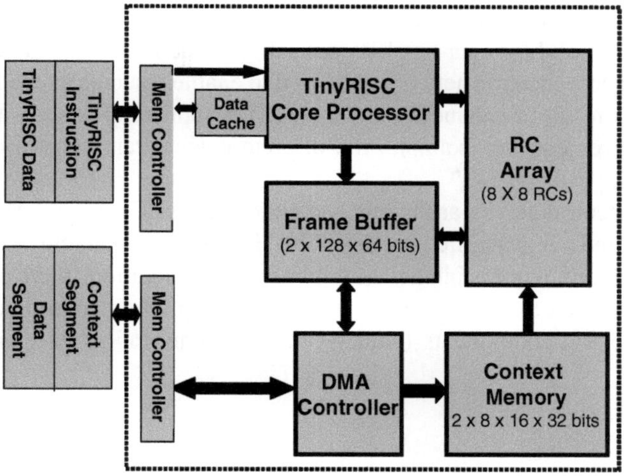

Fig. 4.1 MorphoSys integrated architecture model

4.2.1 RC Array

In the current implementation, the reconfigurable component is an array of RCs or processing elements. Considering that target applications (video compression, etc.) tend to be processed in clusters of 8- by 8 data elements, the RC array has 64 cells in a two-dimensional matrix, as illustrated in Fig. 4.2. This configuration is chosen to maximally utilize the parallelism inherent in an application, which in turn enhances throughput.

The RC array follows the single instruction multiple data (SIMD) model of computation. All RCs in the same row/column share same configuration data

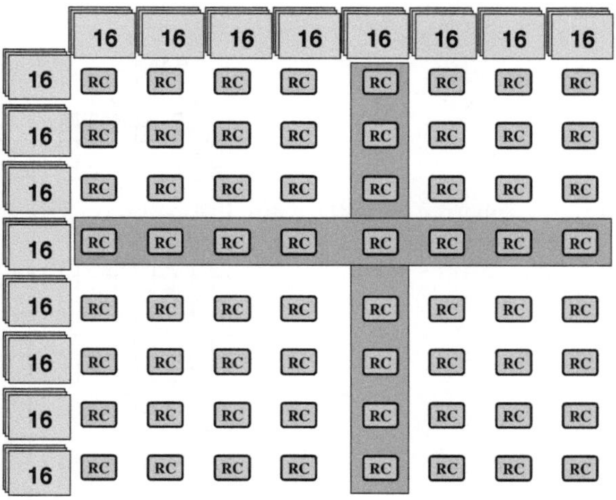

Fig. 4.2 An 8 by 8 RC Array with context memory

(context). However, each RC operates on different data. Sharing the context across a row/column is useful for data-parallel applications. The RC Array has an extensive three-layer interconnection network, designed to enable fast data exchange between the RCs. This result in enhanced performance for application kernels that involve high data movement, for example, the discrete cosine transform (used in video compression).

Each RC incorporates an arithmetic and logic unit (ALU)-multiplier, a shift unit, input muxes, and a register file (see Fig. 4.7). The multiplier is included since many target applications require integer multiplication. In addition, there is a context register that is used to store the current context and provide control/configuration signals to the RC components (namely, the ALU-multiplier, shift unit, and the input muxes).

4.2.2 Frame Buffer and DMA Controller

The potential parallelism of the RC Array would be ineffective if the memory interface is unable to transfer data at an adequate rate. Therefore, a high-speed memory interface consisting of a streaming buffer or Frame Buffer (FB) and a DMA controller (DMAC) is incorporated in the system. The FB has two sets as illustrated in Fig. 4.3. The FB serves as a data cache for the RC Array. It consists of two sets of identical data memory. Each set consists of two banks of memory with each bank having 64 $\times$ 8 bytes of storage. By using these two sets alternatively, the computation of the RC Array is overlapped with the load and store of the Frame Buffer, which makes the memory accesses transparent to the RC Array. MorphoSys performance benefits greatly from the streaming process of this data buffer.

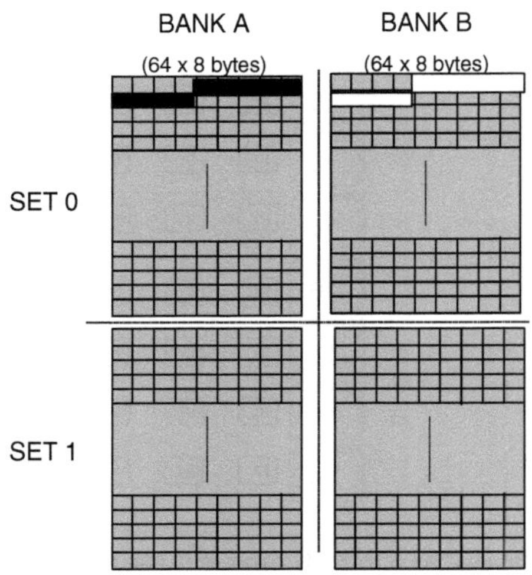

Fig. 4.3 FB block diagram

Frame Buffer is byte addressable. An important feature of the FB is the ability to provide any eight consecutive bytes of data to RC Array in one clock cycle. FB is implemented using an SRAM cell with two read ports and one write port. To access eight consecutive bytes of data, the decoder enables the first read port of the associated decoded row and the second read port of the row next to the decoded row address. Then, these two rows of data are concatenated and a barrel shifter is used to select the desired eight bytes based on the column address. Fig. 4.4 shows the hardware block diagram of the FB.

The DMA controller controls the communication between the FB and main memory. Using the two sets of FB alternatively, the computation of the RC Array and the data overlap the load and store of the FB and, therefore, the memory accesses are virtually transparent to the RC Array.

The DMAC block handles all data/context transfers between Context Memory, Frame Buffer, and main memory. Three TinyRISC instructions for MorphoSys are used to direct the operations of DMAC. The DMAC consists of three components: DMAC state machine, data register unit (DRU), and address generator unit (AGU) (Fig. 4.5). The DRU is used to pack or unpack data since the bus width between main memory and DMAC (32 bits) is different from the bus width between DMAC and Frame Buffer (64 bits). The AGU generates the addresses for the main memory and Frame Buffer when reading or writing Frame Buffer and Context Memory addresses during context loading.

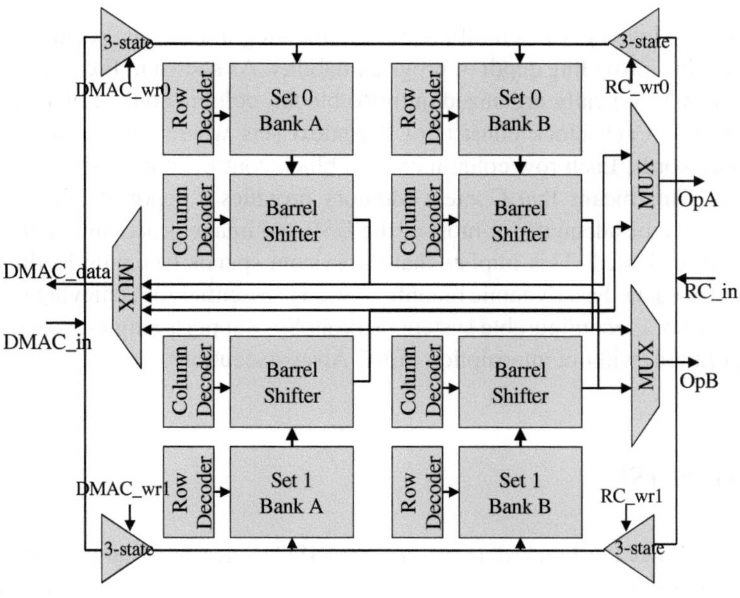

Fig. 4.4 Frame buffer block diagram

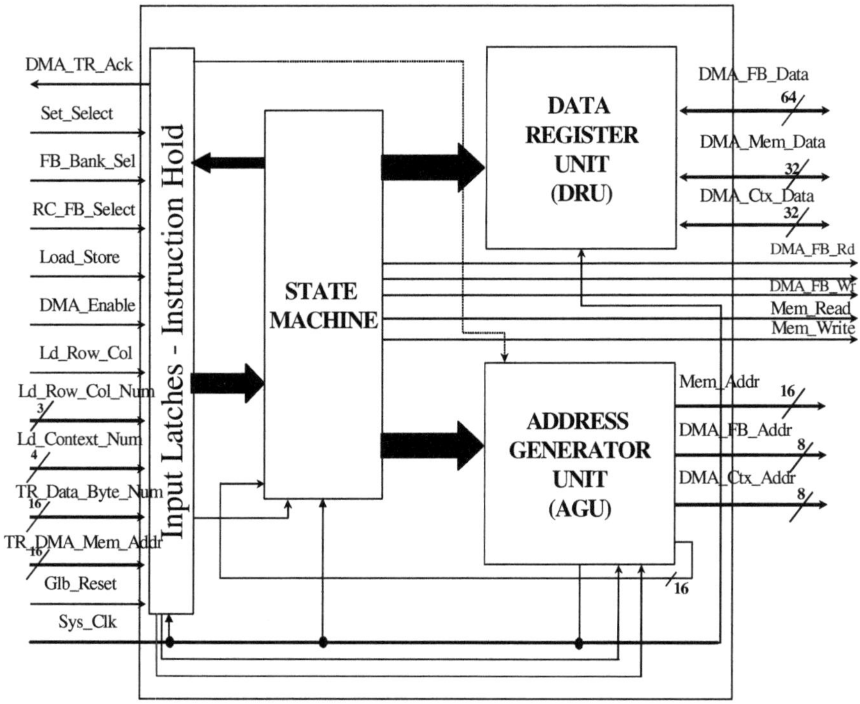

Fig. 4.5 Internal block diagram of DMAC

4.2.3 Context Memory

The Context Memory stores multiple planes of configuration data (context) for the RC Array, thus providing depth of programmability. As shown in Fig. 4.2, the Context Memory is logically organized into two blocks, column context block and row context block. Each block consists of 8 context sets where each context set has 16 context words. Each row/column context block controls one column/row of the RC Array. This means that Context Memory provides a depth of 32, which has been proven to be adequate for most of the DSP and image processing applications we have investigated. This implies that the system spends less time loading fresh configuration data. Fast dynamic reconfiguration is essential for achieving high performance with a reconfigurable system. MorphoSys supports single-cycle dynamic reconfiguration (without interruption of RC Array execution).

4.2.4 TinyRISC

Figure 4.6 shows the block diagram of TinyRISC. Since most target applications involve some sequential processing, a RISC processor, TinyRISC [2], is included in the system. This is a MIPS-like processor with a four-stage scalar pipeline. It

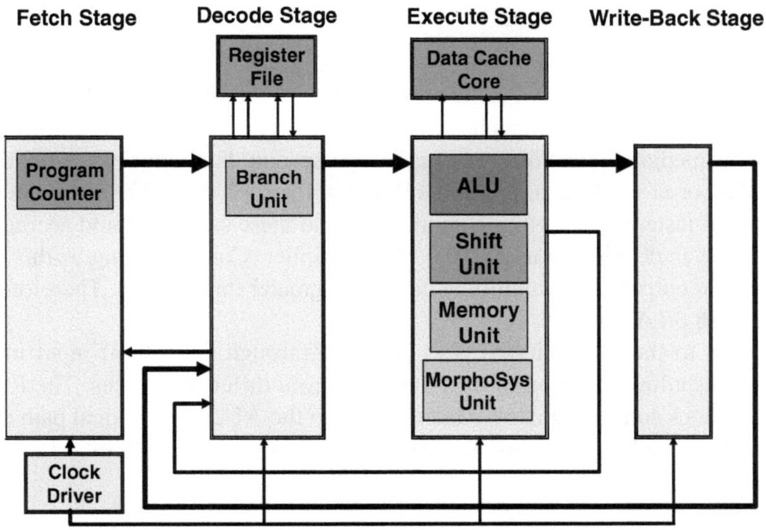

Fig. 4.6 TinyRISC block diagram

has a 32-bit ALU, register file and an on-chip data cache memory. This processor also coordinates system operation and controls its interface with the external world. This is made possible by addition of specific instructions (besides the standard RISC instructions) to the TinyRISC instruction set architecture (ISA). These instructions initiate data transfers between main memory and MorphoSys components, and control execution of the RC Array.

MorphoSys implements a novel control mechanism for the reconfigurable component through the TinyRISC instructions. The TinyRISC ISA has been modified to include several new instructions (see next section) that enable control of different components in the system. These instructions contain fields that directly provide different control signals to the RC Array, DMA controller, FB, and Context Memory. There are two major categories of these new instructions:

- The DMA instructions contain fields that provide the DMA controller with adequate information regarding the direction, destination and source of every data transfer: starting address in main memory, starting address in the FB and Context Memory, number of bytes to be transferred.
- The RC Array instructions have fields that provide the control signals to the RC Array and the Context Memory. This is essential to enable the execution of computations in the RC Array. This information includes the contexts to be executed, the mode of context broadcast (row or column), location of data to be loaded in from the FB, etc.

4.2.5 Reconfigurable Cell

The Reconfigurable Cell (RC) is the basic element of the RC Array, which is the reconfigurable component of MorphoSys. Each RC (Fig. 4.7) has a 16 × 12 multiplier. Most multi-media applications require that the second data input to the multiplier be less than or at most equal to 12 bits. Since a 16 × 12 multiplier is significantly smaller (and faster) than a 16 × 16 multiplier, and these savings would accrue over 64 RCs, it was decided to use a 16 × 12 multiplier. Corresponding to this input data size, the output of the multiplier cannot be greater than 28 bits. Therefore, RC includes a 28 bit ALU.

The data to the multiplier/ALU is provided through two 16-bit input muxes. These muxes allow selection of data operands from different sources. The RC decoder generates control signals for the muxes and the ALU. The critical path of RC consists of the 16 bit input mux, the 16 × 12 bit multiplier, the 28 bit ALU, and a shift unit.

4.2.5.1 RC Array Interconnection Network and Global Routing

The global routing network consists of three parts: RC Array interconnection and data/context bus network, clock tree, and power/ground network. The RC interconnection network is comprised of three hierarchical levels.

There are three levels of RC Array interconnection network. The first level of the RC Array interconnection network is the nearest neighbor layer that connects the Reconfigurable Cells in a 2-D mesh (Fig. 4.8). The second layer of

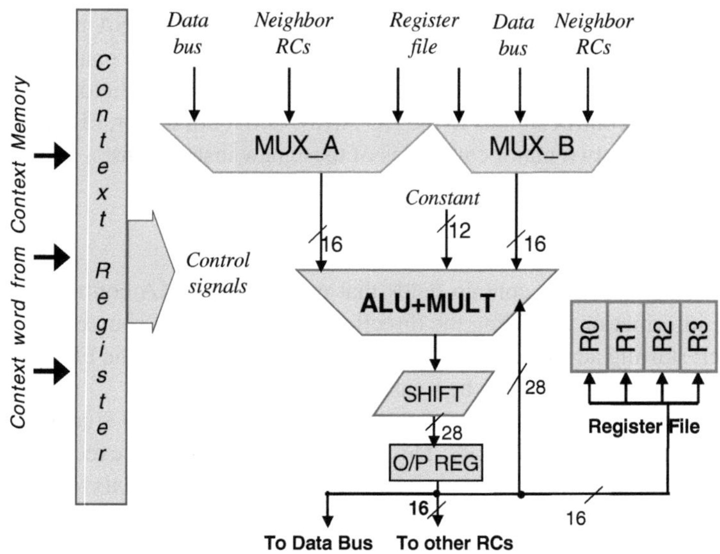

Fig. 4.7 RC Element architecture

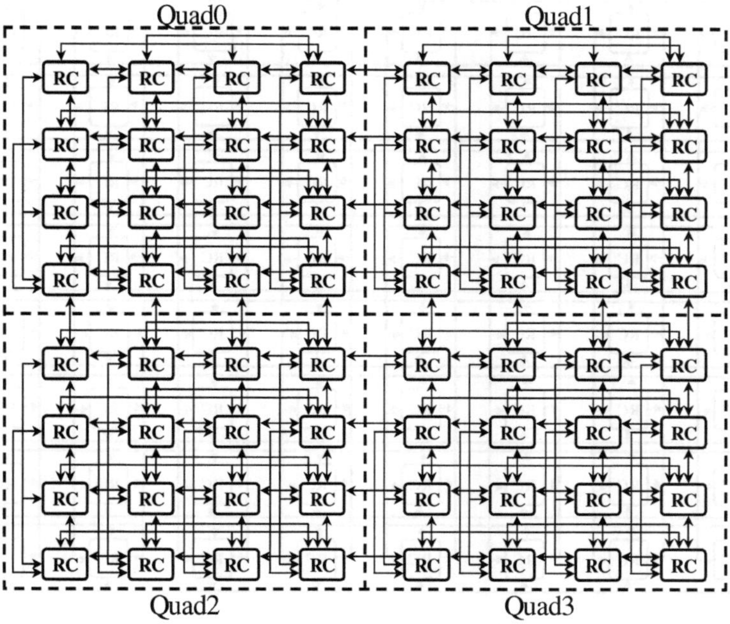

Fig. 4.8 Levels 1 & 2 of interconnection network

connectivity is at the quadrant level (a quadrant is a 4 × 4 RC group), which provides complete row and column connectivity within a quadrant. Therefore, each RC can access data from any other RC in the same row/column within the quadrant. At the highest or global level, there are buses that support inter-quadrant connectivity (Fig. 4.9). These buses are also called express lanes and they run across rows as well as columns. These lanes can supply data from any one Reconfigurable Cell (out of four) in a row (or column) of a quadrant to other Reconfigurable Cells in same row (or column) of the adjacent quadrant. Thus, up to four cells in a row (or column) may access the output value of any one of four cells in the same row (or column) of the adjacent quadrant. The express lanes greatly enhance global connectivity. Even irregular communication patterns, that otherwise require extensive interconnections, can be handled quite efficiently. For example, an eight-point butterfly is accomplished in only three clock cycles.

A 128-bits data bus from the Frame Buffer to RC array is linked to the column elements of the array. As illustrated in Fig. 4.3, each bank of Frame Buffer provides eight consecutive bytes of data. When each Reconfigurable Cell in one column receives two operand data (operand A and operand B), a total of 16 bytes are interleaved and sent to the RC Array through a 128-bits bus. It is possible to load the entire RC Array with same set of 128-bits data in one cycle. To have each RC receive different data, eight cycles are required to load the entire RC Array. The outputs of RC elements of each column are written back to Frame Buffer through a 64-bit data bus.

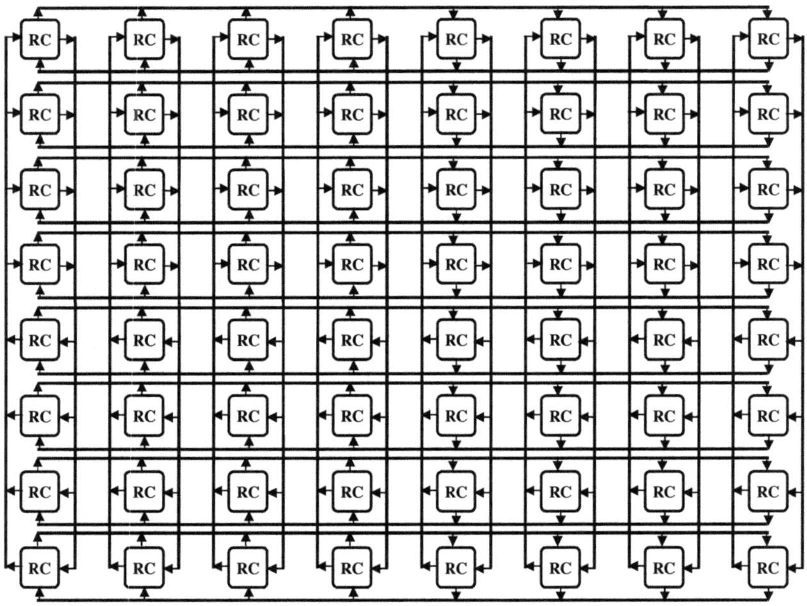

Fig. 4.9 Level 3 of interconnection network

When a TinyRISC instruction specifies that a particular group of context words be executed, these must be distributed to the Context Register in each RC from the Context Memory. The context bus communicates this context data to each RC in a row/column depending upon the broadcast mode. Each context word is 32 bits wide, and there are eight rows (columns), hence the context bus is 256 bits wide.

4.2.5.2 RC Internal Storage Units (RC-RAMs)

Each RC has an internal memory in order to load the intermediate data used by that RC instead of loading them on the Frame Buffer. For applications where each RC has a different set of input (and/or output) data, such as a part of an image, being able to quickly load or unload the internal RAM will have a significant impact on the system performance.

We have also developed a technique to load and unload data from RC's internal RAM quickly by performing DMA transfer between the frame buffer and the internal RAM. To accomplish this operation, Tiny RISC instruction (to manage FB transfers) and a context (to load/store more than one data in RC's RAM) are required. They are different for transfers from RC-RAM to FB or from FB to RC-RAM:

- DMA data Transfer from RC-RAMs to the FB: The RAM initial address must be in one RF register. The register value is loaded in a counter that points to the RAM initial address. The counter is incremented after each read signal received from the DMA controller. The Tiny RISC instruction indicates the frame buffer

address destination, the number of words to load and the column from where these words are going to be transferred.
- DMA Data Transfer from FB to RC's RAM: The RAM starting address must be in one register. The register value is loaded in a counter that points to the RAM initial address. The counter is incremented after each write signal received from the DMA controller. The Tiny RISC instruction indicates the frame buffer starting address, number of words to store and the column where these words are going to be stored.

4.2.6 Summary of MorphoSys Architecture

MorphoSys is a coarse-grain processor designed for data-parallel computation-intensive applications. It has a novel control mechanism and streaming memory interface for the reconfigurable component, and is a complete system-on-a-chip. In summary the most prominent features incorporated are as follows:

- *Integrated model:* MorphoSys has a novel control mechanism for the reconfigurable component that uses a general processor. Except for main memory, MorphoSys is a complete system-on-a-chip.
- *Coarse-grain reconfigurable computing system:* The internal datapath of the RC Array is 16 bits. However, MorphoSys also supports some bit-level operation (e.g., bit-level template matching).
- *Multiple contexts on-chip:* Context Memory can store up to 32 planes of configurations. The users have the option of broadcasting contexts across rows or columns. This feature enables fast single-cycle reconfiguration.
- *Dynamic reconfiguration:* Context data may be loaded into a non active part of Context Memory without interrupting RC Array operation. Context loads and reloads are specified through TinyRISC and are actually done by the DMA controller.
- *On-chip controller:* MorphoSys allows efficient execution of applications that have both serial and parallel tasks.
- *Innovative memory interface:* The reading and writing of the Frame Buffer are controlled by the DMA controller. It supports high data throughput by using a two-set data buffer that allows overlap of computation with data transfer.

4.3 MorphoSys Execution Model

4.3.1 Architectural Point of View

The execution model for MorphoSys is based on partitioning applications into sequential and data-parallel tasks. TinyRISC handles the sequential portion of the application, whereas the data-parallel portions are mapped to the RC Array.

Table 4.1 MorphoSys specific TinyRISC instructions

Mnemonic	Description of Operation
LDCTXT	Load Context from Main Memory to Context Memory.
LDFB STFB	Load (store) data from (into) Main Memory to (from) Frame Buffer.
CBCAST	Context broadcast, no data from Frame Buffer.
DBCBC DBCBR	Column (or row) context broadcast, get data from both banks of Frame Buffer.
DBCB	Context broadcast, get data from both banks of Frame Buffer.
SBCB	Context broadcast, transfer 128 bit data from Frame Buffer.
WFB WFBI	Write the processed data back to Frame Buffer (with indirect or immediate address).
RCRISC	Write one 16-bit data from RC Array to TinyRISC.

TinyRISC initiates all data transfers involving application and configuration data (context). TinyRISC provides various control/address signals for Context Memory, Frame Buffer, and the DMA controller. RC Array execution is enabled through special TinyRISC instructions for context broadcast.

The MorphoSys program flow may be summarized as follows. First, a special TinyRISC instruction, LDCTXT is issued. This initiates loading of context words (configuration data) into the Context Memory through DMA Controller (Table. 4.1). Next, the LDFB instruction causes the TinyRISC to signal the DMA Controller to load application data, such as image frames, from main memory to the Frame Buffer. When both configuration and application data are ready, a TinyRISC instruction for context broadcast, such as CBCAST, SBCB, etc., is issued. This starts execution of the RC Array.

The context broadcast instructions specify the particular context (from among the multiple contexts in Context Memory) to be executed by the RCs. There are two modes of specifying the context: column broadcast and row broadcast. For column (row) broadcast, all RCs in the same column (row) are configured by the same context word. TinyRISC can also selectively enable a row/column, and can access data from selected RC outputs.

During the execution of a context/data broadcast instruction, the intended context and data are taken from the context memory and the FB, respectively, and loaded into the corresponding registers inside the corresponding RCs, so that in the following clock cycle, the data can be processed as instructed by the context. All different context types are executed in one clock cycle.

MorphoSys supports dynamic reconfiguration. Context data may be loaded into a non-active part of the Context Memory without interrupting RC Array operation. Since the Frame Buffer has two sets, it is possible to overlap computation in the RC Array with data transfers between external memory and the Frame Buffer. While the RC Array performs computations on data in one Frame Buffer set, fresh data may be loaded in the other set or the Context Memory may receive new contexts.

4.3.2 Application Point of View

Multi-contexts reconfigurable architectures execution model can be also described from the application point of view. From the previous discussion is inferred that the execution of any application in MorphoSys requires the storage of its data and contexts in the on-chip memories in advance. Then, when a new configuration or a new data is needed, it is loaded from these on-chip memories which are faster than loading from the external memory. However, an application usually has to process an amount of data whose size is larger than the on-chip data memories storage size and it is enough complex to have an amount of different configurations whose size is also greater than CM size. Therefore, a study of the applications usually executed in MorphoSys and their execution model is required.

We can describe the multimedia and DSP applications typically executed in MorphoSys as a group of macro-tasks, which usually contain an important amount of interrelated computations. One example is MPEG [3], a standard for video compression and decompression, whose encoding sequence consists of a set of macro-tasks, kernels, repeatedly executed within a loop. These are motion estimation (ME), motion compensation (MC), discrete cosine transform (DCT), etc. These functions are usually used in a lot of other applications, for example, DCT is widely used in the field of video or image processing.

One of those applications can typically be represented as a loop of N_T iterations of a sequence of kernels (Fig. 4.10a). Then, our execution model can be represented as in Fig. 4.10b. It is a time diagram where the first three rows stand for data and contexts transfers, and the other represents the evolution of kernel processing over data stored in a specific FB set. The superscript j refers to the jth iteration. The computation of the kernels can be overlapped with the data and contexts transfers of the current, previous or next iterations. Then, for example, kernels k_1 and k_2 use data

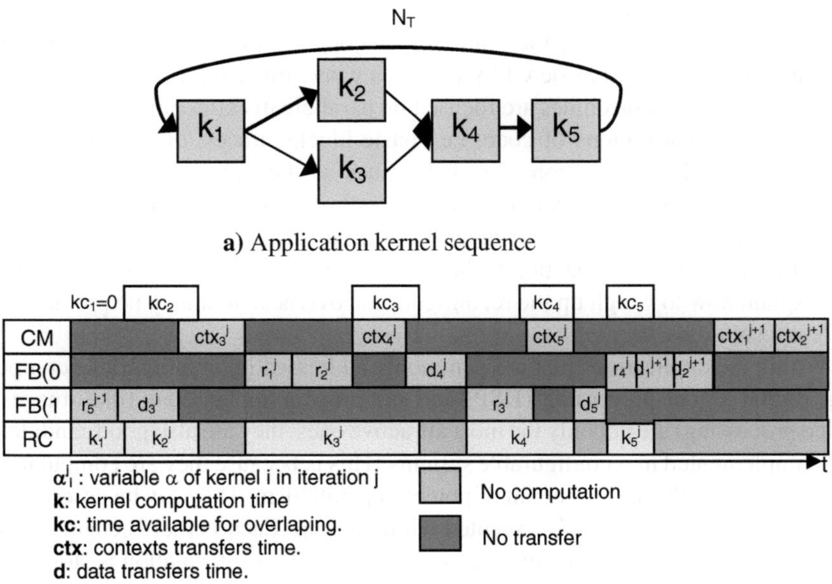

a) Application kernel sequence

b) Execution model illustrating a possible schedule in iteration j

Fig. 4.10 An application kernel sequence and its execution model

from the set 0. At the same time, the results of k_5 from previous iteration ($r_5{}^{j-1}$) and the input data of k_3 for the current iteration ($d_3{}^j$) are transferred between the external memory and the set 1 of the FB. A similar reasoning may be applied to the rest of the kernels. The application execution model imposes that the data and context required for the execution of a kernel must be on the FB and C respectively, before kernel execution. This can cause a hardware stall until the data and contexts are transferred. In order to understand the time diagram shown in Fig. 4.10b we should remember the architectural execution model which allows data and contexts transfer overlapping with computations but not simultaneous transfers of contexts and data. Moreover, contexts loading can overlap with kernel computation, when there is no access from the RC array to the CM, this interval is called *kc* (see Fig. 4.10).

Hereafter, our discussion will be based on this execution model, which establishes the features of the target architecture that can be addressed by the proposed methodology.

4.4 Compilation Framework

The development of computer-aided design (CAD) tools that speed up the design process, exploit the architectural features, and achieve the throughput requirements, becomes extremely necessary. In order to develop an efficient scheduling, the tools have to take into account the particular characteristics of the applications and the architecture. This section briefly describes a CAD framework which produces executable code from a given high-level input description of an application. It considers all possible optimizations that may be carried out at compilation time, through the scheduling of the context/data transfers and their allocation.

From our point of view, there are two major issues that unlike our approach are only superficially considered by previous work. First, the compilers based on research for parallel machines are focused on parallelism extraction, and, therefore, they explore local regions of code, i.e., basic blocks, traces, or hyperblocks [4]. This makes it difficult to consider such issues as the impact of reconfiguration time and data transfers on performance, as well as incremental configuration. All these parameters should be considered from a global point of view. Second, they do not propose any optimization technique regarding reconfiguration management nor explain how to weigh up the reconfiguration overhead that may be produced by different solutions.

Within the applications that are commonly addressed in reconfigurable computing, digital signal processing (DSP) and multimedia applications (like image or video processing) are not only the most attractive ones, they are also good candidates to be implemented in reconfigurable systems. This is because they are computationally intensive with a large amount of potential parallelism. They can be described as a set of kernels consecutively executed within a loop. Each kernel is characterized by its contexts, as well as, its input and output data. This information is stored in a Kernel Library, which is part of the compilation framework.

The general framework that we propose is presented in Fig. 4.11. The application description is provided in C code, and we assume it is written in terms of kernels.

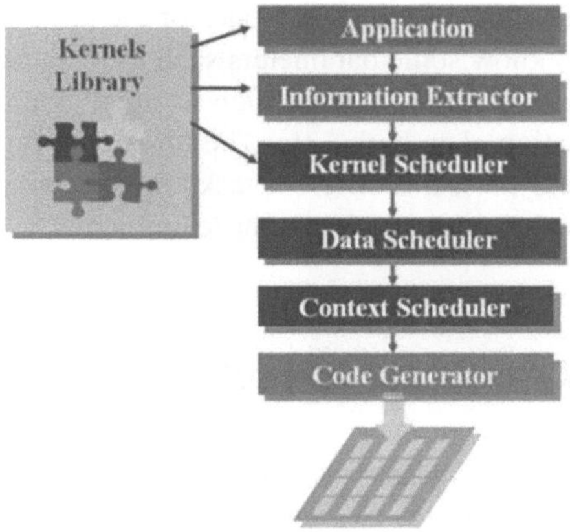

Fig. 4.11 Task flow of the general compilation framework

The kernel library stores the code that specifies not only the functionality, but also the mapping to the reconfigurable units. This approach enables an environment to be supported with the following advantages:

- Reuse of kernel code, which distributes the initial development costs and enables refined optimizations to get optimal results.
- Design modularity, which enables fast application development cycles and easy adaptation to a graphical environment.
- Global optimizations may be exploited more easily above the kernel level.
- Design space manageability, since the number of kernels is low.

The first step of the framework is information extraction, which generates from the input code all the parameters that the following tasks need. This information includes kernel execution time, data size, data dependencies (as data flow graphs) and context size of each kernel. Later, the Kernel Scheduler explores the design space to find a sequence of kernels that minimizes the overall execution time.

At this point, we know the best sequence of kernels, but there are other events that should be carefully scheduled, such as context loading and data transfers. Furthermore, the allocation of contexts and data will strongly influence the memory utilization and so the scheduling and the final execution time. The Data and Context Schedulers decide when and where the data and contexts are transferred trying to reuse the maximum amount of data and contexts.

Finally, the code generator will encode the results of the previous steps so that the optimized version of the code is generated. This includes the code of all the kernels, as well as the scheduling specifications. Then, the executable code is generated by the native compiler. Since an application is always executed as an ordered sequence of configurations and data transfers and the application structure can always be provided in terms of kernels, the proposed compilation framework can be applied to different reconfigurable architectures. However, the specific tasks that compose the compilation framework, as well as the kernel mappings stored in the library must take into account the issues of the target architecture. Thus, it would be necessary to know some parameters such as reconfiguration and data transfer time, and some architectural features such as architecture reconfigurable blocks, memory organization, data buses, etc. The research reported here targets specifically MorphoSys, and thus, the different tasks in the framework can only be applied efficiently to architectures with similar features.

4.4.1 Kernel Scheduler

Each application is described by its Data Flow Graph which imposes some constraints to the execution order. Then, it is possible to have different execution sequences that result in different total execution times. The target of the Kernel Scheduler [5] is to find the sequence of kernels that minimizes the execution time.

We will discuss the scheduling issues with the example in Fig. 4.12. A given application can usually be scheduled in many different ways. Figures 4.12a & b show two extreme schedules of a hypothetical application. If each kernel is executed only once before the execution of the next one (case *a*), each kernel's context has to be loaded as many times as the total number of iterations, N_T. Moreover, if some data that have been used or produced remain in the FB, they can be reused by other kernels, and data reloading is avoided. In the other extreme, if each kernel is executed N_T times before executing the next one (case *b*), each kernel's context has to be loaded only once. However, as the size of the data produced and used by any kernel is likely to exceed the size of the internal memory resources, data reuse may be low. Furthermore, in case *b* data transfers for a kernel can potentially overlap with the computation of any other kernel, and the possibilities of overlapping are maximal. In the former case data/context transfer for a kernel can only overlap with its own computation, and the possibilities of overlapping are minimal. The optimal solution usually remains between these two extreme alternatives (Fig. 4.12c). It can be obtained if the application is partitioned into sets of kernels that can be scheduled independently of the rest of the application. We use the term partition to refer to one of these kernel sets. In these solutions, context and data reuse, as well as the possibilities of overlapping computation and data/context transfers may be possible to some extent between the two extreme sequences, hence the overall effect may produce a better result.

From the above discussion, it is clear that there are three optimization criteria, which are summarized as follows:

- Context reloading should be minimized.
- Data reuse should be maximized.
- Computation and context/data transfer overlapping should be maximized.

The scheduling of kernels within a partition is not just the simple ordering of kernels. In architectures with multiple memory sets such as MorphoSys, the scheduling of kernels can be viewed as an assignment of computations to one of these memory sets. This means finding subsets of kernels and deciding to which memory set are going to be assigned. We use the term partition to refer to one of these subsets. The partitioning and the assignments will influence the reuse of data among kernels, as well as the possible temporal overlapping of computation and data movements. Generally, it is not possible to know in advance which of the feasible solutions is the best. It is necessary to explore the design space, because the above three criteria are mutually conflicting. The solution to the problem is partitioning and scheduling a linear sequence of kernels with the minimal execution time. The scheduling algorithm has to take into consideration the following constraints:

- Loading of input data: before execution.
- Loading of contexts: before execution.
- Write-back /re-use of results: after execution.
- Frame buffer size.
- Context memory size.
- Overlapping of computation and data movements is possible only if they use different sets of the frame buffer.
- When computation is across rows (columns) context, then a new column (row) context may be transferred.

We propose to solve this problem by dividing it into two tasks:

- Exploration Algorithm
- Partitioning of the Application

We use a backtracking technique in order to support the search process in both tasks. This process is guided by some heuristic technique that tries to "explore best candidate solutions first". We also employ bounding heuristics for an early pruning of the search space.

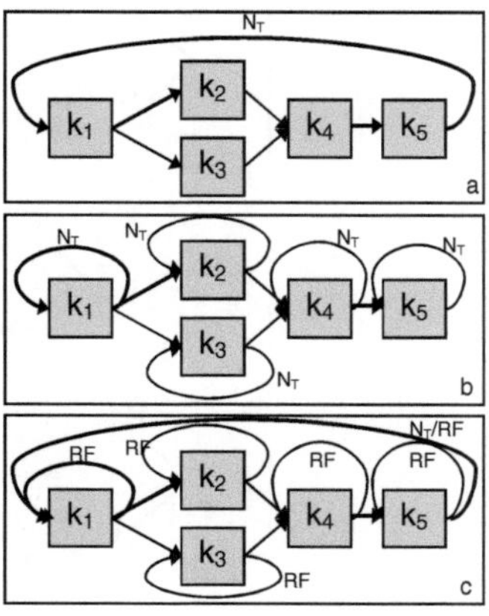

Fig. 4.12 Different possible kernel scheduling

4.4.1.1 Exploration Algorithm

One of the desirable properties of the exploration algorithm is the potential generation of any possible solution. It will ensure that the optimal solution can always be found. Hence, we propose a recursive algorithm with backtracking. The starting set is the whole application to be partitioned. Partitioning of this initial solution will allow all possible partitions (clusters) to be visited. The exploration procedure is implemented as follows. Each edge of the DFG (data flow graph) is numbered in ascending order according to the amount of data reuse between the kernels connected by that edge (Fig. 4.13a) creating the root node for the exploration tree. If data reuse is the same for several edges, any order is valid, but numbers are not repeated. Then, the edges are removed in ascending order generating the branches shown in the exploration tree. For example, in Fig. 4.13b the edge 1 is removed. Other child-nodes are created removing other edges. The procedure continues removing edges to create new child-nodes of the previous one. In a given node of the exploration tree, edge i cannot be removed if previously any edge j such us $j > i$ has been removed. For example, in Fig. 4.13d, edge 2 cannot be removed, as the last removed edge is 3, then Fig. 4.13f is its only possible leaf. Note that if the tree is visited from the top to any leaf, it is not possible to find a descending edge sequence. This fact guarantees that every solution is generated only once. This process results in the formation of groups of kernels that have no joining edges (Fig. 4.13.c–d). Each separated group of kernels forms a potential partition (cluster) whose feasibility has to be checked. For example, in Fig. 4.13c the generated subset $\{k_1, k_3, k_4\}$ is not a partition, since k_4 needs the results from k_2. Every time a new partition appears, a different cover of the DFG is built, and therefore a different solution is explored.

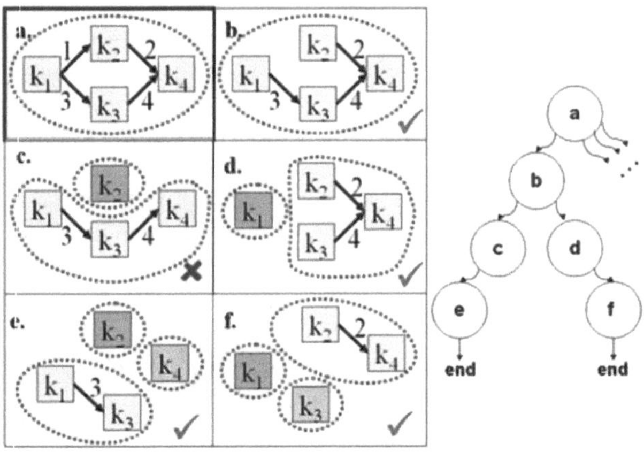

Fig. 4.13 Exploration sequence and tree

4.4.1.2 Partitioning of the Application

From the point of view of performance, the quality of a partition can be estimated without performing scheduling. If we assume that parallel computation and data transfer is always possible, the execution time can be directly estimated through the evaluation of an expression.

Let $P = \{k_1, \ldots, k_n\}$ be a partition, then the estimated execution time is:

$$ET(P) = \begin{cases} MAX\left[\sum_{i=1}^{n} t(k_i), \sum_{i=1}^{n} (t(d_i) + t(r_i)) + t(ctx_i)\right] \; if \sum_{i=1}^{n} t(kc_i) \geq t(ctx_i) \\ MAX\left[\sum_{i=1}^{n} t(k_i) - t(kc_i), \sum_{i=1}^{n} (t(d_i) + t(r_i))\right] + t(ctx_i) \; if \sum_{i=1}^{n} t(kc_i) \leq t(ctx_i) \end{cases} \quad (4.1)$$

$t(k_i)$: the execution time of kernel k_i

$t(ctx_i)$: the time spent to load the contexts of k_i

$t(kc_i)$: the portion of computation time that can overlap with contexts loading for k_i

$t(d_i)$: the time required to transfer the data of k_i from the external memory to the on-chip memory

$t(r_i)$: the time required to transfer the results of k_i from the internal memory to the external memory

ET(P) is graphically illustrated in Fig. 4.14. In the first case of this expression, computation completely overlaps with context transfers, because the system allows enough time to do these transfers (Fig. 4.14a1 and a2). The difference between Figs. 4.14.a1 and 4.14.a2 is the relationship between kernel execution time and kernel data, result and contexts transfers time, which implies different values for *ET(P)*. In the second case (Fig. 4.14.b1 and b2) there is not enough time to do all the contexts transfers. The remaining computation time is overlapped with data transfers. The partition execution time depends not only on the kernels of the partition but also on the associated context and data sizes.

In a general partition the total number of contexts words may be more than the context memory size. In that case contexts transfers cannot be completely overlapped with kernels execution time. However, a subsequent partitioning of this partition may reduce the number of kernels per partition. So the number of contexts to transfer can be overlapped with computation, reducing execution time. Furthermore, the size of data of a general partition may be bigger than the size of one set of the on-chip memory, then data transfers cannot be fully overlapped with kernel execution. The partitioning algorithm continues always that a refinement of the current partition reduces the execution time.

Let $\{P_i, \ldots, P_m\}$ be a refinement of partition $P(P = P_i \mathrm{U} \ldots \mathrm{U} P_m \mid P_i \cap P_j = \emptyset \; (i = j)$. If the new partition fulfils that $ET(P) \leq \sum_{\forall P_i \in P} ET(P_i)$ and its data fit into one set of the FB, no partitioning of P will improve its execution time. Then P represents a sequence of kernels whose data and results are assigned to the same set of the internal memory. We call clusters to this kind of partitions that have the minimum execution time.

The partitioning of the application on clusters give us the execution model shown in Fig. 4.15, for the example of Fig. 4.10. The solution with the minimum execution time is formed by three clusters: $c_1 = \{k_1, k_2\}$; $c_2 = \{k_3\}$; $c_3 = \{k_4, k_5\}$. MorphoSys' computation model allows the overlapping of data and contexts transfers with computation (Fig. 4.15.b) and the Kernel Scheduler takes advantage of this characteristic in order to find the solution than potentially has a minimum execution time. The Data and Contexts Schedulers manage these transfers and their allocation in the on-chip memories in order to reduce the execution time. Moreover, it has also an important impact in the system energy consumption because multimedia applications executed repeatedly a sequence of kernels and deal with a large amount of data that have to be transferred and stored in memory. Both facts consume an important amount of energy. Then the data and contexts schedulers are not only targeted to the reduction of the execution time, but also to minimizing energy consumption.

4.4.2 Data Scheduler

In multimedia applications the energy consumption is clearly dominated by memory accesses [6], [7] and [8]. They deal with a large amount of data, which have to be stored and transferred. The minimization of these transfers from/to external memory achieves an important reduction in the overall execution time as well as a minimization of the power consumption. Then, tools developed for reconfigurable systems must have an efficient data manager for reducing significantly the system overall energy consumption that would be achieved by considering the particularities of multimedia applications and reconfigurable systems. Then the problem could be defined as follows: given an ordered set of macro-tasks, obtained by the kernel scheduler, find the data scheduling that minimizes the energy consumption.

Although multimedia applications require a large amount of memory for the storage of the data structures that they process, most of these applications apply

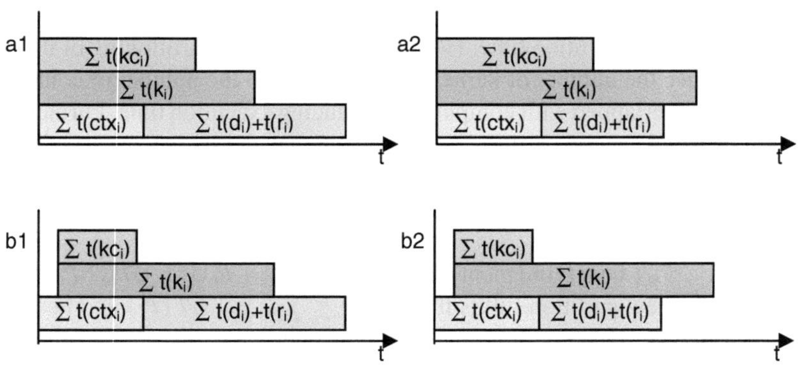

Fig. 4.14 Time diagram with data and contexts overlapping with kernel execution time

block partitioning algorithms to the input data. The partitioning algorithms indicate the data block size processed by the kernel in each iteration (as it was explained before, the sequence of kernels is repeated till all the input data are processed). These blocks are relatively small and can easily fit into reasonably sized on-chip memories. Within these blocks there is a significant data reuse as well as data spatial locality [9]. Particularly, the data framework presented in this paper takes into account the hardware reconfiguration and the memory hierarchy detailed for our target architecture MorphoSys. The energy consumption for one cluster execution is the sum of the energy consumption of data and contexts transfers, memory accesses (FB, CM and RC-RAMs) and reconfigurable hardware. The energy consumption of each memory was estimated using Landman's memory model [10]. Landman has presented a number of black-box capacitance models for architectural energy analysis. The dominant term in memory energy consumption, in this model, comes from charging and discharging the capacitance of the bit lines. Consequently, with small memories a small capacity is switched and less energy is consumed. It implies that the data scheduler should store the most frequently used data in the smallest on-chip memory to reduce energy consumption, which is in the case of MorphoSys the RC-RAM. The context memory has also this energy model and the Context Scheduler is the responsible to find and store in it the most frequently used configurations.

The data scheduler first task processes the application code in order to find the data and results blocks that have a reasonable reuse. The data scheduler finds not only the data and results reused among the kernels of the same cluster, it also finds

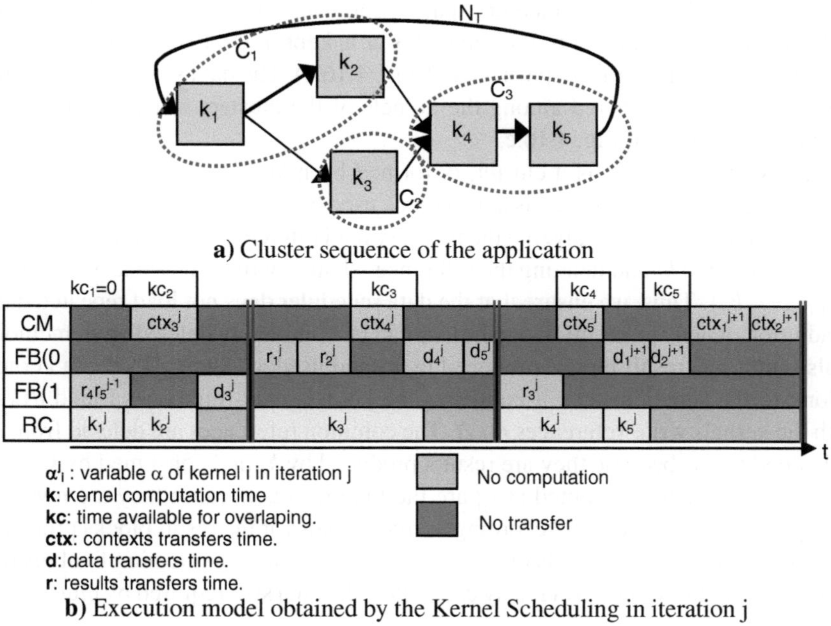

Fig. 4.15 Kernel scheduler results

the data and results shared among kernels that belong to different clusters, promoting their reuse when it is possible. Then, it decides where to store these blocks in order to minimize energy consumption, calculating an energy factor per each block and estimating the energy consumption reduction achieved when a block is stored and kept in one memory of the hierarchy. Based on this factor and the available memory space, the data scheduler decides where to store the block. The third task exploits the possibility of repeating the execution of a given kernel, so reducing context transfers. Finally, an Allocation Algorithm decides the placement of data and results by promoting regularity and simplifying data accesses.

4.4.2.1 Data and Results Blocks

In order to explain the importance of calculating an accurate value of the data or results block sizes used by the kernels, let us use the example described in Fig. 4.16. In this example the Kernel Scheduler has decided to arrange kernels k_1, k_2 and k_3 in a cluster, then their data and results are stored in the same FB set. The kernels' data and results sizes are obtained from the Kernel Library, these values are independent of the final sequence of kernels (Fig. 4.16.a). However, it is highly possible that different kernels of the same application share data. If these shared data are not calculated, the same set of data may be in the FB more than once (case 1 of the Fig. 4.16.b). The Data Scheduler should find the data shared among kernels of the same cluster to obtain the result shown in the case 2 of Fig. 4.16.b, which implies a better usage of the available memory space. Moreover, a result produced by a kernel and used by another kernel of the same cluster is kept in the FB after this kernel is executed taking up memory space (case 1 Fig. 4.16.c). The data scheduler must find the data and results shared among the kernels of the clusters achieving the result shown in the case 2 of Fig. 4.16.c.

Let us suppose there is a cluster, c, formed by n kernels $\{k_1, k_2, \ldots, k_n\}$. The Data Scheduler finds d_i, which is a data block used by kernel k_i and not by any other kernel executed after it, analyzing the memory read references of the part of the code corresponding to k_i and deleting the references shared with kernels executed after it $\{k_{i+1}, \ldots, k_n\}$. Thus, we ensure that the data scheduler does not load repeated data. In addition, when k_i execution finishes, d_i can be completely deleted from memory. It also finds r_{ij}, result blocks produced by k_i and used as data by k_j (both kernels belong to the same cluster). It compares the kernels read-references found $d(k_j)$, with the kernels write-references $r(k_i)$. The common references are deleted from d_j and added to r_{ij}, because they are results produced by k_i and consumed by k_j. The write-references not computed as r_{ij} are the final results because they are not used by any kernel of the cluster. Accordingly, $rout_i$ stands for the result blocks produced by k_i and processed by kernels of others clusters or used as application final results.

Then using the values previously calculates, the data size required by cluster c is, $DS(c)$:

$$DS(c) = \underset{i \in \{1,\ldots n\}}{MAX} \left[\sum_{j=i}^{n} d_j + \sum_{j=1}^{i} \left(rout_j + \sum_{t=i}^{n} r_{jt} \right) \right] \tag{4.2}$$

n: number of kernels of cluster c.
d_j: input data size for kernel k_j of cluster c, candidate to be stored in the RC-RAM.
r_{jt}: intermediate results size of kernel k_j of cluster c which are data for kernel k_t of cluster c and not for any kernel executed after it.
$rout_j$: final results size of kernel k_j of cluster c.

DS(c) calculated the cluster c data size taking into account that:

- All the data used by any kernel of the cluster are in memory before the execution of the first kernel of the cluster.
- When the execution of a kernel k_i ends, the data associated with this kernel, $\{d_i, r_{1i}, r_{2i}, \ldots r_{i-1i}\}$ are released from memory.
- At the end of cluster execution only the kernels final results (*rout*) are in memory.

Then after applying the data scheduler over the sequence of kernels of every cluster of the application, the space required for cluster execution has been minimized, increasing so the free space in the on-chip memory that can be used to store data shared among kernels of different clusters (minimizing data transfers energy penalty), or to store the data required for consecutive kernel executions (minimizing

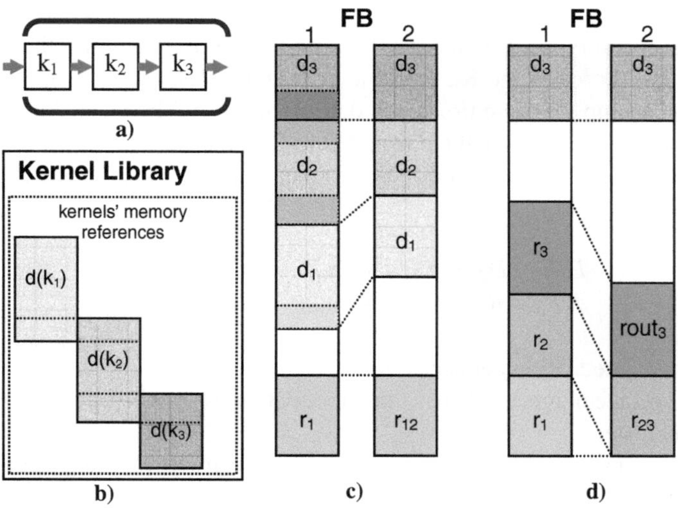

Fig. 4.16 Data memory organization for an example with 3 kernels. a) Kernels data dependencies for the example. b) Kernel Library: data references. c) FB snapshots just after k_1 execution (1: without any scheduling; 2: with data and results reuse). d) FB snapshots just after k_3 execution (1: without any scheduling; 2: with data and results reuse)

the contexts transfers energy penalty). The following sections exploit both penalty reductions.

The algorithm developed can also find data and results shared among clusters analyzing the data and results blocks previously calculated. It finds D_{uv}, which is a data block read by some kernels of clusters u and v. It is composed by the common read-references between u and v. This data block would be located before the execution of the first kernel of cluster u and it would be deleted after execution of the last kernel, which processes it, of cluster v. In a more general example, with three clusters c, u and v, executed in this order, and they could share the same data block, the Data Scheduler will obtain the blocks D_{cu}, D_{cv} and D_{uv}. Among these blocks it will decide, depending on the available memory space and energy reduction, which block to store in memory. Moreover, it finds R_{uv}, result block produced by cluster u and consumed by cluster v, comparing the r_{out} references of cluster u with cluster v read-references.

4.4.2.2 Data and Results Blocks Scheduling

As was stated above memory design in MorphoSys is hierarchical the internal memory within the reconfigurable cell (RC-RAM) being the lowest hierarchical level. It has the lowest energy consumption per access because it is the smallest one and it is within the RC. Moreover, there are specific RISC instructions and contexts, which make efficient transfers between the FB and the RC-RAMs possible. This allows the RC-RAM Scheduler to optimally exploit data storage in RC-RAM to reduce energy consumption.

Data storage in this memory should reduce energy consumption, but this is not always true. There may be an increase in energy consumption because data from the external memory must be loaded first into the FB and then into the RC-RAM. But if the data stored in the RC-RAM are read enough times (n_{min}), the decrease in energy consumption from processed data stored in RC-RAMs is larger than the energy wasted in transferring them from FB to RC-RAM.

EF_{RC} estimates the energy reduced if data or results are kept in RC-RAM:

$$\begin{aligned} EF_{RC}(D) &= E_0 \cdot (N \cdot n_{acc} - n_{min}(d)) \cdot D_{uv} \\ EF_{RC}(R) &= E_0 \cdot (N \cdot n_{acc} - n_{min}(r)) \cdot R_{uv} \end{aligned} \tag{4.3}$$

E_0: Energy reduction per access to the RC-RAM instead FB

nacc: average number of times that those data or results are read by the kernels of a cluster.

nmin(): minimum number of reads required to improve energy consumption (d: data; r: results).

D_{uv}: the size of input data stored in the RC-RAM for cluster u and processed by cluster v.

R_{uv}: the size of results stored in the RC-RAM for cluster u and processed by cluster v.

N: number of clusters that process those data or results.

The energy reduction is positive when the number of accesses is greater than a minimum value experimentally found (n_{min}), that depends on RC-RAM and FB access energy and the FB to RC-RAM transfer energy. Then, a small sized RC-RAM, compared with the FB, produces a small n_{min} value due to its smaller access energy compared with that to the FB. All the blocks that produce an energy reduction should be ideally loaded in the RC-RAMs, but this cannot be usually achieved because of RC-RAMs small size. The Data Scheduler decides which data and results blocks are stored in these memories taking into account the data size stored in the RC memories. The Data Scheduler sorts the data and result blocks candidates to be stored in RC-RAMs according to their energy factor, EF_{RC}. It starts checking if those with the greatest EF_{RC} fit in the RC-RAM. The scheduling continues with data or results with lower EF_{RC}. If the size required to store any block is greater than the size of the RC-RAM, this is not stored. When one block is stored in the RC-RAM the data that it has in common with the remaining blocks are released, the new EF_{RC} value of the remaining blocks is calculated and the scheduling continues until no more blocks fit in the RC-RAMs.

There are also data and results blocks shared among clusters that are not stored in any RC-RAM because there is not enough space in them or because their storage in those memories does not imply an energy reduction. However, if the Data Scheduler keeps them in the FB instead of reloading them in each cluster execution, there is a data transfer reduction between the FB and the external memory that reduces energy consumption.

In order to reduce energy consumption all the blocks shared among clusters should be maintained in the FB instead of reloading them. But it cannot usually store all those blocks due to FB fixed size. The FB Scheduler decides which data and results blocks are kept in this memory taking into account the energy factor (EF_{FB}) and the size of data stored in the FB.

In particular, EF_{FB} estimates the energy reduced if one data or results block is kept in the FB:

$$\begin{aligned} EF_{FB}(D) &= E_0 \cdot D_{uv} \cdot (N-1) \\ EF_{FB}(R) &= E_0 \cdot R_{uv} \cdot (N+1) \end{aligned} \tag{4.4}$$

E_0: Energy reduction per access to the FB instead the external memory.
D_{uv}: the size of input data stored in the FB for cluster u and processed by cluster v.
R_{uv}: the size of results stored in the FB for cluster u and processed by cluster v.
N: number of clusters that process those data or results.

The difference between data and results is due to that a data block used by N clusters has to be transferred to memory once, for the first cluster that uses it, and, as it is maintained in memory, there is an energy reduction because it needs not be transferred for the other $N-1$ clusters. However in the case of a result block used as data by N clusters, there are $N+1$ transfers avoided, because at the end of the execution of the cluster that produces these results, they do not have to be transferred to the external memory.

Following the same algorithm employed with the RC-RAMS the Data Scheduler starts by checking if the block with the highest EF_{FB} fits on FB. The Data Scheduler keeps the largest possible amount of data and results that reduce the energy consumption.

4.4.2.3 Contexts Reuse Scheduling

The Data Scheduler also reduces energy consumed by the reconfiguration, i. e., by transfers to the Context Memory (CM). It is achieved by considering that multimedia applications are composed of a sequence of kernels that are consecutively executed over a part of the input data, until all data are processed. Then if an algorithm needs, to process the total amount of data, execute the kernel sequence of the application N_T times, the contexts of each kernel may be loaded into the CM N_T times. However, loop fission can be applied sometimes to execute a kernel *RF* consecutive times before executing the next one (loading the required data for *RF* consecutive iterations in the on-chip memories). In this case, kernel contexts are reused because they have to be loaded only N_T/RF times, reducing context transfers from the external memory and thus minimizing energy consumption. The number of consecutive kernel executions *RF* (Reuse Factor) is limited by the internal memory sizes. The Data Scheduler finds the maximum *RF* value taking into account memories and data sizes.

As MorphoSys architecture has two levels of on-chip memory with different amount of data stored in each of them. Thus, the CM Scheduler can obtain different *RF* values (Fig. 4.7), RF_{RC} and RF_{FB}, which stand for the reuse factor allowed by the RC-RAMs and by the FB respectively. These values are calculated taking into account that in the memories are stored the data and results blocks chosen by the schedulers to reduce the energy consumption. Moreover, the CM Scheduler calculates $RF_{FB}(max)$ (see Fig. 4.17), that stands for the value obtained when in the FB there are only the data and results required to execute the cluster and not the shared data and results with other clusters. $RF_{FB}(max)$ is clearly the maximum *RF* value allowed by the architecture because the space available in the FB is only used to store the data or results required by the cluster.

Nevertheless, *RF* must be the same for all clusters and reconfigurable cells due to data and results dependencies. Then, the *RF* value that minimizes the energy consumption associated to the configuration transfers could be the maximum common $RF_{FB}(max)$ value for the clusters of the application. However, this value does not guarantee the minimum energy consumption because some data or results blocks could not be stored in on-chip memories because there is not enough space for loading them *RF* times, increasing so the energy consumption. The CM scheduler compares in terms of energy consumption RF_{RC}, RF_{FB} and $RF_{FB}(max)$ computing first the energy consumed by the smallest *RF*. If this value is increased by one, increasing so the contexts reuse, it will require more space in on chip memories, and this space should be obtained through the reduction of the data and results blocks maintained in the on-chip memories with the less energy factor. The CM Scheduler continues increasing the *RF* value until finds the one with the minimum energy consumption.

4.4.2.4 Data Allocation

Data and results placement to efficiently exploit their reusability is no a trivial problem. An allocation algorithm has been developed to minimize fragmentation as well as to promote regularity to simplify data and results accesses when kernels are iteratively executed.

Before cluster c execution starts, the input data that have not been stored into FB yet, have to be loaded from the external memory. Some input data of c are shared with next clusters (D_{ij}). As these data must remain longer in the FB, then they are placed first to minimize fragmentation. They are placed from the upper free addresses *RF* times. After that, the allocation algorithm places the input data of cluster c that are not shared with other clusters (d_i), also from upper free address.

The execution of a kernel produces different results that have to be placed. The data of a kernel shared with other clusters (R_{uv}) are placed then from the lower free addresses. The final results ($rout_i$) are placed from upper addresses. The results shared among kernels of the same cluster (r_{ij}) are also allocated from lower addresses.

When kernel execution finalizes, the Allocation Algorithm releases the space

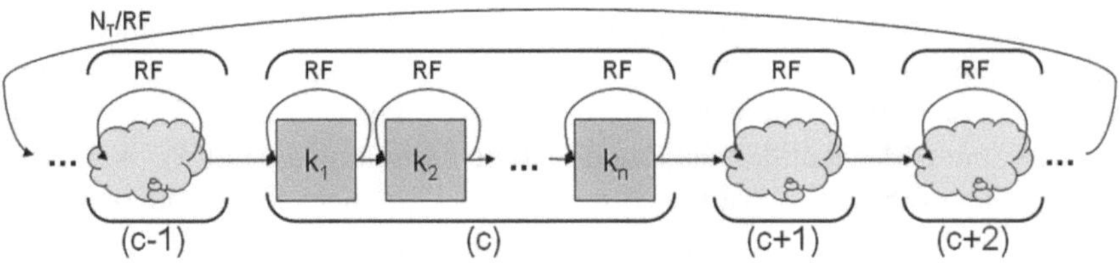

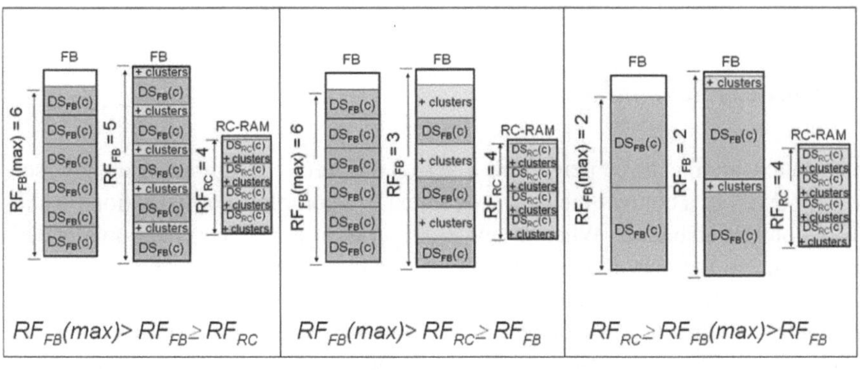

Fig. 4.17 Different $RF_{FB}(max)$, RF_{FB} and RF_{RC} for a synthetic experiment

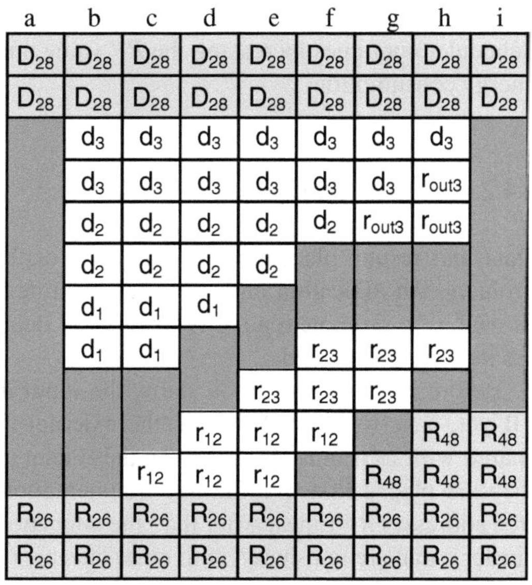

Fig. 4.18 Memory for cluster 4 execution with RF=2 (it is composed of 3 kernels) a) Before cluster 4 execution. b) Before first execution of k_1. c) After first execution of k_1. d) After second execution of k_1. e) After first execution of k_2. f) After second execution of k_2. g) After first execution of k_3. h) After second execution of k_3. i) Before next cluster execution

occupied by data and results that are not going to be used by next kernels of the cluster or next clusters.

In order to preserve regularity, data and results are allocated from the addresses where were placed in the previous iteration (Fig. 4.18). This can achieve a periodic data and result allocation.

In order to illustrate the results obtained by the algorithm described in this subsection in the data memory allocation of any application, Fig. 4.18 presents different memory snapshots for an application composed by 8 clusters. The memory image has been taking during the execution of cluster 4 which is formed by 3 kernels. Moreover, the CM scheduler has found that the optimum value of *RF* is 2. Before the execution of the kernels of the cluster (Fig. 4.17.a) there are on the memory the data and results produced by the previous clusters and consumed by next clusters. In the same way, after cluster 4 execution (Fig. 4.17.b) the memory stores only data and results shared with next clusters.

4.4.3 Context Scheduler

Multi-context architectures may store a set of different configuration planes (so-called contexts) in a context memory. A context is a complete configuration for the entire reconfigurable chip. When a new configuration is needed it is downloaded from the context memory if available. As the context memory is on chip, this operation is much faster than the reconfiguration from an external memory. Therefore, a multi-context architecture is a better choice from the point of view of performance, which is more evident in the case of full reconfiguration. Sometimes, contexts can be loaded in advance to its execution, which will reduce the reconfiguration overhead.

Hence, it is evident that reconfiguration time can be dramatically reduced, but only if the context memory is carefully managed.

In this section, we focus on the possible optimizations related to the context management that can be carried out at compilation time. The key aspect to consider is the minimization of the configuration overhead, which is also influenced by the context memory fragmentation. Then, we propose a methodology to perform context scheduling on a given ordered set of kernels, its goal being to minimize the overall execution time and energy consumption [11]. The input to this problem is a sequence of kernels, which can be obtained through the algorithms previously presented.

In order to understand the problem, we will first provide an intuitive approach to the relevant issues. During the discussion we will consider the example shown in Fig. 4.19. Note that we assume a periodic sequence. The figure represents some CM snapshots taken at different instants of a given kernel sequence: $\{k_1, k_2, k_3, \ldots\}$. Odd CM snapshots stand for the distribution of the context memory, just at the end of the corresponding kernel execution. Even snapshots represent the context memory just at the beginning of a kernel execution. We assume that before the execution of a kernel, its whole context has to be in the CM.

As a kernel always uses the same contexts, it may be possible to keep some of them in the memory and reuse them in the next kernel execution. We will use the term "*static*" to refer to these contexts. On the other hand, the rest of the contexts will have to be loaded before the beginning of the next execution, and they will be stored in the remaining part of the CM. These contexts will be called "*dynamic*". Therefore, dynamic contexts imply loading time, whereas static contexts save loading time. It may seem obvious that we should maximize the number of static contexts, and minimize the dynamic part. However, this is not always true if we consider

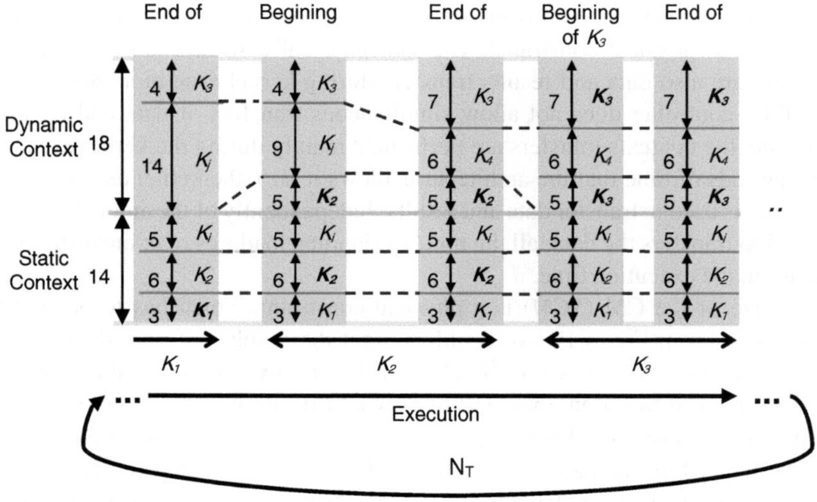

Fig. 4.19 Context memory snapshots

overlapping of context loading and system computation. In this case, we could load some contexts in advance, without affecting latency.

We will say that these loadings are "overlapping loadings" (*Lovp*). Note that overlapping loadings are accomplished between even and odd CM snapshots. For example, in Fig. 4.19 some "overlapping loadings" are carried out between the second and the third snapshot. This will save future loading time, but the amount of computation time available for overlapping is limited, which implies a major drawback: if a context is loaded too soon it will reduce the number of free positions of the CM, which may even decrease the number of static contexts. As shown in Fig. 4.19, k_3 is in the CM since the end of execution of k_1, and these positions cannot be used again until the end of k_3 execution. Furthermore, the loading of k_3 and k_4 removes k_j contexts, which in some cases might be static. In summary, both optimization factors are in conflict. Thus, there is a tradeoff between the number of static contexts and the number of overlapping loadings.

On the other hand, the "*non-overlapping loadings*" are carried out between odd and even snapshots, and they are usually necessary, since the overlapping time is limited. As these non-overlapping loadings increase the latency of the system, they constitute the key optimization aspect to be minimized. In a typical reconfigurable system, the CM cannot be updated from the external memory while it is being accessed for chip reconfiguration. As a kernel execution requires its own contexts, the CM is accessed at some fixed instants within the corresponding kernel execution. Consequently, overlapping loadings can only be performed within some fixed intervals of every kernel execution (kc_i, as explained previously). Moreover, if there is only one external memory, which is used for configuration and data storage, overlapping context loadings are also limited by data transfers.

4.4.3.1 Contexts Loading Minimization

In order to minimize the non-overlapping context loading we have to take in consideration the physical constrains. The first physical constrain is due to the fact that there are also data and results transfers during kernel execution, and the system DMA controller does not allow simultaneous transfers of data and contexts. Therefore, the contexts transfers are performed mainly during the kernel execution in the periods of time that the architecture allows it (*kc*), the kernel execution time remainder is used to transfer data and results. Independently of the amount of cycles required to transfers the data, all the non-overlapping loadings of contexts imply an increase in the execution time.

The size of the CM (*SCM*) is a physical constraint, which has to be fulfilled in all cases. From Fig. 4.19, we could say that the number of both dynamic and static contexts cannot exceed *SCM*. Then, if the non-overlapping context loadings are accomplished too soon, they will occupy CM positions, which will reduce the number of free positions. Therefore, any context loading regardless of the type has to be performed *as late as possible (ALAP)*. Overlapping loadings do not increase system latency, so they can be done just before the corresponding kernel execution.

When a context is loaded to the CM, it has to remain there until it has been used.

Consequently, if a context has been loaded before or during k_i execution, and is used after or during k_i execution, this context will be in the CM during k_i execution. For example, in Fig. 4.19 some k_3 contexts have been loaded before and during k_2 execution. Note that the terms *before* and *after* have to consider the periodic nature of the problem.

Let us consider a cluster formed by $\{k_1, k_2, k_3\}$ and the corresponding numbers of context words: $ctx_1 = 18$; $ctx_2 = 14$; $ctx_3 = 10$. The total number of context words is 42, and having a size of the CM of 32 contexts, it is necessary to reload some of them.

We present our method through a matrix called *PREP_EX*. The "*ith*" column represents the number of context words that are in the CM corresponding to kernel k_i. Different rows stand for different instants. This matrix is split into PREP (odd rows) and EX (even rows) which group the rows concerning *preparation* (before execution), and those ones related to *execution* itself. Since during the execution of a kernel its whole contexts have to be in the CM, the number corresponding to EX_{ii} always equals ctx_i. Moreover, $MAX(PREP_{i+1,j} - EX_{ij})$ is the number of context words of k_j, that have been loaded without any overlap with execution, just after the execution of k_i. Thus the goal of the context scheduler is to minimize the sum of these terms for all the kernels in the cluster. Fig. 4.20 represents the matrix formed with the 3 kernels example. If a kernel has just been executed, its context words have the highest probability of being replaced since before its next execution all other kernels have to be executed. If there are not enough locations, the previous kernel in the execution flow will also be used. The rest of the context words remain in the CM. Consequently, we can compute the row i, of EX from the row i of PREP: $EX_{i,j} = MIN[ctx_j;\ PREP_{i,j} + \delta]$. Where δ is the portion of the overlapping time that has not yet been used. The same reasoning is applied if the context loading does not overlap with computation. However, all the CM locations are then candidates. The first locations replaced belong to the kernel that has just been executed and the previous ones are used if necessary. Consequently, we can compute the row i of PREP from the row $i-1$ of EX: $PREP_{i,j} = MIN\{EX_{i-1};\ S\}$ where S is the number of locations in CM that are currently not being used. Both expressions are used within two iterative procedures, so that one EX row and one PREP row are generated. The matrix for the example, with a Context Memory of 32 contexts is shown in Fig. 4.21. Notice that each row is generated from the previous row. Therefore, it is necessary to produce a starting row. The process is repeated until a periodical solution is found or, otherwise, it stops after a number of iterations and selects the best solution found so far.

contexts / time	k_1	k_2	k_3	
	18	12	2	PREP **k_1**
	18	12	2	EX **k_1**
	16	14	2	PREP **k_2**
	14	14	4	EX **k_2**
	10	12	10	PREP **k_3**
	10	12	10	EX **k_3**

Fig. 4.20 Context configuration array

Fig. 4.21 Scheduled contexts configuration array

$$\begin{matrix} TAOC_1=2 \\ \\ TAOC_2=2 \\ \\ TAOC_3=0 \\ \\ \end{matrix} \begin{pmatrix} 18 & 9 & 5 \\ 18 & \overset{(2)}{11} & 3 \\ 15 & 14 & 3 \\ 13 & 14 & \overset{(2)}{5} \\ 13 & 9 & 10 \\ 13 & 9 & 10 \end{pmatrix} \Rightarrow \begin{pmatrix} 5 & 0 & 2 \\ 5 & 2 & 0 \\ 4 & 5 & 0 \\ 0 & 5 & 2 \\ 0 & 0 & 7 \\ 0 & 0 & 7 \end{pmatrix} \begin{matrix} \text{PREP } \mathbf{k_1} \\ \text{EX } \mathbf{k_1} \\ \text{PREP } \mathbf{k_2} \\ \text{EX } \mathbf{k_2} \\ \text{PREP } \mathbf{k_3} \\ \text{EX } \mathbf{k_3} \end{matrix}$$

$$Static + DynamicPREP\_EX \rightarrow DynamicPREP\_EX$$

In order to find the contexts that do not overlap with computation, dynamic contexts, which have a negative effect in the system performance, we need to calculate the number of possible static contexts and those that can be overlapped with computation (*TAOC: Total computation time Available for Overlapping with Context loading*). *TAOC* is limited by the kernel overlapping available time *kc* and by the free space in the contexts memory. This means, that the scheduling algorithm can decide to transfer $TAOC_i$ contexts for the next execution of kernel k_i when this number is lower than kc_{i-1} and the free space in the CM. For the example of Fig. 4.20 we have obtained the *Static+Dinamic PREP_EX* matrix of Fig. 4.21, which fulfils that the total number of contexts stored in CM is lower than or equal to the context memory size. Furthermore, during the k_i execution, represented by the EX_i row, the number of context loaded is $TAOC_i$. Then we can easily find the *Dynamic PREP_EX* subtracting to the previous one the minimum value of each column that represents the static contexts (Fig. 4.21), so we find the dynamic contexts.

4.4.3.2 Contexts Allocation

The context loading minimization process obtains a solution that provides information about the optimal configuration of context words in the memory (periodical "in time"), as well as whether the context loads overlap with computation or not. The next step is to decide where to place each context word so that memory fragmentation is minimized, while providing a periodical solution "in space".

In the context selection solution, there are some context words that remain in the CM (static context words) during all the iterations and some others that are reloaded (dynamic). We can always choose to allocate the dynamic positions in a single block. Therefore, the problem is reduced to studying the allocation of the dynamic context words. We have created a *Dynamic PREP_EX* matrix for the dynamic contexts word. The specification of a context selection configuration includes complete information about which context words have to be replaced and which ones have to be placed. This fact constrains the context allocation task. Once a kernel has been executed, all its dynamic positions are free and the context allocation task decides which positions to use. The allocation process now has to avoid the fragmentation of the context blocks (contexts placed or replaced at the same time) that are loaded as a whole in one step. Thus, a graph (Fig. 4.22) representing the dependencies among the context blocks is generated, so that all possibilities can be explored in an orderly manner. A node in the graph represents a dynamic block. An edge is added between the nodes if the source node disappears from the set of dynamic PREP EX and the destination node appears later.

Let us consider the set of dynamic PREP EX array in Fig. 4.21. If we suppose that the context loadings that are necessary in each row are loaded consecutively, but independently of the other rows, there are five context blocks. These blocks are identified with letters from "a" to "e" and the number indicates how many context words the block has. (Fig. 4.22).

The process is now divided into two steps. First, we look for a periodical solution, because it is the most important condition, the contexts loads must be independent of the application iteration. Then, we try to place the context blocks in consecutive positions beginning with the first context block, called a, and following the first-fit algorithm the other dynamic blocks are placed. The context words that belong to the same block should be in consecutive positions. All the blocks, except "c", are placed in consecutive positions for solution 1 (Fig. 4.22). We can apply backtracking, changing the first-fit placement chosen, to find a better solution, as solution 2 probe (Fig. 4.22), which does not have fragmentation. Moreover, if the initial context memory values have not been the adequate this process finds more static contexts, as contexts 2c or 2b in the Fig. 4.22 solution 1 and 2 respectively demonstrate, since these contexts are in the same position for all the kernels they can be stored at the beginning of the application as static contexts.

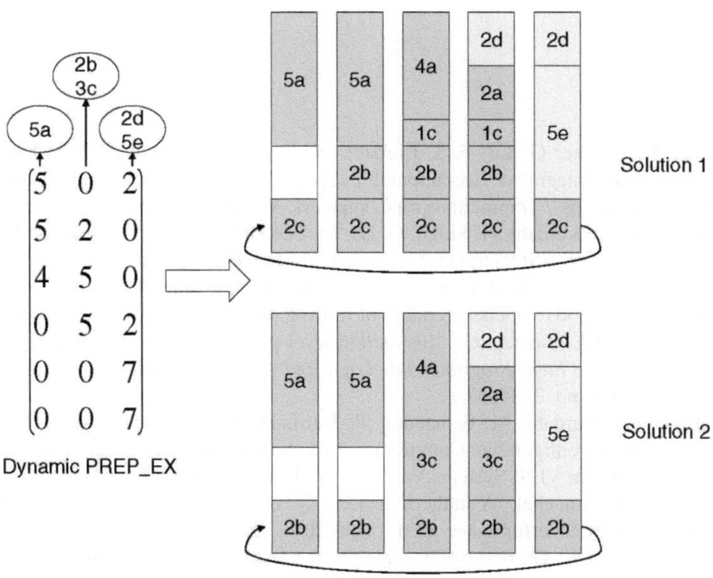

Fig. 4.22 Context memory final allocation

4.4.4 Summary of the Compilation Framework

Reconfigurable computing is emerging as a viable design alternative to implement a wide range of computationally intensive applications. The scheduling problem becomes a really critical issue in achieving the high performance that these kind of applications demands. These sections have described the different aspects regarding the scheduling problem in a reconfigurable architecture.

- Kernel Scheduler: it finds the sequence of kernels that have a minimum execution time arranging them in clusters. It partitions the application in clusters trying that the kernels of the same clusters have the maximum data overlapping and a minimum reconfiguration penalty (context loading). It creates the seed for the following compilations tasks.
- Data Scheduler: it finds the data and results shared among kernels and clusters, and tries to load them in their corresponding memory such that it produces an important energy consumption reduction. It also calculates the number of times that a kernel can be consecutively executed to reduced the energy consumption associated with the system reconfiguration.
- Context Scheduler: it schedules the contexts transfers and decides which of them should be kept in the CM memory (static contexts) and which must be reloaded in each kernel execution in order to clearly reduce power consumption and execution time.

These three tasks resolve one of the important problems in reconfigurable computing which is the complete scheduling of a group of kernels that constitute a complex application for minimum execution time and power consumption.

References

1. H. Singh, M.-H. Lee, G. Lu, F. J. Kurdahi, N. Bagherzadeh, and E. M. Chaves Filho, "MorphoSys: An Integrated Reconfigurable System for Data-Parallel and Computation-Intensive Applications". Transactions on Computers, Vol. 49, No. 5, pp. 465–481.
2. N. Bagherzadeh, F. Kurdahi, H. Singh, G. Lu, M. Lee and E. Filho, "Design and Implementation of the MorphoSys Reconfigurable Computing Processor", Journal of VLSI and Signal Processing-Systems for Signal, Image and Video Technology. Mar 2000, pp 147–164.
3. ISO/IEC JTC1 CD 13818. Generic coding of moving pictures, 1994 (MPEG-2 standard)
4. T. J. Callahan and J. Wawrzynek, "Instruction-level parallelism for reconfigurable computing," in *Proc. FPL'98 Field-Programmable Logic and Applications 8th Int. Workshop*, Tallinn, Estonia, Sept. 1998, pp. 248–257.
5. R. Maestre, F. J. Kurdahi, M. Fernandez, R. Hermida, N. Bagherzadeh, and H. Singh "A Framework for Reconfigurable Computing: Task Scheduling and Context Management", IEEE Transactions on VLSI Systems,Vol. 9, No. 6, Dec. 2001, pp. 858–873.
6. S. Sohoni, and R. Min, et al. "A study of memory system performance of multimedia applications". SIGMETRICS Performance 2001, pages 206–215.
7. M. B. Kamble, and K. Ghose (1997). "Analytical energy dissipation models for low power caches". *Proceedings of the ACM/IEEE International Symposium on Microarchitecture*, pages 184–193.

8. F. Catthoor, S. Wuytack, and F. Balasa, et al. (1998). “Custom Memory Management Methodology. Exploration of Memory Organization for Embedded Multimedia System Design”. Kluwer Academic Publishers.
9. M. Kaul, and R. Vemury, et. al (1999). “An automated temporal partitioning and loop fission approach for FPGA based reconfigurable synthesis of DSP applications”. *Proceedings 36th Design Automation Conference*, pages 616–622.
10. P. Landman (1994). Low-Power architectural design methodologies. Doctoral Dissertation, U. C. Berkeley.
11. R. Maestre, F. J. Kurdahi, M. Fernandez, R. Hermida, N. Bagherzadeh, and H. Singh “A Formal Approach to Context Scheduling for Multi-Context Reconfigurable Architectures”, *IEEE Transactions on VLSI Systems*, Special Issue on Reconfigurable and Adaptive VLSI Systems, Vol. 9, No. 1, Feb. 2001, pp. 173–185.

Chapter 5
Polymorphic Instruction Set Computers

Georgi Kuzmanov and Stamatis Vassiliadis

Abstract The Polymorphic Instruction Set Computers (PISC), also referred to as polymorphic processors, are a class of reconfigurable machines based on the co-processor architectural paradigm. The PISC paradigm, in difference to RISC/CISC, introduces hardware extended functionality of a general purpose processor provided on explicit programmer's demand without the need of ISA extensions. We motivate the necessity of PISCs through an example, which arises several design challenges unsolvable by traditional architectures and fixed hardware designs. More specifically, we address: architectural and microarchitectural concepts; a programming paradigm allowing hardware and software to coexist in a program; and particular spacial compilation techniques. This chapter illustrates the theoretical performance boundaries and efficiency of the PISC paradigm utilizing established evaluation metrics such as potential zero execution (PZE) and the Amdahl's law. Overall, the PISC paradigm allows designers to ride the Amdahl's curve easily by considering the specific features of the reconfigurable technology and the general purpose processors in the context of application specific execution scenarios.

5.1 Motivation and Design Challenges

Overall performance measurements in terms of Millions Instructions Per Cycle (MIPS) or Cycles Per Instruction (CPI) depend greatly on the CPU implementation. Potential performance improvements due to the parallel/concurrent execution of instructions, independent of technology or implementations, can be measured by the number of instructions which may be executed in zero time, denoted by PZE (potential zero-cycle execution) [Vassiliadis et al., 1994]. The rationale behind this measurement, as described in [Vassiliadis et al., 1994] for compound instruction sets is:

"If one instruction in a compound instruction pair executes in n cycles and the other instruction executes in $m \leq n$ cycles, the instruction taking m cycles to execute appears to execute in zero time. Because factors such as cache size and branch prediction accuracy vary from one implementation to the next, PZE measures the potential, not the actual, rate of zero-cycle execution. Additionally, note that zero-cycle instruction execution does not translate directly to cycles per instruction (CPI)

S. Vassiliadis, D. Soudris (eds.), *Fine- and Coarse-grain Reconfigurable Computing.*

because all instructions do not require the same number of cycles for their execution. The PZE measure simply indicates the number of instructions that potentially have been "removed" from the instruction stream during the execution of a program."

Consequently, PZE is a measurement that indicates maximum speedup attainable when parallelism/concurrency mechanisms are applied. The main advantage of PZE is that given a base machine design the benefits of proposed mechanisms can be measured and compared. We can thus evaluate the efficiency of a real design design expressed as a percentage of the potentially maximum attainable speedup indicated by PZE. An example is illustrated in Fig. 5.1. Four instructions, executing in a pipelined machine are considered. The instructions from the example are parallelized applying different techniques, such as instruction level parallelism (ILP), pipelining, technological advances, etc., as depicted in Fig. 5.1. Timewise, the result is that the execution of 4 instructions is equivalent to the execution of 2 instructions, which corresponds to a seeming code reduction by 50%, i.e., *2 out of 4 instructions potentially have been removed from the instruction stream during the program execution.* It means that the maximum theoretically attainable speedup (i.e., again potentially) in such a scenario is a factor of 2. In the particular example from Fig. 5.1, the execution cycles count for 4 instructions is reduced from 8 to 6 cycles, allowing 1.33 times speedup, which compared to the maximum speedup of 2, suggests efficiency of 65%. The above example suggests that PZE allows to measure the efficiency of a real machine implementation by comparing to a theoretical base machine, i.e., PZE gives an indication of how close a practical implementation performs to the theoretically attainable best performance boundaries. These theoretical boundaries are described by Amdahl's law [Amdahl, 1967].

5.1.1 Amdahl's Law and the Polymorphic Paradigm

The maximum theoretically attainable (i.e., the potentially maximum) speedup, considered for the PZE, with respect to the parallelizable portion of the program code,

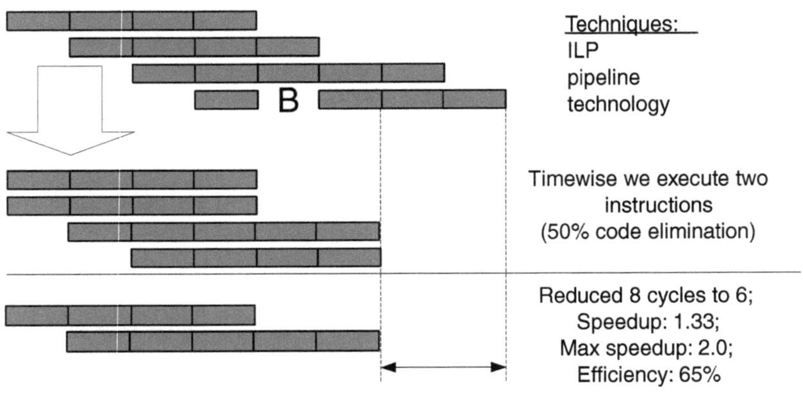

Fig. 5.1 PZE example

is determined by Amdahl's law. Amdahl's curve, graphically illustrated in Fig. 5.2, suggests that if, say half of an application program is parallelized and that its entire parallel fraction is assumingly executed in zero time, the speedup would potentially be 2. Moreover, the Amdahl's curve suggests that to achieve an order of magnitude speedup, a designer should parallelize at least 90% of the application execution. In such cases, when over 90% of the application workload is considered for parallelization, it is practical to create an ASIC, rather than utilizing programmable GPP. The design cycle of an ASIC, however, is extremely inflexible and very expensive. Therefore, ASICs may not appear to be an efficient solution when we consider smaller portions (i.e., less than 90%) of an algorithm for hardware acceleration. Obviously, there exist potentials for hardware proposals that perform better than GPPs and are more flexible alternative to design and operate than ASICs.

In this chapter, we introduce an architectural paradigm targeting the existing gap between GPPs and ASICs in terms of flexibility and performance. This paradigm exploits specific features of the reconfigurable hardware technologies. In consistence with the classical RISC and CISC paradigms [Patterson et al., 1980, Bhandarkar et al., 1991], we refer to the architectural paradigm as to a Polymorphic Instruction Set Computer (PISC). The practically significant scope of PISC covers between 50% and 90% application parallelization illustrated with the Amdahl's curve in Fig. 5.2. This interval provides a designer with potentials to benefit from the best of two worlds, i.e., with a synergism between purely programmable solutions on GPPs and reconfigurable hardware. That is, the infinite flexibility of the programmable GPPs combined with reconfigurable accelerators results into a PISC - a programmable system that substantially outperforms GPP. Therefore, the gap between GPP and ASIC, illustrated in Fig. 5.2, belongs to PISC. In the following discussion we will motivate the need for PISC and will introduce in more details their most important properties.

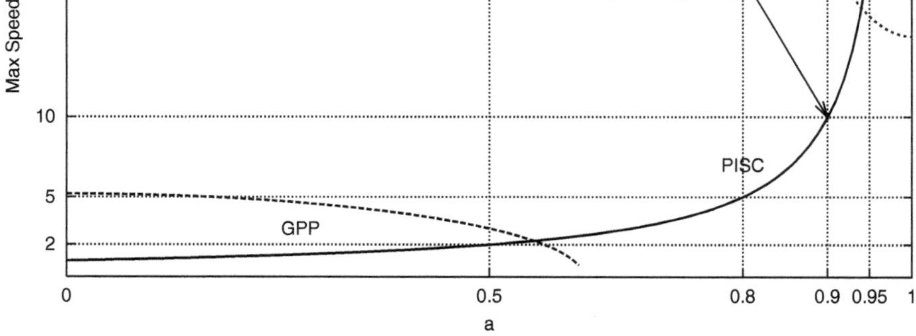

Fig. 5.2 The Amdahl's curve and PISC

5.1.2 Motivating Example

To illustrate the necessity of the PISC computing paradigm, we present a motivating example based on the Portable Network Graphics (PNG) standard [Roelofs, 1999]. PNG is a popular standard for image compression and decompression, it is a native standard for the graphics implemented in *Microsoft Office* as well as in a number of other applications. We consider the piece of C-code presented in Fig. 5.3, which is extracted from an implementation of the PNG standard. This code fragment implements an important stage of the PNG coding process. It computes the Paeth prediction for each pixel d of the current row, starting from the second pixel. The Paeth prediction scheme, illustrated in Fig. 5.4, selects from the 3 neighboring pixels a, b, and c, that surround d, the pixel that differs the least from the value $p = a + b - c$ (which is called the initial prediction). The selected pixel is called the *Paeth prediction* for d. If the pixel rows contained *length + 1* elements, *length* prediction values are produced. This prediction scheme is used during the image filtering stage of the image coding and decoding. Fig. 5.5 presents an implementation of the code fragment in pseudocode derived from the AltiVec assembly. In this figure, the general-purpose register GPRi of the underlying ISA is denoted by `ri`, `vri` denotes the *i*-th vector register of AltiVec.

Analysis of the motivating example presented above suggests the following. If the Paeth predictor must be computed for a row of 1024 pixels, the complete AltiVec code presented in Fig. 5.5 will result in a dynamic instruction count of $8(prologue) + 64 \cdot 3(load) + 6(unpack) + 76(compute) + 1(pack) + 1(store) + 2(miscellaneous) + 3(pointerupdate) + 3(loopcontrol) = 8 + 6464 \cdot 95 = 6088$ *instructions*. This high instruction count, which limits the performance, is caused by

```
void Paeth_predict_row(char *prev_row, char *curr_row, char *predict_row, int
length)
{char *bptr, *dptr, *predptr;
char a, b, c, d;
short p, pa, pb, pc;

bptr = prev_row+1;
dptr = curr_row+1;
predptr= predict_row+1;

for(i=1; i<length; i++)
{c = *(bptr-1); b = *bptr;
a = *(dptr-1); d = *dptr;
p = a + b - c; /* this is the initial prediction */
pa = abs( p - a ); /* distance of each member */
pb = abs( p - b ); /* to the */
pc = abs( p - c ); /* initial estimate */
if ((pa≤pb)&&(pa≤pc) *predptr = a;
else if (pb≤pc) *predptr = b;
else *predptr = c;

bptr++; dptr++; predptr++; } }
```

Fig. 5.3 The Paeth prediction routine according to PNG specification [Roelofs, 1999]

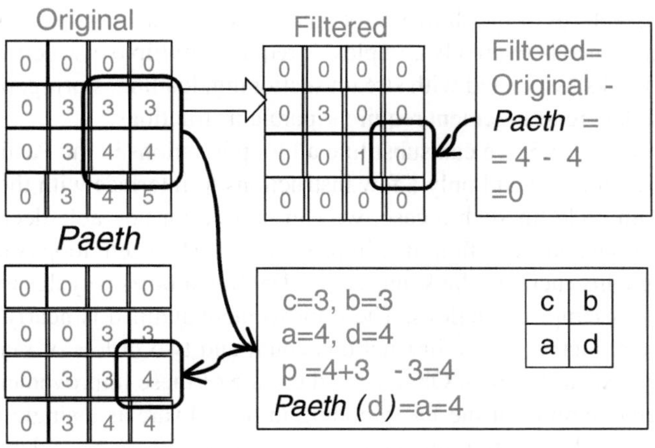

Fig. 5.4 The Paeth prediction scheme

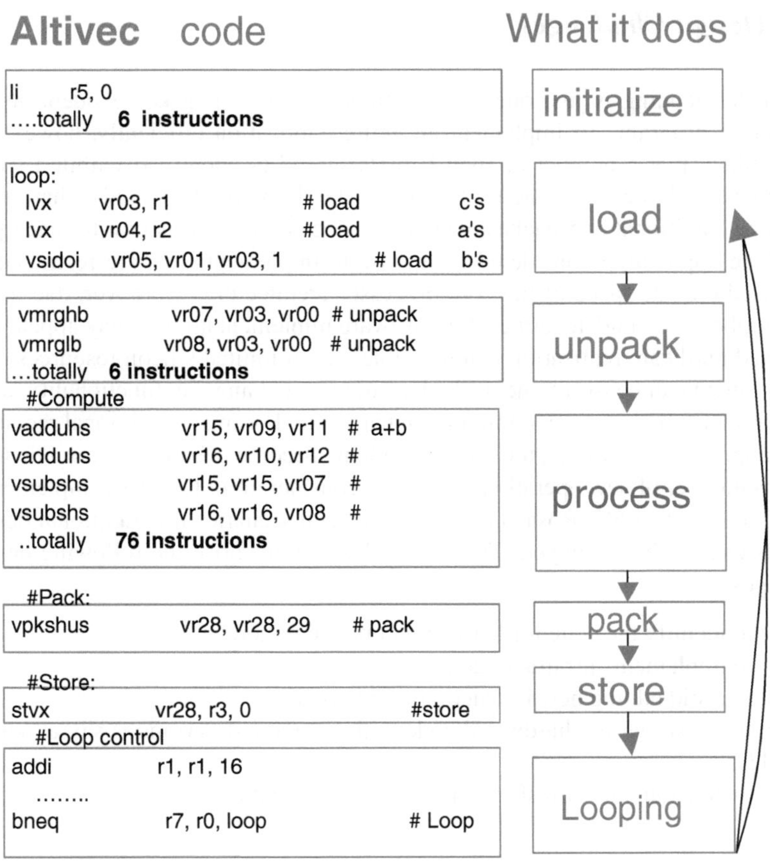

Fig. 5.5 AltiVec code for the *Paeth predict* kernel

the following features of the short-vector media extensions. First, if the main operation to be performed is relatively complex, it requires multiple instructions. Second, the overhead tasks associated with stream sectioning, loading, storing, packing, unpacking, and data rearrangement require separate instructions.

Considering Fig. 5.5, we can substitute all loop iterations in the Paeth code with a single instruction and add only a few instructions to interface with the remainder of the program code. In such a case, we can expect considerable decrease of the instructions count and execution time improvements. The Paeth loop is transformed now in a single instruction [Hakkennes et al.,1999] that takes 5 cycles to complete[1] and requires 20 setup instructions. The improvement attained is nearly two orders of magnitude reduction of the instructions count and two orders of magnitude reduction of the execution time. Obviously, the scale of these improvements depends on the implementability of the Paeth coding into hardware as a single instruction. An efficient Paeth hardware implementation comprises 24 32-bit adders allowing a throughput of 16 pixels/cycle (i.e., 6 8-bit adders per pixel).

5.1.3 Design Challenges

The Paeth encoding is just one computationally demanding kernel identified in a particular program. To implement an entire application efficiently, however, it is very likely that a number of such kernels should be identified within a single program execution context and each of them should be implemented in hardware. Therefore, traditional approaches, which introduce a new instruction for each portion of the application considered for hardware implementation, are restricted by the unused opcode space of the core processor architecture. Moreover, due to the large number of candidate kernels for hardware implementation, it may appear that their fixed hardware realization is impossible within limited silicon resources. The latter problem can be overcome, if the hardware can change its functionality at the designer's wish, i.e., using reconfigurable hardware. For many traditional reconfigurable approaches, however, the above problems become even more dramatic if an arbitrary number of new operations should be considered for hardware implementation [Hauck et al., 1997, Rosa et al., 2003]. In such scenarios,the traditional design methods can not be employed. The above observations arise the following design challenges:

1. How to identify the code for hardware implementation?
2. How to implement "arbitrary" code?
3. How to avoid adding new instructions per kernel?
4. How to substitute the hardwired code with SW/HW descriptions say at source level?
5. How to generate the "transformed" program automatically?

[1] One cycle is the duration of a single ALU operation

To meet the above design challenges, the following engineering topics should be adequately addressed:

1. HW/SW co-design tools
2. Microarchitecture design.
3. Processor architecture (behavior and logical structure).
4. Programming paradigm allowing HW and SW to coexist in a program.
5. Compilation supporting the programming paradigm

5.2 General Approach

To meet all design challenges stated in the previous section, probably the most sound solution is to employ the specific synergism between a general-purpose processor (GPP) and a reconfigurable processor (RP) referred to as PISC. In the discussion to follow, we present the general concept of transforming an existing program to one that can be executed on a polymorphic computing platform and hints to the mechanisms, intended to improve existing approaches. The conceptual view of how program P (intended to execute only on the GPP) is transformed into program P' (executing on both the GPP core and the reconfigurable hardware) is depicted in Fig. 5.6. The purpose is to obtain a functionally equivalent program P' from program

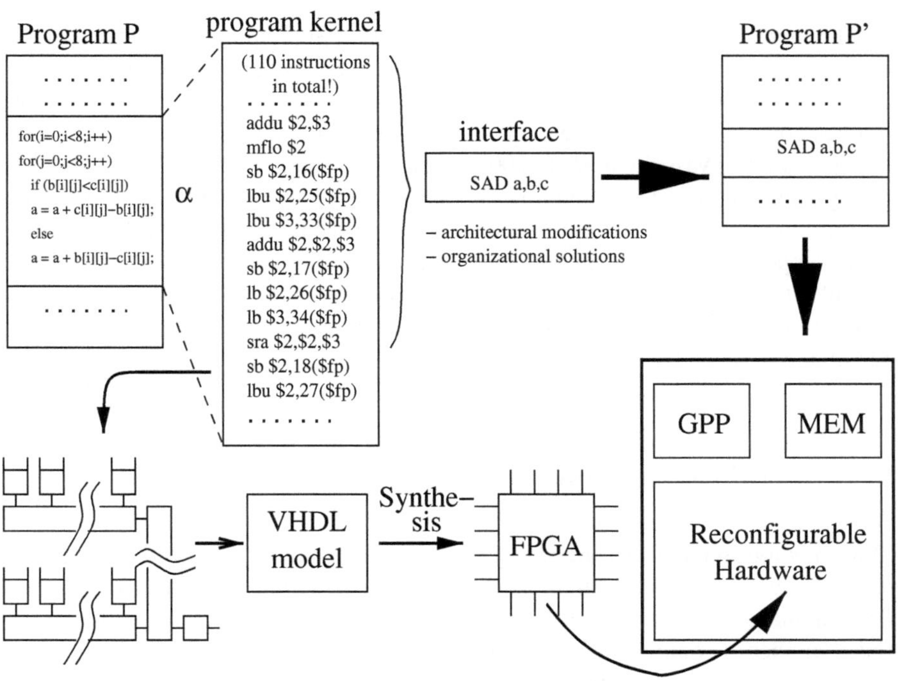

Fig. 5.6 The general Molen approach: Program transformation example

P which (using specialized instructions) can initiate both the configuration and execution processes on the reconfigurable hardware. The sum of absolute differences (SAD) calculation, a well known multimedia operation, is considered as an example in Fig. 5.6. The steps involved in this transformation are the following:

1. **Identify** pieces of software code "α" in program P to be mapped in reconfigurable hardware that can be possibly benefited from hardware implementation (see piece α corresponding to the SW code of SAD in Fig. 5.6).
2. **Design** a hardware unit performing the functionality of the extracted program kernel "α" and describe the design in Hardware Description Language (HDL). Show that "α" can be implemented in hardware in an existing technology, e.g., FPGA, and map "α" onto reconfigurable hardware.
3. **Eliminate** the identified code "α" from program P. Insert an equivalent code A (e.g., SAD a,b,c), which calls the hardware through a preestablished SW/HW calling interface. This interface reflects the architectural and organizational modifications of the original GPP and comprises:
   - Parameters and results communication between the GPP and the reconfigurable processor.
   - Configuration code, inserted to configure the reconfigurable hardware.
   - Emulation code, used to perform the functionality of the hardware accelerated kernel "α".
4. **Compile and execute** program P' with original code plus code having functionality A (equivalent to "α", i.e., SAD a,b,c) on the GPP/reconfigurable processor.

The mentioned steps illustrate the need for a programming paradigm in which both software and hardware descriptions are present in the same program. It should also be noted that the only constraint on "α" is implementability, which possibly implies complex hardware. Consequently, the microarchitecture may have to support emulation [Vassiliadis et al., 2003a] via microcode. We have termed this reconfigurable microcode ($\rho\mu$-code) as it is different from the traditional microcode. The difference is that such microcode does not execute on fixed hardware facilities. It operates on facilities that the $\rho\mu$-code itself "designs" to operate upon. We refer to such facilities as to configurable computing units (CCU). A processor supporting $\rho\mu$-code is referred to as a $\rho\mu$-coded processor [Vassiliadis et al., 2001, Vassiliadis et al., 2004].

5.3 General Design Flow

In the previous sections, we argued that a PISC organization can potentially improve the GPP performance by configuring customizable computing resources (e.g., Field-Programmable Gate Arrays - FPGA) on a per-application basis. Thus, virtually infinite number of complex hardware functions can be emulated within the otherwise limited reconfigurable resources. In this section, we present a general approach for hardware-software co-design of polymorphic processors. A general methodology, used to fit a given (media) application into a GPP augmented with reconfigurable hardware is sketched in Fig. 5.7. The design process is performed in several interactive stages. First, an analysis of the application algorithms is performed. This stage requires extensive profiling and software-hardware (SW/HW) partitioning of the application. Candidate functions or kernels for hardware implementation are identified through the SW/HW partitioning and considered for further hardware design. SW/HW interface solutions have to be made at this initial design stage as well, and later considered for program code annotation and hardware implementation. The remaining design stages are performed in two separate tracks, interacting with each other – one in software and the other in hardware.

5.3.1 Software Track

Consider Fig. 5.7. The original application code is first modified according to the SW/HW partitioning and the interface solutions made in the preceding stage. Usually, these modifications include code annotations utilizing either high level programming language techniques (e.g.,in C, Java, etc.) or a lower assembler level language. The modified/annotated application code is then compiled and linked for the targeted GPP architecture. If code annotations are made in high level programming language manner, an accordingly modified retargetable compiler has to be used. In the case of lower assembler level annotations, the native compiler for the GPP architecture can be employed. The result of the compilation

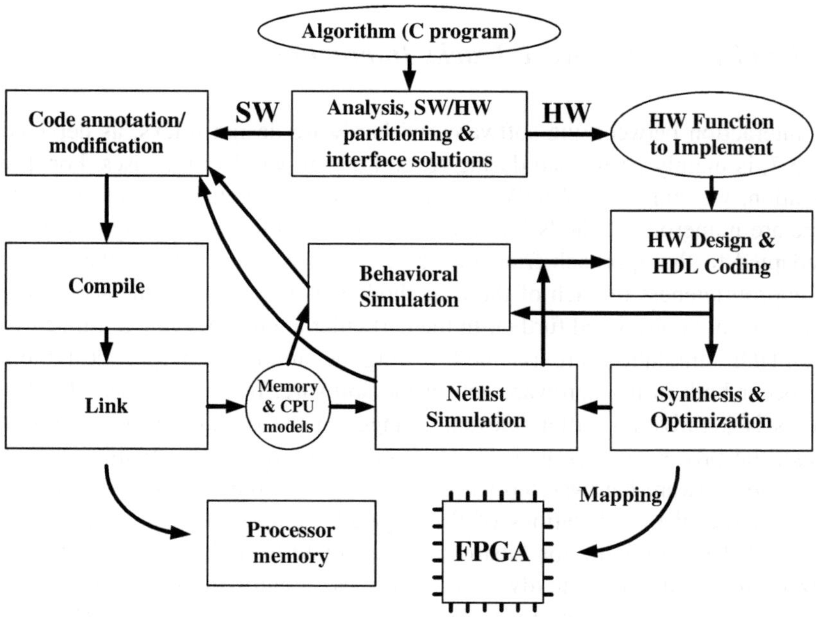

Fig. 5.7 Make applications fit - a typical reconfigurable design flow

and link processes is a single or a number of binary sequences (codes), each of them dedicated for a certain location in the target memory organization. The generated binary codes are loaded into corresponding memory models for SW/HW co-simulation.

5.3.2 Hardware Track

Consider Fig. 5.7. Hardware units supporting the functions extracted for HW implementation are designed and coded in hardware description language HDL. The HDL models are simulated at behavioral level to validate the functional correctness of the designs. Behavioral simulations may be performed over stand-alone models of the units. It is far more essential, however, behavioral simulations to be performed over a model of the entire reconfigurable processor, i.e., including the compiled application programs. The results of these simulations may impose changes in the initial hardware design as well as some changes in the program code annotations. After the reconfigurable design is validated at behavioral level, the HDL codes of the hardware units are synthesized and optimized. Once again, the resulting netlist design description is co-simulated with the software to detect possible design errors. Performed at lower level of abstraction, this simulation is the final validation of the entire reconfigurable design before its physical implementation. Finally, the synthesized and optimized design is mapped onto the targeted reconfigurable device (FPGA) and a configuration bitstream is generated.

5.3.3 Software-Hardware Tracks Interaction

The interaction between the software and hardware design tracks, as depicted in Fig. 5.7, is mainly performed during several design validation phases. For design validation, we consider SW/HW co-simulations at different levels of abstraction. There are numerous methods for simulating the reconfigurable design, which can be adapted to the approach from Fig. 5.7. A discussion regarding the relevance and appropriateness of each of these methods would be outside the scope of this thesis, thus not considered further. In the particular approach, cycle accurate event-driven HDL simulations are assumed. During the simulation phases, design errors may occur both in the hardware and in the software. In these phases, the design process is iterative and after relevant changes in the designs (resp. their source codes), the process is repeated. Once an error free design is obtained and validated, the software-hardware co-simulation is considered complete. Next, binary codes for the distinct locations of the targeted memory organization are generated. An FPGA configuration bitstream is generated from the synthesized and optimized HDL code of the hardware design track. Finally, the linked binary codes of the application software are loaded into the physical memories of the processor and the generated FPGA bitstream is loaded into the targeted reconfigurable device(s).

5.4 The Polymorphic Instruction Set

In this section, we present the polymorphic programming paradigm [Vassiliadis et al., 2003b], the instruction set architecture (ISA) that supports it, and the program sequencing required to implement this programming paradigm.

The polymorphic programming paradigm is a sequential consistency paradigm targeting the previously described organization, which allows parallel and concurrent hardware execution. Further in our discussion and experiments, we will assume that the programming paradigm is intended for single program execution, which is not its general limitation, however. The polymorphic programming paradigm requires only a one-time architectural extension of few instructions to provide a large user reconfigurable operation space. The complete list of the eight required instructions, denoted as polymorphic ($\pi o\lambda\upsilon\mu o\rho\phi\iota\kappa\acute{o}$) **I**nstruction **S**et **A**rchitecture (πISA), is as follows:

Six instructions are required for controlling the reconfigurable hardware:

- Two **set** instructions: these instructions initiate the configurations of the CCU. When assuming partial reconfigurable hardware, two instructions for such purpose are provided, namely:
  - the partial set (*p*-**set** <*address*>) instruction performs those configurations that cover common and frequently used functions of an application or set of applications. In this manner, a considerable number of reconfigurable blocks in the CCU can be preconfigured.
  - the complete set (*c*-**set** <*address*>) instruction performs the configurations of the remaining blocks of the CCU (not covered by the *p-set*). This *completes* the CCU functionality by enabling it to perform the less frequently used functions. Due to the reduced amount of blocks to configure, reconfiguration latencies can be reduced.

  We must note that in case no partial reconfigurable hardware is present, the *c*-**set** instruction alone can be utilized to perform all configurations.
- **execute** <*address*>: this instruction controls the execution of the operations implemented on the CCU. These implementations are configured onto the CCU by the *set* instructions.
- **set prefetch** <*address*>: this instruction prefetches the needed microcode responsible for CCU reconfigurations into a local on-chip storage facility (the $\rho\mu$-code unit) in order to possibly diminish microcode loading times.
- **execute prefetch** <*address*>: the same reasoning as for the **set prefetch** instruction holds, but now relating to microcode responsible for CCU executions.
- **break**: this instruction is utilized to facilitate the parallel execution of both the reconfigurable processor and the core processor. More precisely, it is utilized as a synchronization mechanism to complete the parallel execution.

Two *move* **instructions** for passing values between the register file and exchange registers (XREGs) since the reconfigurable processor is not allowed direct access to the general-purpose register file:

- **movtx** $XREG_a \leftarrow R_b$: (move to XREG) used to move the content of general-purpose register R_b to $XREG_a$.
- **movfx** $R_a \leftarrow XREG_b$: (move from XREG) used to move the content of exchange register $XREG_b$ to general-purpose register R_a.

The <*address*> field in the instructions introduced above denotes the location[2] of the reconfigurable microcode responsible for the configuration and execution processes, described in Section 5.5. It must be noted that a single address space is provided with at least $2^{(n-op)}$ addressable functions, where n represents the instruction length and op the opcode length. Code fragments of contiguous statements (as they are represented in high-level programming languages) can be isolated as generally implementable functions (that is code with multiple identifiable input/output values). The parameters are passed via the exchange registers. In order to maintain correct program semantics, the code is annotated and a hardware description file provides the compiler with implementation specific information such as the addresses where the reconfigurable microcodes are to be stored, the number of exchange registers, etc.

It should be noted that it is not imperative to include all instructions when designing a polymorphic architecture. The programmer/implementor can opt for different ISA extensions depending on the required performance to be achieved and the available technology. There are basically three distinctive πISA possibilities with respect to the polymorphic instructions introduced earlier - the *minimal*, the *preferred* and the *complete* πISA extension. In more detail, they are:

- **the minimal πISA:** This is essentially the smallest set of polymorphic instructions needed to provide a working scenario. The four basic instructions needed are **set** (more precisely: *c*-**set**), **execute**, **movtx** and **movfx**. By implementing the first two instructions (**set/execute**) any suitable CCU implementation can be loaded and executed in the RP. Furthermore, reconfiguration latencies can be hidden by scheduling the **set** instruction considerably earlier than the **execute** instruction. The **movtx** and **movfx** instructions are needed to provide the input/output interface between the RP targeted code and the remainder application code.
- **the preferred πISA:** The minimal set provides the basic support, but it may suffer from time-consuming reconfiguration latencies, which could not be hidden, and that can become prohibitive for some real-time applications. In order to address this issue, two **set** (*p*-**set** and *c*-**set**) instructions are utilized to distinguish among frequently and less frequently used CCU functions. In this manner, the *c*-**set** instruction only configures a smaller portion of the CCU and thereby requiring less reconfiguration time. As the reconfiguration latencies are substantially (or completely) hidden by the previously discussed mechanisms, the loading time of microcode will play an increasingly important role. In these cases, the two **prefetch** instructions (**set prefetch** and **execute prefetch**) provide a way to diminish the microcode loading times by scheduling them well ahead of the

[2] Indirect pointing could be required in order to extend the $\rho\mu$-code addressing space.

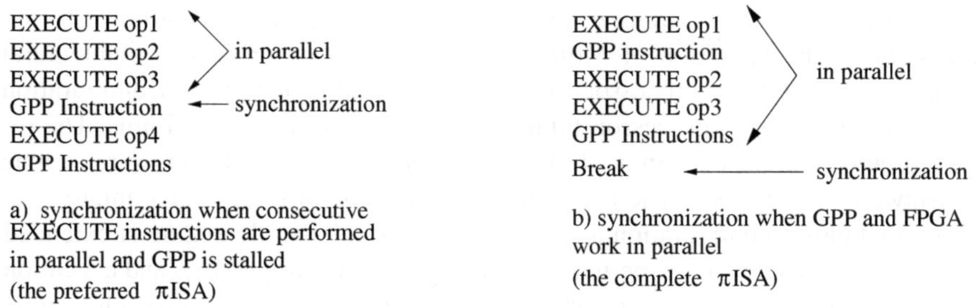

Fig. 5.8 Parallel execution and models of synchronization

moment that the microcode is needed. Parallel execution is initiated by a πISA **set/execute** instruction and ended by a general-purpose instruction as described in Fig. 5.8(a).

- **the complete πISA:** This scenario involves all πISA instructions including the **break** instruction. In some applications, it might be beneficial performance-wise to execute instructions on the core processor and the reconfigurable processor in parallel. In order to facilitate this parallel execution, the preferred ISA is further extended with the **break** instruction. The **break** instruction provides a mechanism to synchronize the parallel execution of instructions by halting the execution of instructions following the **break** instruction. The sequence of instructions performed in parallel is initiated by an **execute** instruction. The end of the parallel execution is marked by the **break** instruction. It indicates where the parallel execution stops (see Fig. 5.8 (b)).

5.4.1 Parallel Execution

Parallel execution, for all πISA modifications is initiated by a **set/execute** instruction. For both minimal and preferred πISA, a parallel execution is ended by a general-purpose instruction as described in Fig. 5.8(a). When a complete πISA is implemented and a sequence of instructions is performed in parallel, the end of the parallel execution is marked by the **break** instruction. It indicates where the parallel execution stops (see Fig. 5.8 (b)).

5.5 The Molen Polymorphic Processor

An example of a PISC is the Molen $\rho\mu$-coded processor introduced in [Vassiliadis et al., 2001]. More details on the Molen microarchitecture have been published in [Vassiliadis et al., 2004]. In this section, we briefly present the Molen polymorphic processor simply referred to as Molen in the remainder of the presentation. The general proposal is: *by displaying means to maintain the reconfiguration at architectural level, to achieve a high flexibility in tuning the system for the specific application.* The operation of Molen is based on the co-processor architectural paradigm. A GPP (core processor) controls the execution and the (re)configuration of a reconfigurable co-processor, tuning the latter for specific algorithms. The reconfiguration and the execution of the code on the reconfigurable hardware is done in firmware via $\rho\mu$-code. The $\rho\mu$-code is an extension of the classical microcode, which includes reconfiguration and execution. The microcode engine is extended with mechanisms that allow permanent and pageable reconfiguration and execution microcode to coexist. Details regarding the Molen architecture, microarchitecture, organization and implementation are presented in the remainder of this section.

5.5.1 The Molen Organization

The two main components in the Molen machine organization (depicted in Fig. 5.9) are the 'Core Processor', which is a general-purpose processor (GPP), and the 'Reconfigurable Processor' (RP). Instructions are issued to either processors by the 'Arbiter' by means of a partial decoding of the instructions received from the instruction fetch unit. Data are fetched (stored) by the 'Data Fetch' unit from(to) the main memory. The 'Memory MUX' unit is responsible for distributing(collecting) data to(from) either the reconfigurable or the core processor. The reconfigurable processor is further subdivided into the $\rho\mu$-code unit and the *custom configured unit* (CCU). The CCU consists of reconfigurable hardware, e.g., an FPGA, and memory. Essentially, the CCU is intended to support additional and future functionalities that are not implemented in the core processor. Pieces of application code can be implemented on the CCU in order to speed up the execution of the overall application

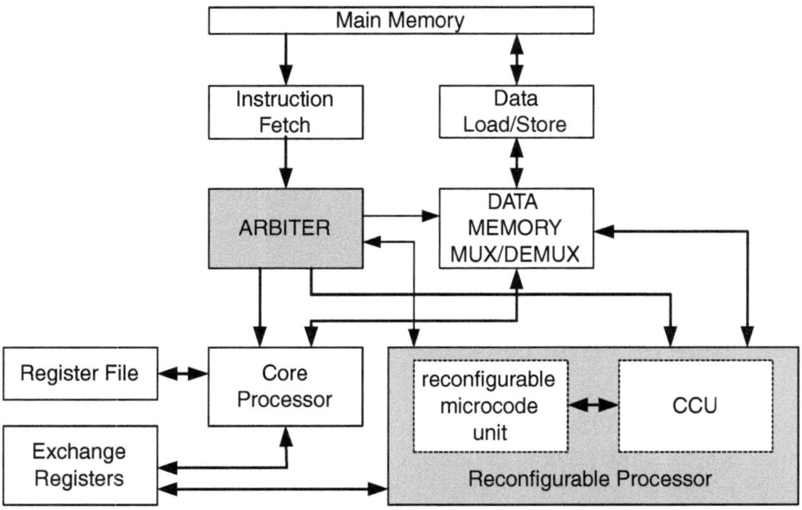

Fig. 5.9 The Molen machine organization

code. A clear distinction exists between code that is executed on the reconfigurable unit (the RP targeted code) and code that is executed on the core processor (remaining code). Data must be transferred across the code boundaries in order for the overall application code to be meaningful. Such data includes predefined parameters (or pointers to such parameters) or results (or pointers to such results). The parameter and result passing is performed through a mechanism utilizing so-called exchange registers (XREGs) depicted in Fig. 5.9.

The support of operations [3] by the reconfigurable processor can be initially divided into two distinct phases: set and execute. In the set phase, the CCU is configured to perform the supported operations. Subsequently, in the execute phase the actual execution of the operations is performed. This decoupling allows the set phase to be scheduled well ahead of the execute phase and thereby hiding the reconfiguration latency. As no actual execution is performed in the set phase, it can be even scheduled upwards across the code boundary in the code preceding the RP targeted code. Furthermore, no specific instructions are associated with specific operations to configure and execute on the CCU as this will greatly reduce the opcode space. Instead, pointers to *reconfigurable microcode* ($\rho\mu$-code), which emulates both the configuration and the execution of programs, are used. Consequently, two types of $\rho\mu$-code are distinguished:

- reconfiguration microcode that controls the configuration of the CCU;
- execution microcode that controls the execution of the implementation configured on the CCU.

5.5.2 The Molen Microarchitecture

Experienced microcode designers will recognize that for performance reasons, there is a necessity of having microcode that resides permanently in the control store and microcode that is pageable. To represent this difference, a bit from the instruction word is dedicated to implement resident/pageable microcode.In the instruction format, depicted in Fig. 5.10, the location of the microcode is indicated by the resident/pageable-bit (R/P-bit) which implicitly determines the interpretation of the address field, i.e., as a memory address α (R/P=1) or as a ρ-control store ad-

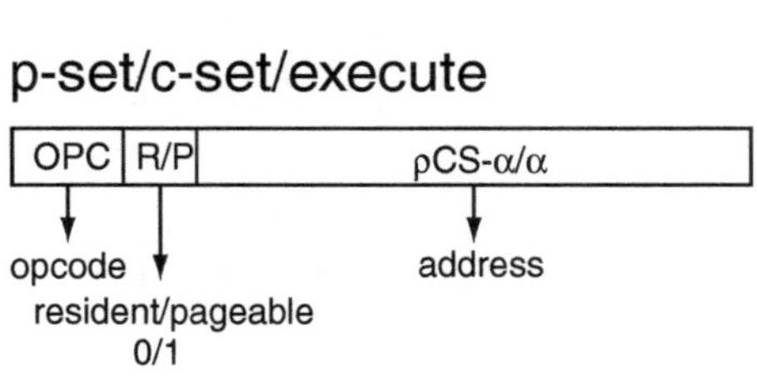

Fig. 5.10 The *p*-**set**, *c*-**set**, and **execute** instruction format

[3] An operation can be as simple as a single instruction or as complex as a piece of code.

dress ρCS-α (R/P $= 0$) indicating a location within the $\rho\mu$-code unit. This location contains the first instruction of the microcode which must always be terminated by a dedicated microinstruction, e.g., *end_op*.

5.5.3 The $\rho\mu$-Code Unit

The internal organization of the $\rho\mu$-code unit is depicted in Fig. 5.11. The $\rho\mu$-code unit comprises three main parts: the sequencer, the ρ-control store, and the $\rho\mu$-code loading unit. The sequencer mainly determines the microcode execution sequence.

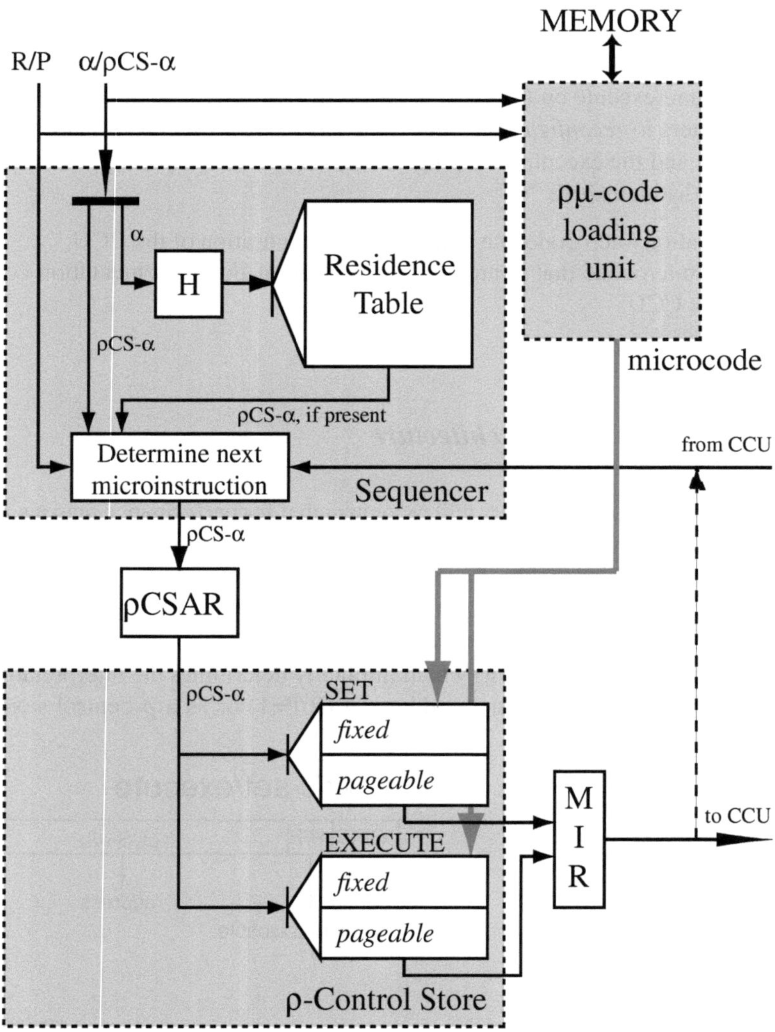

Fig. 5.11 $\rho\mu$-code unit internal organization

The ρ-control store is used as a storage facility for microcode. The $\rho\mu$-code loading unit, as its name suggests, is responsible for the loading of reconfigurable microcode from the memory. The execution of microcode starts with the sequencer receiving an address from the arbiter (see Fig. 5.9) and interpreting it according to the R/P-bit. When receiving a memory address, it must be determined whether the microcode is already cached in the ρ-control store or not. This is done by checking the residence table (see Fig. 5.12) which stores the most frequently used translations of memory addresses into ρ-control store addresses and keeps track of the validity of these translations. It can also store other information: least recently used (LRU) and possibly additional information, e.g., required for virtual addressing[4] support. In the case that a memory address is received and the associated microcode is not present in the ρ-control store, the $\rho\mu$-code unit initiates the loading of microcode from the memory into the ρ-control store. In the case a ρCS-α is received or a valid translation into a ρCS-α is found, the ρCS-α is transferred to the 'determine next microinstruction'-block. This block determines the next microinstruction to be executed:

- When receiving the address of the first microinstruction: Depending on the R/P-bit, the correct ρCS-α is selected, i.e., from the instruction field or from the residence table.
- When already executing microcode: Depending on previous microinstruction(s) and/or results from the CCU, the next microinstruction address is determined.

The resulting ρCS-α is stored in the ρ-control store address register (ρCSAR) before entering the ρ-control store. Using the ρCS-α, a microinstruction is fetched from the ρ-control store and then stored in the microinstruction register (MIR) before it controls the CCU reconfiguration or before it is executed by the CCU. The ρ-control store comprises two sections, namely a *set* section and an *execute* section. Both sections can be identical, probably differing in microinstruction word

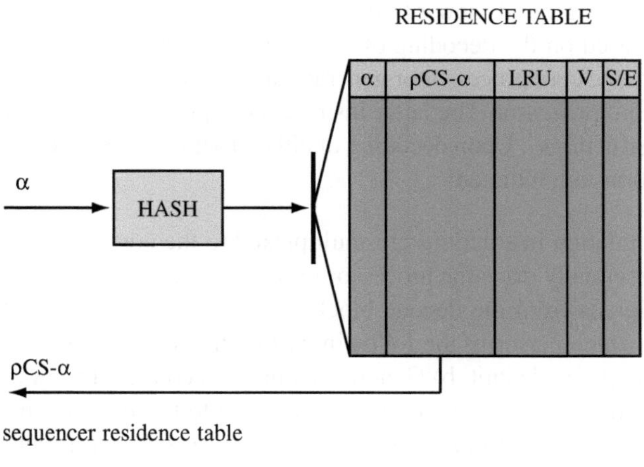

Fig. 5.12 The sequencer residence table

[4] For the simplicity of the discussion, we assume that the system only allows real addressing.

sizes only. Each sections is further divided into a *fixed* part and a *pageable* part. The fixed part stores the resident reconfiguration and execution microcode of the *set* and *execute* phases, respectively. Resident microcode is commonly used by several invocations (including reconfigurations) and it is stored in the fixed part so that the performance of the *set* and *execute* phases is possibly enhanced. Which microcode resides in the fixed part of the ρ-control store is determined by performance analysis of various applications and by considering various software and hardware parameters. Other microcode is stored in memory and the pageable part of the ρ-control store acts like a cache to provide temporal storage. Cache mechanisms are incorporated into the design to ensure the proper substitution and access of the microcode in the ρ-control store. This is exactly what is provided by the residence table which invalidates entries when microcode has been replaced (utilizing the valid (V) bit) or substitutes the least recently used (LRU) entries with new ones. Finally, the residence table can be separate or common for both the set and execute pageable ρ-control store sections. In assuming a common table implementation, an additional bit needs to be added to determine which part of the pageable ρ-control store is addressed (depicted as the S/E-bit in Fig. 5.12).

5.5.4 The Arbiter - General Requirements

The Molen arbiter performs partial decoding of instructions in order to determine where instructions should be issued. Its organization is of great importance for the time efficient operation of the entire Molen organization. The arbiter controls the proper co-processing of the core processor and the reconfigurable processor (see Fig. 5.9) by directing instructions to either of these processors. It arbitrates the data memory access of the reconfigurable and core processors and it distributes control signals and the starting microcode address to the $\rho\mu$-code unit.

In Fig. 5.13, a general view of an arbiter organization is depicted. The arbiter operation is based on the decoding of the incoming instructions and either directs instructions to the core processor or generates an instruction sequence to control the state of the core processor. The latter instruction sequence is referred to as "arbiter emulation instructions". Upon decoding of either a **set** or an **execute** instruction, the following actions are initiated:

1. Arbiter emulation instructions are multiplexed to the core processor instruction bus and essentially drive the processor into a wait state.
2. Control signals from the decode block are transmitted to the control block in Fig. 5.13, which performs the following: a) Redirect the microcode location address to the $\rho\mu$-code unit. b) Generate an internal code representing either a **set** or **execute** instruction (Ex/Set) and delivering it to the $\rho\mu$-code unit. c) Initiate the reconfigurable operation by generating *'start reconf. operation'* signal to the $\rho\mu$-code unit. d) Reserve the data memory control for the $\rho\mu$-code unit by generating a *memory occupy* signal to the (data) memory controller. e) Enter a wait state until the signal *'end of reconf. operation'* arrives.

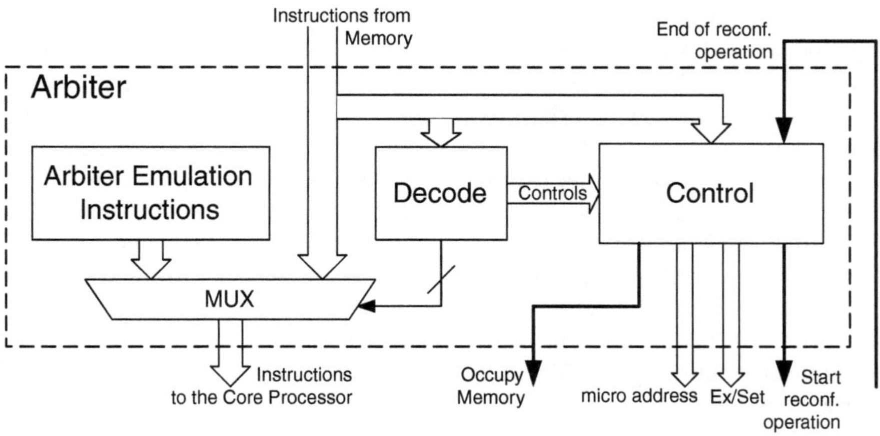

Fig. 5.13 General arbiter organization

An active *'end of reconf. operation'* signal initiates the following actions: 1) Data memory control is released back to the core processor. 2) An instruction sequence is generated to ensure proper exiting of the core processor from the wait state. 3) After exiting the wait state, the program execution continues with the instruction immediately following the last executed reconfigurable processor instruction.

5.5.5 The Exchange Registers

The exchange registers (XREGs) are used for passing operation parameters to the reconfigurable hardware and returning the computed values after operation execution. In order to avoid dependencies between the reconfigurable processor and the core processor, the needed parameters are moved from the register file to the XREGs (**movtx**) and the results stored back in the register file (**movfx**). During the execute phase, the defined $\rho\mu$-code is responsible for taking the parameters of its associated operation from the XREGs and returning the result(s). A single **execute** instruction does not pose any specific challenge, because the complete set of exchange registers is available. When executing multiple **execute** instructions simultaneously, overlapping utilization of the available XREGs has to be avoided. This assumes an agreement on the conventions for the XREGs allocation.

5.5.6 Parameter Exchange, Parallelism and Modularity

As shown earlier, the exchange registers solve the limitation on the number of parameters as present in other reconfigurable computing approaches (e.g., [Campi et al., 2003, Ye et al., 2000]). If the parameters do not exceed the number of XREGs, parameters are passed by value, otherwise - by reference. This allows an arbitrary number of parameters to be exchanged between the calling (software) and called (hardware) functions, where only the hardware resources determine the upper bound. The Molen architecture also addresses an additional shortcoming of other reconfigurable computing approaches concerning parallel execution. In case that two or more functions considered for CCU implementation do not have any true dependencies, they can be executed in parallel. There is always a physical maximum of how many operations can be executed in parallel on the CCU. This is, however, an implementation dependent issue, e.g., reconfigurable hardware size, number of XREGs, etc. and can not be considered as a limitation of the Molen architecture. In addition, it should be emphasized that the Molen hardware/software (HW/SW) division ability is not limited to functions only. In case the targeted kernel is part of a function, e.g., a highly computational demanding loop, it can be appropriately transformed for use in the Molen programming paradigm by defining a clear set of interface parameters and passing them via the XREGs (as values or references) to the CCU implementation of the kernel.

The Molen paradigm facilitates modular system design. For instance, hardware implementations described in an HDL (VHDL, Verilog or System-C) language are mappable to any FPGA technology, e.g., Xilinx or Altera, in a straightforward manner. The only requirement is to satisfy the Molen set and execute interface. In addition, a wide set of functionally similar CCU designs (from different providers), e.g. sum of absolute differences (SAD) or IDCT, can be collected in a database allowing easy design space explorations.

5.5.7 Interrupts and Miscellaneous Considerations

The Molen approach is based on the GPP co-processor paradigm (see for example [Padegs et al., 1988, Buchholz, 1986, Moudgill et al., 1996]). Consequently, all known co-processor interrupt techniques are applicable. In order to support the core processor interrupts properly, the following parts are essential for any Molen implementation:

1. Hardware to detect interrupts and terminate the execution before the state of the machine is changed are assumed to be implemented in both core processor and reconfigurable processor.
2. Interrupts are handled by the core processor. Consequently, hardware to communicate interrupts to the core processor is implemented in CCU.
3. Initialization (via the core processor) of the appropriate routines for interrupt handling.

It is assumed that the implementor of a reconfigurable hardware follows a co-processor type of configuration. With respect to the GPP paradigm, the FPGA co-processor facility can be viewed as an extension of the core processor architecture. This is identical with the way co-processors, such as floating point, vector facilities, etc., have been viewed in the conventional architectures.

5.5.8 Molen- Summary

Many current reconfigurable proposals fall short of expectations due to a number of shortcomings, the most essential of which are opcode space explosion, lack of ISA compatibility, technology dependence, no design modularity, limited number of processing parameters, and no support for parallel reconfigurable execution. All these critical issues have been addressed and successfully solved by the Molen polymorphic processor paradigm. *The basis of the Molen processor is established on the capability to control program execution and hardware reconfiguration allowing intermingling of program code and hardware description and by utilizing emulation microcode.* This special microcode is termed as reconfigurable microcode ($\rho\mu$-code) as it is different from the traditional one. The difference is that instead of executing on fixed hardware facilities, the $\rho\mu$-code itself "designs" the facilities to operate upon. The main advantages of the Molen approach can be summarized as follows:

- **Compact ISA extension.** For a given ISA, a single architectural extension comprising 4 to 8 additional instructions provides unlimited number of reconfigurable functionalities per single programming space. This realization is application independent and resolves the opcode space explosion problem as well as provides ISA compatibility and portability of reconfigurable programs.
- **Technology independent and modular design.** The design concept is not bound to any particular reconfigurable technology. It allows reconfigurable modules designed by a third party to be ported easily into the Molen organization.
- **Arbitrary number of parameters and parallel processing.** The Molen processor organization and the programming paradigm based on sequential consistency allow an arbitrary number of parameters as well as parallel executions of no data dependent operations.

5.6 Compiling for PISC

The specific PISC compiling techniques will be illustrated with examples from the Molen compiler [Moscu-Panainte et al., 2003]. Currently, the Molen compiler relies on the Stanford SUIF2 [SUIF] (Stanford University Intermediate Format) Compiler Infrastructure for the front-end and for the back-end on the Harvard Machine SUIF [Machine SUIF] framework. The following essential features for a compiler targeting custom computing machines (CCM) have currently been implemented:

- Code identification: for the identification of the code mapped on the reconfigurable hardware, we added a special pass in the SUIF front-end. This identification is based on code annotation with special pragma directives (similar to [Gokhale et al., 1998]). In this pass, all the calls of the recognized functions are marked for further modification.
- Instruction set extension: the the compiler takes into account the instruction set extension and inserts the appropriate **set/execute** instructions both at the

medium intermediate representation level and at low intermediate representation (LIR) level.

- Register file extension: the register file set has been extended with the exchange registers. The register allocation algorithm allocates the XREGs in a distinct pass applied before the register allocation; it is introduced in Machine SUIF, at LIR level. The conventions introduced for the XREGs are implemented in this pass.
- Code generation: code generation for the reconfigurable hardware (as previously presented) is performed when translating SUIF to Machine SUIF intermediate representation, and affects the function calls marked in the front-end. The code generation schedules the **set** instructions to hide the reconfiguration latency and to guarantee that the functions can be mapped on the available area [Moscu-Panainte et al., 2006].

An example of the code generated by the extended compiler for the Molen programming paradigm is presented in Fig. 5.14. On the left, the C code is depicted. The function implemented in reconfigurable hardware is annotated with a pragma directive named *call_fpga*. It has incorporated the operation name, *op1* as specified in the hardware description file. In the middle, the code generated by the original compiler for the C code is depicted. The pragma annotation is ignored and a normal function call is included. On the right, the code generated by the compiler extended for the Molen programming paradigm is depicted; the function call is replaced with the appropriate instructions for sending parameters to the reconfigurable hardware in XREGs, hardware reconfiguration, preparing the fixed XREG for the microcode of the **execute** instruction, execution of the operation and the transfer of the result back to the general-purpose register file. The presented code is at medium intermediate representation level in which the register allocation pass has not been applied yet.

The compiler extracts from a hardware description file the information about the target architecture such as the microcode address of the **set** and **execute** instructions for each operation implemented in the reconfigurable hardware, the number of XREGs, the fixed XREG associated with each operation, etc. The compiler may also decide not to use a reconfigurable hardware function and to include a pure software based execution.

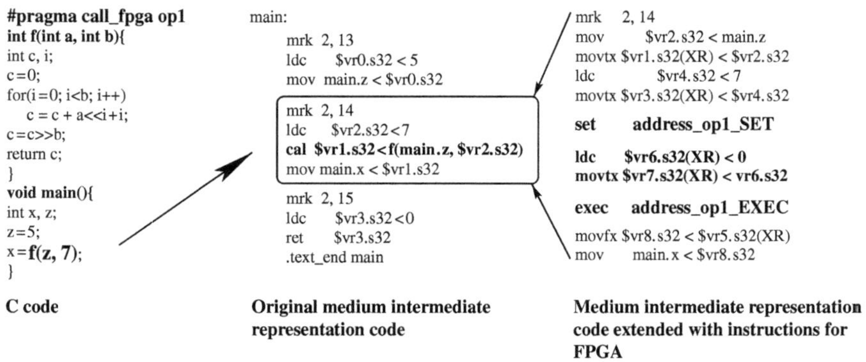

Fig. 5.14 Medium intermediate representation code

5.7 Performance Evaluation

Assume a sequential software benchmark, where T is the execution time of the original program and T_{SEi} - the time to execute its kernel i in software. Further assume $T_{\rho i}$ is the execution time for a reconfigurable hardware implementation of the same kernel i. The overall speed-up of the benchmark program with respect to the reconfigurable hardware implementation of kernel i is:

$$S_i = \frac{T}{T - T_{SEi} + T_{\rho i}} = \frac{1}{(1 - a_i) + \frac{a_i}{s_i}} \tag{5.1}$$

Where a_i is the fraction taken by kernel i from the total time of the sequential software execution, and s_i is the (local) speedup of kernel i in reconfigurable execution scenario. That is, $a_i = \frac{T_{SEi}}{T}$, $0 < a_i \leq 1$ and $s_i = \frac{T_{SEi}}{T_{\rho i}}$, $s_i > 0$. Identically, assume that totally all considered kernels take a fraction of a from the application execution time, i.e., $a = \sum_i a_i$, $0 < a \leq 1$. Then, all considered kernels will speedup the benchmark by:

$$S = \frac{T}{T - \sum_i T_{SEi} + \sum_i T_{\rho i}} = \frac{1}{(1 - a) + \sum_i \frac{a_i}{s_i}} \tag{5.2}$$

5.7.1 Theoretical Analysis

Consider Equation (5.2). We are interested to establish the theoretical boundaries of the execution speedup that can be achieved. Clearly, if we assume that a reconfigurable kernel implementation executes for time 0, the local speedup with respect to its software execution will approach its maximum, i.e., infinity:

$$s_i^{max} = \lim_{T_{\rho i} \to 0} \frac{T_{SEi}}{T_{\rho i}} = \infty \tag{5.3}$$

Assume that all considered kernels execute for time 0 in reconfigurable scenario. That is, $\forall i, s_i \rightarrow \infty$ (from Equation (5.3)) and considering Equation (5.2), we devise:

$$S_{max} = \lim_{\forall i, s_i \to \infty} S = \frac{1}{1 - a} \tag{5.4}$$

Where S_{max} is the theoretical maximum of the achievable speed-up, given the fraction a, which corresponds to a certain kernel partitioning of the benchmark application. Obviously, the bigger a, the bigger the speedup. That is, the more sequential portions of the execution time of a program we accelerate in reconfigurable hardware, the faster the overall speedup. A careful analysis of Equation (5.4) suggests that an order of magnitude acceleration is attainable only if more than 90% (a = 0.9) of the application are accelerated. Therefore, *speedups of sequential algorithms in orders of magnitude are impractical to achieve, unless virtually the entire application execution time is accelerated by dedicated hardware.* One more support for this statement is the fact that two orders of magnitude accelerations can be achieved only if 99% (a = 0.99) of the execution time is reduced to 0. Obviously, for orders of magnitude speedups, a GPP is virtually unnecessary and the entire application should be implemented (if possible) in dedicated hardware. Generally, such a solution will increase the silicon cost of a design and will bring all disadvantages related to the hardwired designs, mainly losing flexibility and even implementability. On the other hand, in the general purpose computing society, even accelerations of 10% are considered spectacular. In this field, if we are capable to achieve accelerations between 50% and 1000% by implementing only a few selected kernels in (reconfigurable) hardware, we can safely claim a considerable speedup.

5.7.2 Graphical Interpretations of Amdahl's Law

Fig. 5.15 illustrates Equation (5.4) giving the dependency of the maximum speedup theoretically attainable with respect to the portion of the application execution time considered for acceleration. Obviously, this dependency is not linear.

If half of the execution time of a program ($a = 0.5$) is reduced to a minimum, the maximum attainable theoretical speedup will be 2. If, however, 80% of the application are accelerated, the theoretical limit of the speedup is already $5\times$. An order of magnitude acceleration is attainable only if more than 90% of the application are accelerated, as explicitly indicated in Fig. 5.15.

Figure 5.16 illustrates how the overall application speedup depends on the local speedup of a kernel. The dependency is depicted for values of a between 0.5 and 0.9

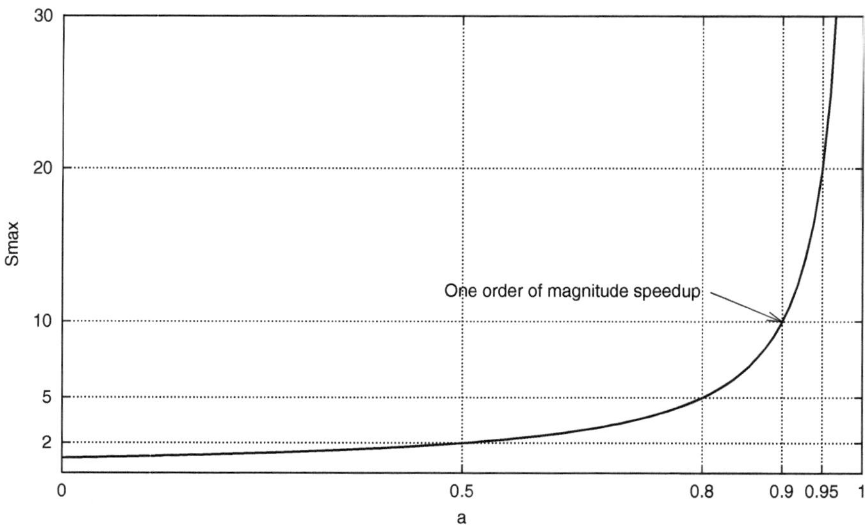

Fig. 5.15 Theoretically maximum attainable speedup, $S_{max} = \frac{1}{1-a}$

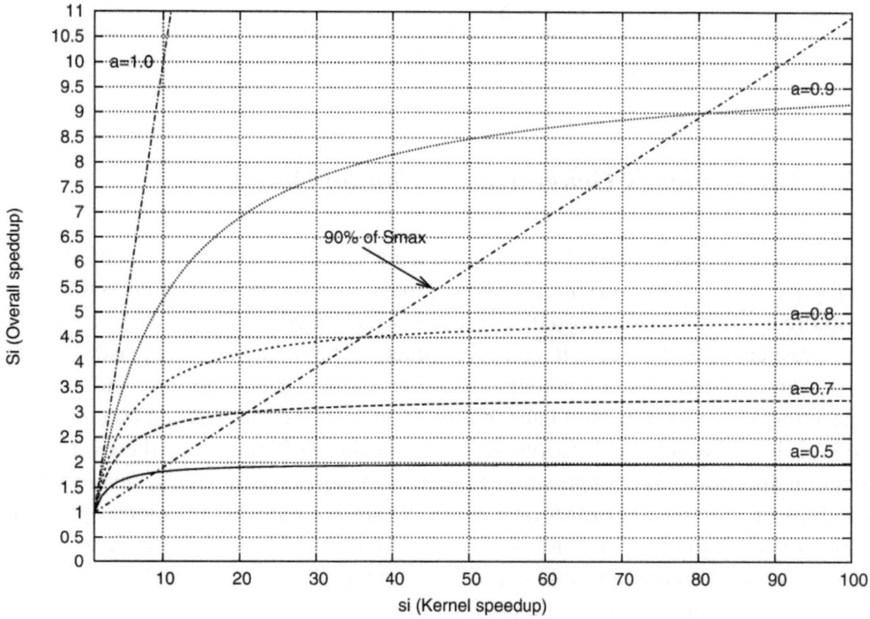

Fig. 5.16 Overall speedup dependance on the kernel speedup (different a)

after Equation (5.1), assuming only one kernel, i.e., $a = a_i$ (a $= 1.0$ is given only as a reference). Assume a practical value of the overall speedup to be 90% of the theoretically attainable maximum in acceleration, i.e., $S_i = 0.9 \times S_{max}$. Figure 5.16 suggests that *to achieve 90% of S_{max}, a kernel that consumes between 50% and 90% of the entire application execution time, must be accelerated locally between 10 and 80 times, respectively.* In Fig. 5.16, this is illustrated by the intersecting points between the speedup curves for different a and the curve presenting 90% of S_{max}. We make this analysis more clear by the following example:

Example: Consider a kernel, which consumes 80% of the entire application execution time. According to Fig. 5.15, the maximum speedup that can be theoretically attained is 5 times. Figure 5.16 suggests that 90% of this theoretical speedup can be achieved if the kernel is speeded up locally by a factor of about 36. Thus, the overall speedup of the application will be $0.9 \times 5 = 4.5$.

Similar analysis can be done for overall speedups less than 90% of S_{max}. In Fig. 5.16, the curves denoting less than 90% of S_{max} occupy some space to the left of the depicted curve for 90% of S_{max}, i.e., less local kernel speedups will be required.

5.7.3 Efficiency Evaluation

The performance efficiency of a polymorphic processor (denoted as Eff_s) can be quantified by the relation between the real speedup (5.2) and the theoretically attainable maximum speedup (5.4), that is:

$$Eff_s = \frac{S}{S_{max}} = \frac{1-a}{(1-a) + \sum_i \frac{a_i}{s_i}} \quad (5.5)$$

5.7.4 Automatically Generated vs. Manual Designs

During the discussion related to Fig. 5.16 we argue that in order to obtain 90% of the top theoretical speedup (S_{max}), a kernel that consumes between 50% and 90% of the entire application execution time, must be accelerated locally between 10 and 80 times, respectively. This conclusion gives practical boundaries of the feasibility and the design requirements of a potential reconfigurable implementation. Additionally, experimental results, reported later in this chapter, prove that beyond certain point, growing local kernel accelerations do not contribute efficiently to the overall application speedups. The implication is that in many practical cases, severe kernel speedups of, say, several orders of magnitude, are not required for their supporting hardware accelerators. Thus, not necessarily the most efficient kernel accelerator is the most efficient reconfigurable implementation application-wide. This conclusion opens a design gap, which can be filled by automatically generated hardware.

Generally speaking, automatically generated designs are far from the optimal hardware as they are neither faster, nor more cost-effective in silicon area compared to the manually designed units. In the reconfigurable devices, however, the customizable hardware gates are constantly increasing in number, which releases the requirement for cost-effective area designs. Moreover, as far as power consumption is not concerned, expanding the hardware within the available reconfigurable resources may not be considered as implementation prohibitive. On the other hand, our analysis suggests that a Molen machine organization can speed up application processing significantly, without implementing the most time efficient CCU designs but rather relatively less efficient ones.

Yet, in certain cases, where some severe design requirements have to be met, the manual design of CCUs is indispensable. Very often, the manual design can make the difference between implementability and non-implementability within a given set of requirements (e.g., limited availability of reconfigurable resources). In addition, the low maturity of recent tools for automated hardware generation as well as the vast number of non-trivial and irregular hardware operations, often makes the manual approach the only option for a CCU design. In the experiments described below, we employed both the automated and the manual approaches for the designs of the considered CCUs.

5.7.5 Experimental Evaluation

For experimental evaluation purposes, two popular media applications are considered: MJPEG and MPEG-2. We evaluated the performance gains on a Xilinx Virtex II Pro Molen prototype and using a dedicated Molen ANSI-C compiler.

Molen prototype design bitstreams and a web-based Molen ANSI-C compiler are available for free download and experiments from `http://ce.et.tudelft.nl/MOLEN/Prototype/`. Some design bitstreams and examples are also provided with the CD-ROM appended to this book.

5.7.6 Prototype Design Features

The referred Molen prototype implementation has the following features:

- Platform FPGAs: xc2vp20/xc2vp30 (Xilinx Virtex II Pro)
- Program memory: 64KB in BRAM.
- Data memory: 64KB in BRAM.
- XREGs: 512 $\times$ 32-bit.
- Microcode word length: 64 bits.
- Logical data memory segment for microprograms: 4M x 64-bits (22-bit address).
- ρ-control store size: 8KB organized in 64-bit words.
- PowerPC clock: 300 MHz.
- Memory clock: 100 MHz.

5.7.7 Synthesis Data for the Designs Considered

For the experiments considered, we utilized the Compaan [Turjan et al., 2003], [Kienhuis et al., 2000] and Laura [Zissulescu et al., 2003] tools to generate automatically a CCU, which supports four computationally demanding operations of the MJPEG encoding algorithm altogether. The obtained synthesis results are reported together with the experiment description. Regarding the MPEG-2 application, the Sum-of-Absolute-Differences (SAD) operation, the Discrete-Cosine Transform (DCT) and its inverse transform (IDCT) are considered for CCU implementations. Since these three operations have been extensively investigated in the literature, we are not focusing on their organizational details. Interested readers are referred to the sources, describing the devices we implemented for our experiments. In the following, synthesis data for the considered designs are presented. Table 5.1 displays synthesis results considering the Xilinx xc2vp20/50 FPGA devices. For the SAD function, we implemented the organization proposed in [Vassiliadis et al., 1998]. The super-pipelined 16-byte version of this SAD organization (SAD16) is capable of processing one 16-pixel line (1 pixel is 1 byte) of a macroblock in 17 cycles at over 300 MHz. The 128-byte version (SAD128) processes eight macroblock lines in 23 cycles, and the 256-byte version (SAD256), processes an entire 16 x 16-pixel macroblock in 25 cycles. SAD256 requires more resources than available in the xc2vp20 chip, therefore we consider it for an implementation on a larger FPGA (e.g., xc2vp50). To support the DCT and IDCT kernels, we synthesized the 2-D DCT and 2D-IDCT v.2.0 cores available as IPs in the Xilinx Core Generator Tool. The parameters for their synthesis are

Table 5.1 Synthesis results per CCU implementation

Device xc2vp20 Speed Grade -5	SAD16	SAD128	SAD256 (xc2vp50)	DCT	IDCT	Available Resources
Slices	831	6807	13613*	4314	5436	10304
Slice Flip Flops	1448	11862	23724*	7964	9876	20608
4 input LUTs	1390	11379	22757*	6832	8624	20608
BRAMs	N.A.	N.A.	N.A. *	2	2	112
F_{max} [MHz]	310	310	310*	96	96	N.A.

*Results for xc2vp50 FPGA

Table 5.2 Synthesis parameters for the Core Generator™ IPs

Parameter	2-D DCT	2-D IDCT
Data width [bits]	16 (signed)	16 (signed)
Coeff. width [bits]	24	24
Result width [bits]	16 (rounded)	16 (rounded)
cycles/input sample	6	8
Internal latency [cyc]	94	97

presented in Table 5.2. Considering the implemented clock domains and synthesis results (from Table 5.1) in our experiments, we have run the DCT and IDCT functions at mem_clk frequency (100MHz). The SAD designs were clocked at 300MHz.

5.7.8 Mapping MJPEG on the Molen Prototype

An MJPEG (Motion JPEG) encoder processes the frames in a video sequence as a series of JPEG images. We consider an MJPEG object-oriented source code written in C. An automated process is employed to synthesize a CCU, supporting four computationally demanding operations of the MJPEG encoding algorithm altogether. Figure 5.17 illustrates how the MJPEG encoder is mapped on the Molen Virtex II Pro prototype.

The four operations, considered for automatic hardware generation are *block input*, *pre-shift*, *2D-DCT* and *block output*. These three operations are embedded in a single hardware design generated with the Compaan [Turjan et al., 2003, Kienhuis et al., 2000] and the Laura [Zissulescu et al., 2003] tool-sets, which utilize Kahn Process Networks (KPN) [Kahn, 1974] for intermediate modelling format. The output of the Laura tool-set is a synthesizable VHDL code. Since the Laura tool-set does not consider the Molen CCU interface, we manually designed a Molen consistent wrapping interface to embed the generated hardware unit as a CCU. The core functionality among the considered four is the 2D-DCT, therefore, we refer to all four operations as to a single kernel denoted by DCT*. The DCT* CCU is synthesized with the Xilinx tools and mapped on the Virtex II Pro FPGA. Synthesis results are reported in Table 5.3. Column 2 contains the resource utilization for the automatically generated unit without the Molen specific interface implemented. In Column 3, synthesis data for the Molen interface wrapper (see Fig. 5.17) is also considered.

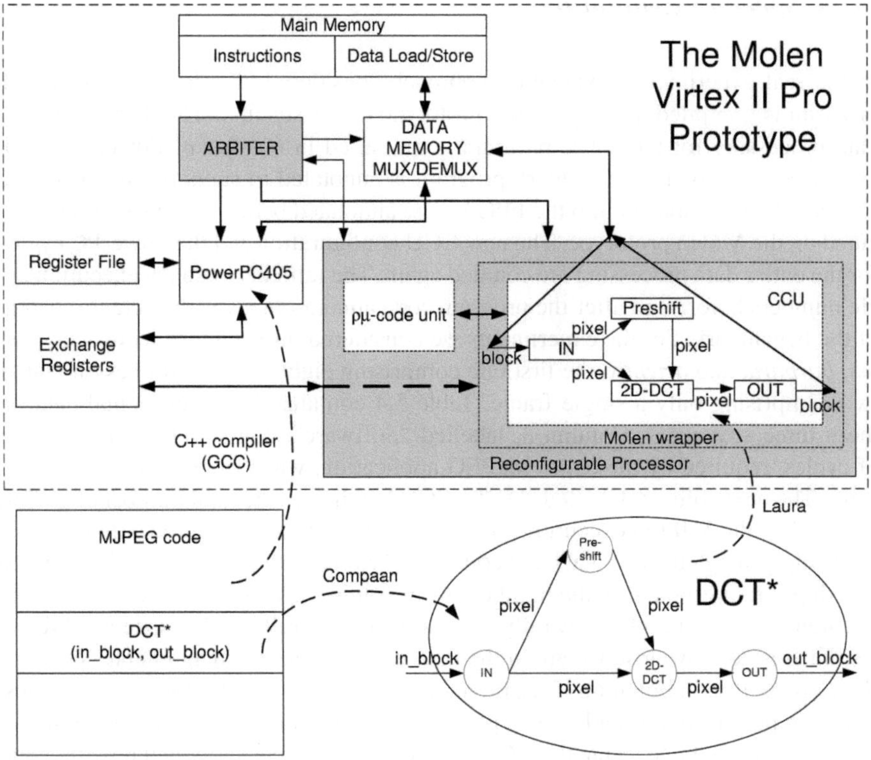

Fig. 5.17 Mapping MJPEG onto the Virtex II Pro Molen prototype

The Molen program code supporting the DCT* CCU , as well as the rest of the MJPEG C code (i.e., not including the DCT* kernel) are considered for PowerPC execution, more precisely they are mapped into the main memory. For the experiments, we considered an image size of 48 × 48 and 4:2:2 YUV macroblock format, i.e., a 16 × 16-pixel macroblock comprises four 8 × 8 Y (luminance) blocks and four 8 × 8 chrominance blocks (two U and two V blocks). The DCT* kernel processes "half" macroblocks at a time, i.e., two luminance and two (U and V) chrominance blocks.

Table 5.3 Synthesis results for the automatically generated DCT* CCU

Device xc2vp20 Speed Grade -5	CCU	CCU+ wrapper	Available Resources
Slices	1804	1975	10304
Slice Flip Flops	2271	2388	20608
4 input LUTs	2014	2228	20608
BRAMs	4	4	112
Multipliers18x18	8	8	112
F_{max} [MHz]	100	100	N.A.

5.7.9 Speedup Evaluation

In the real experimental evaluation approach considered, the original benchmark program is compiled and run on the prototype processor first. The duration of the data processing for the entire program is measured in number of PowerPC clock cycles. Separately, the benchmark program is annotated to support the considered CCUs, which are loaded into the FPGA. The annotated benchmark program is executed on the Molen prototype. The new CCU configuration and the PowerPC cycles for the entire data processing are counted again. The ratio between the execution cycle numbers before and after the program code annotation gives the actual speedup of the benchmark. For the experiments we considered three picture sequences: *tennis*, *barbara*, and *artemis*, the first one comprising eight 48×48 frames, the latter two comprising only a single frame. Table 5.4 contains the experimental data on these three sequences. Column 3, labelled "Software", contains the total number of cycles, required by the entire MJPEG application, when running as a pure software. The next column, labelled "DCT* CCU", indicates the total number of cycles, required by the MJPEG execution on the Molen prototype configured with the automatically generated DCT* CCU. In column 6 ("exper."), the experimentally attained speedups are presented, as the numbers from column 2 are divided by the numbers in column 4, thus straight-forwardly calculating the overall MJPEG speedup. We are also interested how close we are to the theoretically maximum attainable speedups S_{max}, as devised in Equation (5.4), therefore, we carried out additional experiments, to obtain parameters a_i and s_i. For parameter a_i, we simply measured the number of cycles required to execute the DCT* kernel in software and divided it by the total number of MJPEG execution cycles. Thus, we calculated, that the DCT* software kernel constitutes roughly 61% of the total execution time of the pure software MJPEG encoding algorithm. Employing Equation (5.4) with the calculated exact values of a_i results to the theoretical speedups, reported in column 6 of Table 5.4. Finally, in column 7, we estimate how close the experimentally measured speedups are to the theoretical maximum.

Table 5.4 Overall MJPEG speedup by the DCT* Molen CCU implementation

sequence	frame No	Total MJPEG execution [cycles]		Overall Speedup S_i for $a_i = 0.61, s_i = 6.22$		
		Software	DCT* CCU	exper.	S_{max}	% of S_{max}
tennis	1	84556800	40307208	2.10	2.57	81.69
	2	84615272	40393200	2.09	2.56	81.67
	3	84689544	40462000	2.09	2.56	81.69
	4	84629288	40439904	2.09	2.56	81.59
	5	84615808	40436592	2.09	2.57	81.57
	6	84594184	40409512	2.09	2.57	81.58
	7	84471640	40308680	2.10	2.57	81.50
	8	84434216	40263576	2.10	2.57	81.49
barbara	1	85371112	41131512	2.08	2.53	81.94
artemis	1	85577112	41354208	2.07	2.52	82.01

5.7.10 MPEG-2 Experimentally Projected Evaluation

We target the Berkeley implementation of the MPEG-2 encoder and decoder included in libmpeg2. First, we profile the application after running the pure software code on a PowerPC based system. The profiling data are used to identify and design performance critical kernels as CCU implementations. We run the extracted kernels on the prototype Molen processor and directly measure the performance gains. Using these measurements, the profiling data, and Amdahl's law, we estimate the projected overall MPEG-2 speedup.

5.7.11 Software Profiling Results

The first phase of the experimentation is to identify the kernels, which consume most of the application execution time. These kernels will be considered as candidates for reconfigurable hardware implementations. The input data comprised a set of four popular video sequences, namely *carphone*, *claire*, *container* and *tennis*. Profiling results for each considered function and its descendants (obtained with the GNU profiler **gprof**) are presented in Table 5.5 per sequence. For the MPEG-2 encoder, the total execution time spent in SAD, DCT and IDCT operations (Table 5.5, column 6) emphasizes that these functions require around 2/3 of the total application time. Note, that although the IDCT function in MPEG-2 encoder takes only around 1% of the total encoding time (Table 5.5, column 5), in the MPEG-2 decoder it requires on average around 42% of the total decoding. Also note, that the profiling results are data dependent and slightly vary per data sequence. Consequently, all three considered functions are good candidates for hardware implementations although their individual part of the total execution time vary per sequence and per (encoder or decoder) application.

5.7.12 Local Kernel Speedups

We have embedded the considered CCU implementations of SAD, DCT and IDCT within the Virtex II Pro Molen prototype and carried out experiments in two stages: *Stage 1.* Extract the kernels of interest from the original MPEG-2 application source code used in the profiling phase without any further code modifications. Compile these software kernels for the original PowerPC ISA and run them on one of the

Table 5.5 MPEG-2 profiling results for the considered functions

MPEG-2 application		encoder				decoder
sequence	# frames@Res.	SAD	DCT	IDCT	Total	IDCT
carphone	96@176x144	51.1 %	12.5 %	1.3 %	64.9 %	50.4 %
claire	168@360x288	53.8 %	11.8 %	1.0 %	66.6 %	37.6 %
container	300@352x288	56.2 %	10.7 %	1.0 %	67.9 %	40.4 %
tennis	112@352x240	60.0 %	9.5 %	0.8 %	70.3 %	40.5 %

embedded PowerPC405 processors. Obtain the number of PowerPC cycles consumed per kernel execution.
Stage 2. Substitute the kernel software code with a new piece of code to support πISA. Compile the new code. Load the CCU configuration supporting the corresponding kernel into the reconfigurable processor. Run the newly compiled CCU-enabled kernel on the Virtex II Pro Molen prototype and obtain the number of PowerPC cycles consumed during its execution.

For our experiments, we considered the same data sequences as used in the profiling phase. In both stages, the PowerPC timers are initialized before a kernel is executed and are read immediately after the kernel execution has completed. Thus, the exact number of PowerPC cycles, required for the entire kernel execution is obtained. Fig. 5.18 depicts in *logarithmic scale* the measured cycles obtained in the two experimentation stages for each of the three kernels, considered in the experiments. It has been noted that the SAD, the DCT and the IDCT software implementations are slightly data dependent. Therefore, there are four chart groups illustrated in Fig. 5.18, which depict the cycle numbers consumed in the software execution of each testbench sequence. On the contrary, the CCU implementations of all three kernels are data independent. This implies that the same processing cycles are required for the same amount of data, regardless the contents of the benchmark data sequence. Therefore, only one chart group (the last one in Fig. 5.18) presents the cycle numbers, consumed by the Molen prototype. In Fig. 5.18, only results for fixed $\rho\mu$-code CCU implementations are depicted. Additionally, we have considered both fixed and pageable $\rho\mu$-code implementations for SAD16 and SAD128 and the obtained execution cycle numbers are reported in Table 5.6. The cycle numbers of the right-most chart group and in Table 5.6 include all kernel related XREG transfers, memory transfers and data processing. The pagable $\rho\mu$-code total cycle numbers in Table 5.6 include the transfers from the main memory to the ρ-control store, as well. After obtaining the execution cycle numbers for each kernel both on PowerPC and on the Molen prototype, the kernel speedup is calculated for all data sequences with respect to each CCU implementation. Table 5.7 presents the calculated kernel speedups.

5.7.13 Overall Application Speedup

Consider Equation 5.1. Further consider parameters a_i to be the profiling results reported in Table 5.5 and parameters s_i- the local kernel speedups in Table 5.7. Thus, employing Equation 5.1, the overall speedup figures for the entire MPEG-2 encoder and MPEG-2 decoder are estimated per kernel contribution and results are reported in Table 5.8.

The data related to the three considered SAD implementations and reported in Tables 5.5, 5.7, and 5.8 are summarized and illustrated in Fig. 5.19.

The figure depicts the overall MPEG-2 encoder speedup depending on the local speedup of the SAD operation for the four considered data sequences. The speedup, attained by each of the proposed SAD configurations is marked with different symbols. Figure 5.19 suggests that the SAD16 configuration alone (empty squares)

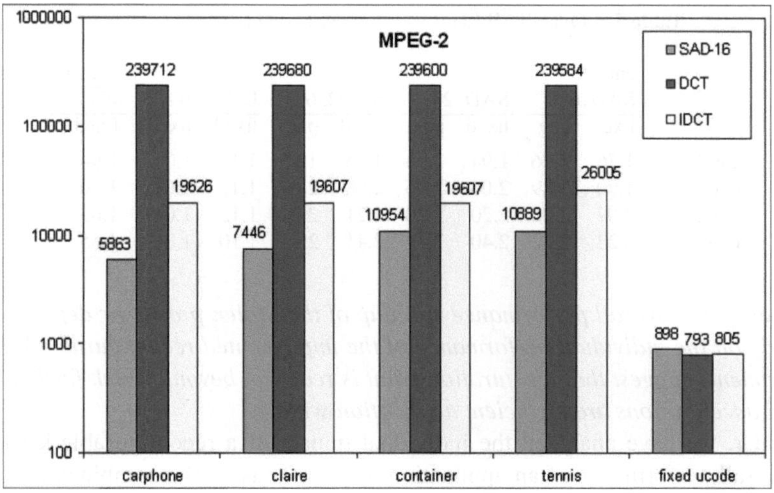

Fig. 5.18 Kernels execution cycles for PowerPC ISA and fixed $\rho\mu$-code

Table 5.6 Cycle numbers for different SAD implementations

	SAD16	SAD128	SAD256
fixed $\rho\mu$-code	898	311	264
peageable $\rho\mu$-code	914	331	284

can speedup the entire MPEG-2 encoder by less or around 90% of the maximum theoretically attainable speedup. Obviously, SAD128 (solid squares) and SAD256 (circles) CCU implementations outperform SAD16 allowing more than 90% of the theoretical limit to be attained in speedup, which is due to their parallel processing organization. Though SAD256 clearly outperforms SAD128 in processing the SAD kernel alone, the overall impact of this processing superiority over the entire MPEG-2 decoder is negligible. Due to the fact that both SAD128 and SAD256 configurations are in the saturation zone of the overall performance curves in Fig. 5.19, both of them perform with almost identical overall efficiency. Therefore, we can conclude that from the three proposed SAD configurations, SAD128 is the optimal one, because it severely outperforms SAD16. On the other hand, the hardware complexity of SAD128 is dramatically (twice) lower compared to SAD256 at the cost of a slight performance decrease. Similar analysis can be carried out for the DCT and IDCT implementations, as well as for an arbitrary design, considered for a CCU implementation.

Table 5.7 Local speedup for the MPEG-2 kernels considered ($s_i = \frac{T_{SEi}}{T_{\rho i}}$)

	SAD16		SAD128		SAD256		DCT	IDCT
	fixed	pag.	fixed	pag.	fixed	pag.	fixed	fixed
carphone	6.5	6.4	18.9	17.7	22.2	20.6	302.3	24.4
claire	8.3	8.1	23.9	22.5	28.2	26.2	302.2	24.4
container	12.2	12.0	35.2	33.1	41.5	38.6	302.1	24.4
tennis	12.1	11.9	35.0	32.9	41.2	38.3	302.1	32.3

Table 5.8 Overall MPEG-2 speedup per kernel ($Si = \frac{1}{1-(a_i - \frac{a_i}{s_i})}$)

	encode								decode
	SAD16		SAD128		SAD256		DCT	IDCT	IDCT
	fixed	pag.	fixed	pag.	fixed	pag.	fixed	fixed	fixed
carphone	1.76	1.76	1.94	1.93	1.95	1.95	1.14	1.01	1.94
claire	1.90	1.89	2.06	2.06	2.08	2.07	1.13	1.01	1.56
container	2.07	2.06	2.20	2.20	2.21	2.21	1.12	1.01	1.63
tennis	2.22	2.22	2.40	2.39	2.41	2.41	1.10	1.01	1.65

That is, *the overall performance speedup of the Molen processor depends non-linearly on the individual performance of the implemented reconfigurable kernels. Experiments suggest that a saturation point is reached, beyond which further local kernel accelerations are inefficient application-wide.*

So far, we have analyzed the individual impact of a reconfigurable kernel on the overall performance of an application. Let us focus on the combined influence of several reconfigurable kernels on the overall performance. Consider Equation (5.2) and the experimental results in Tables 5.5 and 5.7. We calculated the projected overall speedup figures for the entire MPEG-2 encoder and MPEG-2 decoder applications and report them in Table 5.9. Columns labelled "theory" present the theoretically attainable maximum speedup (S_{max}) calculated with respect to Equation (5.4) and illustrated in Fig. 5.15. Columns labelled with "impl." contain data for the projected speedups with respect to the considered Molen implementation and Equation (5.2). For the MPEG-2 encoder, the simultaneous configuration of the SAD128, DCT, and IDCT operations employing fixed microcode implementations has been

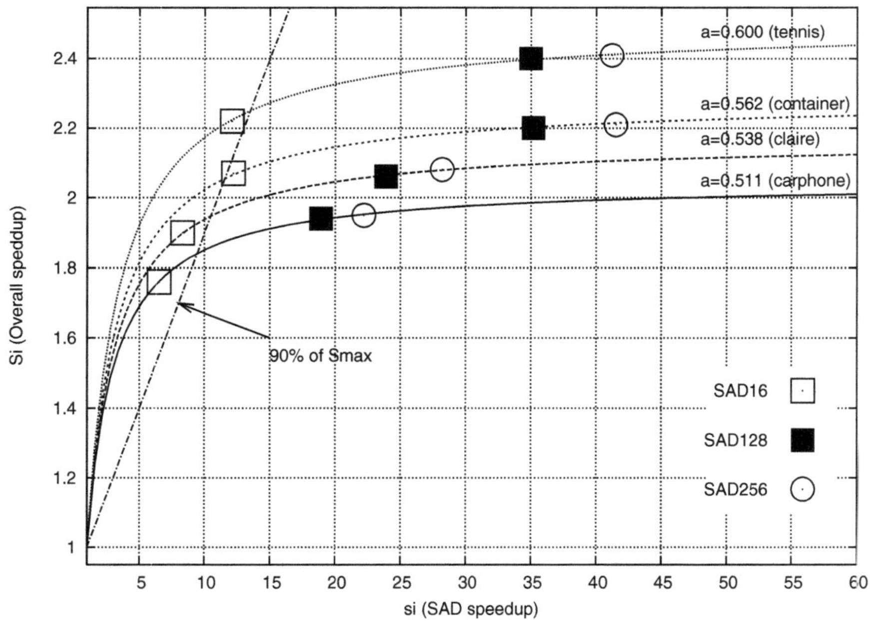

Fig. 5.19 Overall MPEG-2 encoder speedup with three SAD configurations

Table 5.9 Overall speedup estimations for the entire MPEG-2

	MPEG2 encoder*			MPEG2 decoder		
	theory	impl.	impl./th.	theory	impl.	impl./th.
carphone	2.85	2.64	93%	2.02	1.94	96%
claire	2.99	2.80	94%	1.60	1.56	98%
container	3.12	2.96	95%	1.68	1.63	97%
tennis	3.37	3.18	94%	1.68	1.65	98%

* fixed $\rho\mu$-code SAD128 + DCT + IDCT

considered. For the MPEG-2 decoder, only the IDCT reconfigurable implementation has been employed. Columns with label "imp./th." in Table 5.9 indicate (in %) how close the real speedup is to the theoretically attainable one. Reported results strongly suggest that the actual speedups of the MPEG-2 encoder and decoder obtained during our practical experimentation very closely approach the theoretically estimated maximum possible speedups, which is graphically illustrated in Fig. 5.20.

5.7.14 Speedup Amplification Effect:

Fig. 5.21 illustrates how the nonlinearity of the speedup curve influences the overall MPEG-2 encoder speedup. On the left-hand side of the figure, experimental results for the individual CCU implementations of IDCT, DCT and SAD are depicted. Note that when DCT is implemented alone, just 12,5% from the application are accelerated yielding a total of 14% overall acceleration. When SAD is implemented alone, the speedup is 1.94×, i.e., an acceleration of 94%. If both SAD and DCT are implemented altogether, the speedup is 2.55, which is 31.4% acceleration with respect to the SAD alone implementation. Thus, the contribution of the DCT

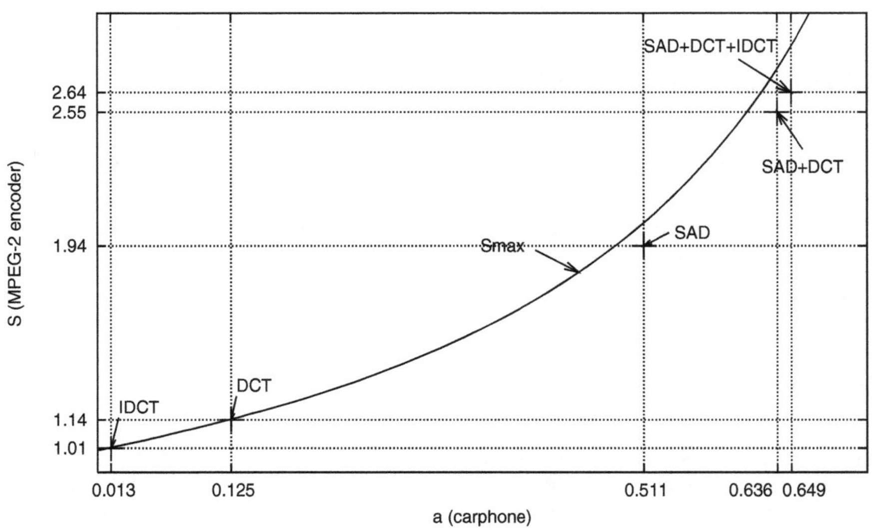

Fig. 5.20 Influence of nonlinearity on the overall MPEG-2 encoder speedup

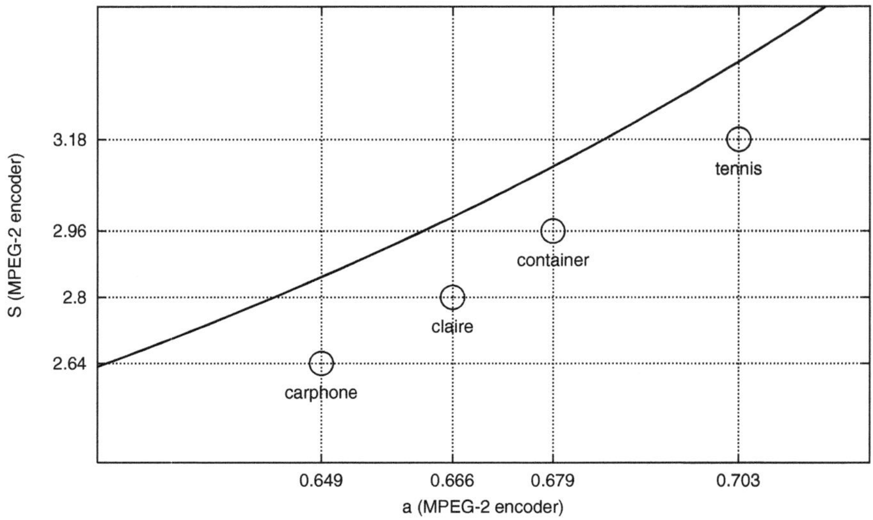

Fig. 5.21 Experimental versus theoretical speedups

CCU implementation to the overall speedup is amplified more than twice in the SAD+DCT configuration (i.e., 14% vs. 31.4% DCT-affected overall acceleration). The more to the right on the overall speedup curve a unit operates, the stronger the *speedup amplification effect*. This is proved once again by the SAD + DCT + IDCT configuration in Fig. 5.21 where a speedup of 2.64 is attained versus 2.55 for the SAD+DCT configuration. It is a 3.5% acceleration caused by the IDCT which, due to its small part of the total execution time, contributes with only around 1% to the overall acceleration if implemented alone (i.e., when operating in the left-most part of the overall speedup curve).

5.7.15 Summary of the MPEG-2 Experiments:

The MPEG-2 application was accelerated very closely to its theoretical limits by implementing SAD, DCT and IDCT as reconfigurable co-processors in the Molen Virtex II Pro prototype. The MPEG-2 encoder overall speedup was in the range between 2.64 and 3.18 while the speedup of the MPEG-2 decoder varies between 1.65 and 1.94.

References

[Vassiliadis et al., 1994] S. Vassiliadis, B. Blaner, and R. J. Eickemeyer, *SCISM: A scalable compound instruction set machine.* IBM J. Res. Develop. Vol. 38, No. 2, January 1994, pp. 59–78.

[Amdahl, 1967] G. M. Amdahl, Validity of the single processor approach to achieving large scale computing capabilities, in *Proc. AFIPS 1967 Spring Joint Computer Conference*, 1967, pp. 483–485.

[Patterson et al., 1980] D. A. Patterson and D. R. Ditzel, *The case for the reduced instruction set computer*, SIGARCH Comput. Archit. News, Vol. 8, No. 6, Oct 1980, pp. 25–33.

[Bhandarkar et al., 1991] D. Bhandarkar and D. W. Clark. *Performance from Architecture: Comparing a RISC and a CISC with Similar Hardware Organization.* Communications of the ACM, Sep 1991, pp. 310–319.

[Roelofs, 1999] G. Roelofs. *PNG: The Definitive Guide.* O'Reilly and Associates, 1999.

[Hakkennes et al.,1999] E. A. Hakkennes and S. Vassiliadis, Hardwired Paeth codec for portable network graphics (PNG), *Euromicro 99*, September 1999, pp. 318–325,

[Hauck et al., 1997] S. Hauck, T. Fry, M. Hosler, and J. Kao, The Chimaera Reconfigurable Functional Unit, in *Proc. IEEE Symp. on Field-Programmable Custom Computing Machines*, 1997, pp. 87–96.

[Rosa et al., 2003] A. L. Rosa, L. Lavagno, and C. Passerone, Hardware/Software Design Space Exploration for a Reconfigurable Processor, in *Proc. Design, Automation and Test in Europe 2003 (DATE 2003)*, 2003, pp. 570–575.

[Vassiliadis et al., 2003a] S. Vassiliadis, S. Wong, and S. Cotofana, "Microcode Processing: Positioning and Directions," *IEEE Micro*, vol. 23, no. 4, pp. 21–30, July/August 2003.

[Vassiliadis et al., 2003b] S. Vassiliadis, G. Gaydadjiev, K. Bertels, and E. Moscu Panainte, "The Molen Programming Paradigm," in *Proceedings of the Third International Workshop on Systems, Architectures, Modeling, and Simulation*, Samos, Greece, July 2003, pp. 1–7.

[Vassiliadis et al., 2001] S. Vassiliadis, S. Wong, and S. Cotofana, "The MOLEN $\rho\mu$-Coded Processor," in *11th International Conference on Field Programmable Logic and Applications (FPL)*, vol. 2147. Belfast, UK: Springer-Verlag Lecture Notes in Computer Science (LNCS), Aug 2001, pp. 275–285.

[Vassiliadis et al., 2004] S. Vassiliadis, S. Wong, G. N. Gaydadjiev, K. Bertels, G. Kuzmanov, and E. M. Panainte, "The Molen Polymorphic Processor," *IEEE Transactions on Computers*, vol. 53, pp. 1363–1375, November 2004.

[Campi et al., 2003] F. Campi, M. Toma, A. Lodi, A. Cappelli, R. Canegallo, and R. Guerrieri, "A VLIW Processor with Reconfigurable Instruction Set for Embedded Applications," in *In ISSCC Digest of Technical Papers*, Feb 2003, pp. 250–251.

[Ye et al., 2000] A. Ye, N. Shenoy, and P. Banerjee, "A C Compiler for a Processor with a Reconfigurable Functional Unit," in *ACM/SIGDA Symposium on FPGAs*, Montery, California, USA, 2000, pp. 95–100.

[Padegs et al., 1988] A. Padegs, B. Moore, R. Smith, and W. Buchholz, "The IBM System/370 Vector Architecture: Design Considerations," *IEEE Transactions on Computers*, vol. 37, pp. 509–520, 1988.

[Buchholz, 1986] W. Buchholz, "The IBM System/370 Vector Architecture," *IBM Systems Journal*, vol. 25, no. 1, pp. 51–62, 1986.

[Moudgill et al., 1996] M. Moudgill and S. Vassiliadis, "Precise Interrupts," *IEEE Micro*, vol. 16, no. 1, pp. 58–67, January 1996.

[Moscu-Panainte et al., 2003] E. Moscu-Panainte, K. Bertels, and S. Vassiliadis, "Compiling for the Molen Programming Paradigm," in *13th International Conference on Field Programmable Logic and Applications (FPL)*, vol. 2778. Lisbon, Portugal: Springer-Verlag Lecture Notes in Computer Science (LNCS), Sep 2003, pp. 900–910.

[SUIF] http://suif.stanford.edu/suif/suif2.

[Machine SUIF] http://www.eecs.hardvard.edu/hube/research/machsuif.html.

[Gokhale et al., 1998] M. Gokhale and J. Stone, "Napa C: Compiling for a Hybrid RISC/FPGA Architecture," in *Proc. IEEE Symp. on Field-Programmable Custom Computing Machines*, Napa, California, April 1998, pp. 126–135.

[Moscu-Panainte et al., 2006] E. Moscu-Panainte, K. Bertels, and S. Vassiliadis, Compiler-driven FPGA-area Allocation for Reconfigurable Computing, in *Proceedings of Design, Automation and Test in Europe 2006 (DATE 06)*, March 2006.

[Turjan et al., 2003] A. Turjan, T. Stefanov, B. Kienhuis, and E. Deprettere, "The Compaan Tool Chain: Converting Matlab into Process Networks," in *Designer's Forum of DATE 2002*, pp. 258–264, 2003.

[Kienhuis et al., 2000] B. Kienhuis, E. Rypkema, and E. Deprettere, "Compaan: deriving process networks from Matlab for embedded signal processing architectures," in *Proceedings of the 8th International Workshop on Hardware/Software Codesign (CODES)*, pp. 13–17, May 2000.

[Zissulescu et al., 2003] C. Zissulescu, T. Stefanov, B. Kienhuis, and E. Deprettere, "Laura: Leiden Architecture Research and Exploration Tool," in *13th International Conference on Field Programmable Logic and Applications (FPL 2003)*, pp. 911–920, LNCS 2778, September 2003.

[Kahn, 1974] G. Kahn, "The Semantics of a Simple Language for Parallel Programming," in *Proceedings of the IFIP Congress '74*, pp. 471–475, August 5-10 1974.

[Vassiliadis et al., 1998] S. Vassiliadis, E. Hakkennes, S. Wong, and G. Pechanek, "The Sum-of-Absolute-Difference Motion Estimation Accelerator," in *Proceedings of the 24th Euromicro Conference*, August 1998, pp. 559–566.

Chapter 6
ADRES & DRESC: Architecture and Compiler for Coarse-Grain Reconfigurable Processors

B. Mei, M. Berekovic and J-Y. Mignolet

Abstract Nowadays, a typical embedded system requires high performance to perform tasks such as video encoding/decoding at run-time. It should consume little energy to work hours or even days using a lightweight battery. It should be flexible enough to integrate multiple applications and standards in one single device. Coarse-grained reconfigurable architectures (CGRAs) are emerging as potential candidates to meet the above challenges. Many of them were proposed in recent years. However, existing CGRAs have yet been widely adopted mainly because of programming difficulty for such complex architecture.

In this chapter, a novel CGRA, ADRES (architecture for dynamically reconfigurable embedded systems), and a compiler framework, DRESC (dynamically reconfigurable embedded system compiler), are presented to address issues of existing CGRAs. Our approach possesses several unique features. First, the ADRES architecture tightly couples a very-long instruction word (VLIW) processor and a coarse-grained array by providing two functional views on the same physical resources. It brings advantages such as high performance, low communication overhead and easiness of programming. Second, the DRESC framework introduces software-like design experience. An application written in C can be quickly mapped onto an ADRES instance. The key technology behind the DRESC framework is a novel modulo scheduling algorithm, which can pipeline a loop onto the partially interconnected array to achieve high parallelism. Finally, ADRES is a template instead of a concrete architecture. With the retargetable compilation support from DRESC, architectural exploration becomes possible to discover better architectures or design domain-specific architectures.

A number of multimedia and telecommunication kernels are mapped. The instruction-per-cycle (IPC) up to 42.7 has been observed on an 8×8 ADRES instance. A multimedia application, MPEG-2 decoder, is mapped within one week starting from a software implementation. The speed-up over an 8-issue VLIW are about 12 times for kernels and 5 times for the entire application.

S. Vassiliadis, D. Soudris (eds.), *Fine- and Coarse-Grain Reconfigurable Computing.*

6.1 Introduction

Today's embedded systems and applications are very different from those 10 years ago. Now a typical embedded system such as a 3G mobile phone or a personal digital assistant (PDA) requires very high performance to encode and decode video streams at run-time and to handle the high-speed wireless data communication. It should consume little energy to work hours or even days using a lightweight battery. Designing embedded systems involves many trade-offs among different design metrics, e.g. performance, power, design costs, time-to-market, manufacturing costs. Fundamentally, there is one most important trade-off between *flexibility* and *efficiency*. Flexibility dictates design costs, time-to-market, non-recurring engineering (NRE) costs, etc., whereas efficiency determines performance, power dissipation, and silicon costs.

Coarse-grained reconfigurable architectures (CGRAs) are emerging as a solution to combine both high flexibilty and efficiency. A CGRA normally comprises an array of basic computational and storage resources. The compuational resources are functional units (FUs), which are fully customized designs capable of executing word-level or subword-level operations such as *add* and *sub*. The storage resources include memory blocks and register files (RFs). These components are connected by a partially connected interconnection network, which provides a scalable communication channel. Many CGRAs are proposed in past decade [11, 29, 32, 31, 22]. They vary a lot in architectural aspects, computational models and design tools. They clearly show the great potential of CGRAs, However, CGRAs haven't been used in mainstream applications mainly due to diffculty of using such complex architectures and lack of automated tool support.

Our approach addresses the limitations of existing CGRAs from both architecture and compiler sides. A novel coarse-grained architecture, ADRES (architecture for dynamically reconfigurable embedded systems), and a compilation-based design framework, DRESC (dynamically reconfigurable embedded system compiler), are presented. The goal is to provide a high-performance and low-power reconfigurable platform for future embedded systems with software-like design experience. Unlike other approaches, we develop our CGRA in a top-down style. The DRESC design framework was developed first, including many compilation techniques for generic array-like CGRA. Then based on the obtained compiler knowledge, we carefully devised the ADRES architecture to enable the automatic compilation by adding many compiler-friendly features.

The rest of this chapter is organized as follows: Section 6.2 describes te ADRES architecture template; Section 6.3 summarizes the DRESC compilation flow, while Section 6.4 digs into the details of the modulo scheduling alogrithm of the compiler; Section 6.5 depicts the simulation methodology and Section 6.6 presents a case study of the mapping of an application on ADRES; Section 6.7 explains some preliminary architecture exploration experiments; finally, Section 6.8 draws conclusions and adresses future work.

6.2 ADRES Architecture Template

This section describes the ADRES architecture template.

6.2.1 Overall Architecture

The ADRES architecture template is shown in Fig. 6.1. It consists of an array of basic components, including FUs, register files (RFs) and routing resources. At the top level, it tightly couples a VLIW processor and a reconfigurable array in the same physical entity. The execution model is that of a processor with coprocessor.

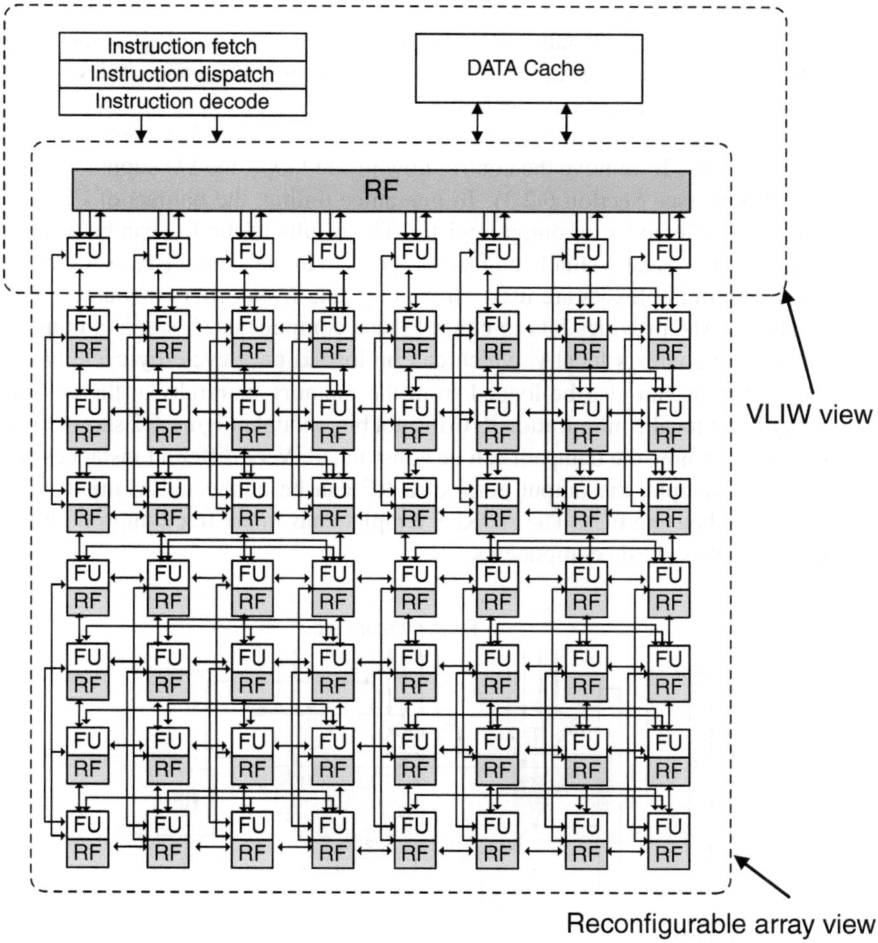

Fig. 6.1 ADRES architecture template

The identified computation-intensive kernels, typically loops, are mapped onto the reconfigurable array, whereas the remaining code is mapped onto the VLIW processor. The data communication between the VLIW processor and the reconfigurable array is performed through the shared RF and shared memory access. ADRES is a flexible template specified by an XML-based architecture specification language, which is integrated into the DRESC design framework.

Inside the ADRES array (Fig. 6.1), we find many basic components, including computational resources, storage resources and routing resources. The computational resources are functional units (FUs), which are capable of executing a set of operations. The storage resources mainly refer to the register file (RFs) and memory blocks, which can store intermediate data. The routing resources include wires, multiplexors and busses. Basically, computational resources and storage resources are connected by the routing resources in the ADRES array. This is similar to other CGRAs. The ADRES array is a flexible template instead of a concrete instance. Figure 6.1 only shows one instance of the ADRES array with a topology resembling the MorphoSys architecture [32]. An XML-based description language is developed to specify ADRES instances (see Section 6.2.7).

Figure 6.2 shows an example of the detailed datapath. The FU performs *coarse-grained* operations. To remove the control flow inside loops, the FU supports predicated operations (see Section 6.2.3). To guarantee timing, the outputs of FUs are required to be buffered by an output register. The results of the FU can be written to the RF, which is usually small and has less ports than the shared RF, or routed to other FUs. The multiplexors are used for routing data from different sources. The configuration RAM provides bits to control these components. It stores a number of configuration contexts locally, which can be loaded on a cycle-by-cycle basis. The configurations can also be loaded from the memory hierarchy at the cost of extra delay if the local configuration RAM is not big enough. Figure 6.2 shows only one possibility of how the datapath can be constructed. Very different instances are possible. For example, the output ports of a RF can be connected to input ports of several neighbouring FUs. The ADRES template has much freedom to build an instance out of these basic components.

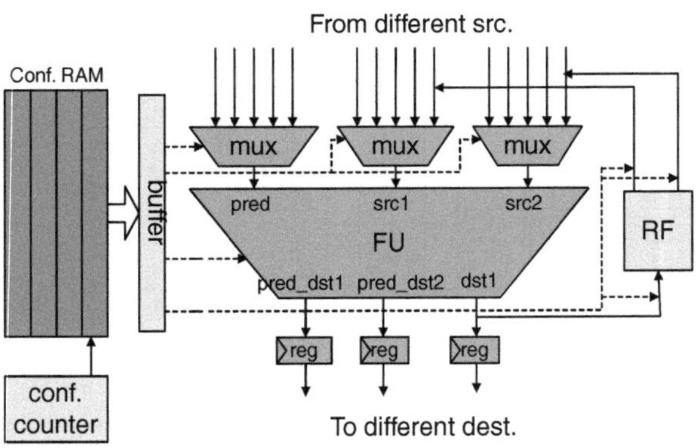

Fig. 6.2 An example of detailed datapath

6.2.2 Execution and Configuration Model

The most important feature of the ADRES architecture is the tight coupling between a VLIW processor and a coarse-grained reconfigurable array. Since VLIW processors and CGRAs use similar components like FUs and RFs, a natural thought is to make them share those components though the FUs and RFs in the VLIW are typically more complex and powerful. The whole ADRES architecture has two virtual functional views: a VLIW processor and a reconfigurable array. These two virtual views share some physical resources because their executions will never overlap with each other thanks to the processor/co-processor execution model. For the VLIW processor, several FUs are allocated and connected together through one multi-port register file. The FUs used in VLIW are generally more powerful. For example, some of them have to support the branch and subroutine call operations. The instructions of the VLIW processor are loaded from the main instruction memory hierarchy. This requires typical steps like instruction fetching, dispatching and decoding. For the reconfigurable array part, all the resources, including the RF and FUs of the VLIW processor, form a big 2D array. The array is connected by partial routing resources. Dataflow like kernels are mapped to the array in a pipelined way to exploit high parallelism. The FUs and RFs of the array are simpler than those of the VLIW processor. The communication between these two virtual views is through the shared VLIW register file and memory access. The sharing is in the time dimension so that it does not increase the hardware cost. For example, it does not require more ports in the VLIW RF.

In the VLIW mode, the configuration is performed as in all other VLIW processors: in each cycle, an instruction is fetched and executed in each cycle from the instruction memory hierarchy. In the array mode, the configuration contexts are fetched from the on-chip configuration memory. Each kernel may use one or more consecutive contexts. If the configuration memory is big enough to accommodate all the kernels, all these kernels only need to be loaded once at start-time. Afterward, the reconfiguration can be done on cycle-by-cycle basis. If the configuration memory is not big enough for all the kernels, one or more existing kernel has to be carefully chosen and discarded in order to load a new kernel from main instruction-memory hierarchy. This is known as the kernel scheduling problem [17]. Proper algorithms can minimize the reconfiguration overhead [17]. However, this aspect is not addressed in this chapter.

6.2.3 Functional Units

An FU can perform a set of operations. In ADRES, only fixed-point operations are supported because they are considered sufficient for typical telecommunication and multimedia applications. All FUs are fully pipelined so that one instruction can be issued at each cycle even when the latency of that instruction is bigger than one cycle. Different implementations may lead to different latency, which can be specified in the architecture description (Section 6.2.7) and is supported by the compiler.

The supported instruction set is very close to that of a standard RISC processor. Typical instructions supported include arithmetic opeations (ADD, SUB, MUL), logic operations (AND, OR, etc.), and compare operations (CMP, PRED). The VLIW FUs, typically the first row of FUs, also support load/store and control operaitions (LOAD, STORE, Branch, JMP).

Unlike most other CGRAs, predicated execution is introduced in the FUs in order to remove control-flow and do other transformations. Basically, it has three source operands: *pred*, *src1* and *src2*. *pred* is a 1-bit signal. If it is 1, the operation is executed; otherwise, the operation is nullified. *src1* and *src2* are normal data source operands. Some operations may only use one of them. To enhance the routability of the ADRES array, the FU is augmented with swapping logic for *src1* and *src2* operands. Therefore, all the operations are commutative, increasing the scheduling freedom. *src2* can also take a constant, normally less than full data width, as input. The FU also has three destination operands: *pred_dst1*, *pred_dst2* and *dst*. *pred_dst1* and *pred_dst2* are complementary 1-bit predicates holding the results of special comparison operations. By using *if-conversion*, each one of them guards the execution of one branch of the if-else construct. *dst* is the normal output operand. The operation formats are shown below for normal operations and predicate-defining comparison operations respectively.

$$\begin{array}{r} < pred > \; Opcode \; [src1, src2] \; [dst] \\ < pred > \; Opcode \; [src1, src2] \left[pred\_dst1, pred\_dst2\right] \end{array} \tag{6.0}$$

The FUs are enhanced for routing purposes with routing operations. SEL1 and SEL2 copy data from *src1* or *src2* to *dst*, similar to MOV operation. CON1, CON2 and CON12 take care of the predicate part by copying a predicate from *pred* to *pred_dst1*, *pred_dst2* or both. To perform normalized static-single assignment (SSA) transformation, FUs are also enhanced with PHI operations. A PHI operation actually works like a multiplexor selected by the *pred* operand. The RPHI operation works just in an opposite way.

6.2.4 Register Files

The register files (RFs) are used to store temporary data. There are two types of RFs: predicate and data RFs. The predicate RFs are 1-bit to store the predicate signal and the data RFs have the same data width as FUs.

The modulo scheduling used for pipelining kernels imposes special requirements on the register file. In the pipelined loops, different iterations are overlapped. Therefore, the life-time of the same variable may overlap over different iterations (Fig. 6.3). To accommodate this situation, each of the simultaneously live instances needs its own register. Furthermore, the name of the used register has to be clearly identified, either in software or in hardware.

Two traditional methods used to support register naming are *modulo variable expansion* (MVE) [13] and *rotating register file* (RRF) [25]. MVE is a software-based

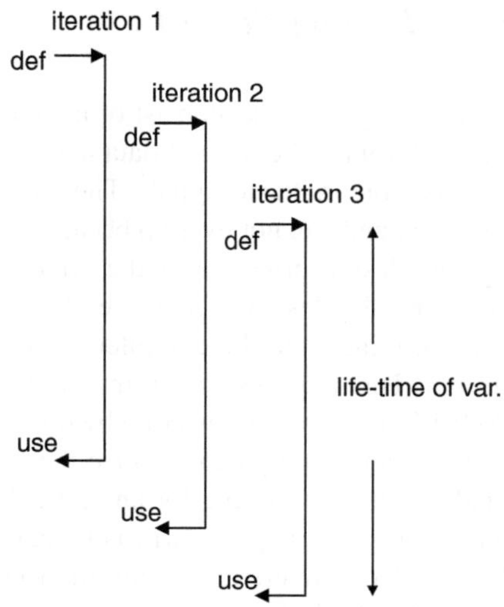

Fig. 6.3 Overlapped life-time of a variable

technique that unrolls the loop body and renames the register access to insure there is no name clash. The disadvantage of the MVE approach is that it expands the loop body significantly. In a coarse-grained array, it translates to many more reconfiguration contexts. RRF is a hardware-based register renaming method. Each physical RF address is calculated by adding a virtual RF address and a value from the iteration counter (Fig. 6.4). Hence, the different iterations of the same variable are actually assigned to different physical registers to avoid name clashes. In ADRES, the RRF approach is adopted. It increases hardware cost by requiring an additional counter and adder for each port of the register file. However, since the distributed RF is very small. An 8-entry RRF only needs 3-bit adders and counters. Hence the extra hardware costs are limited.

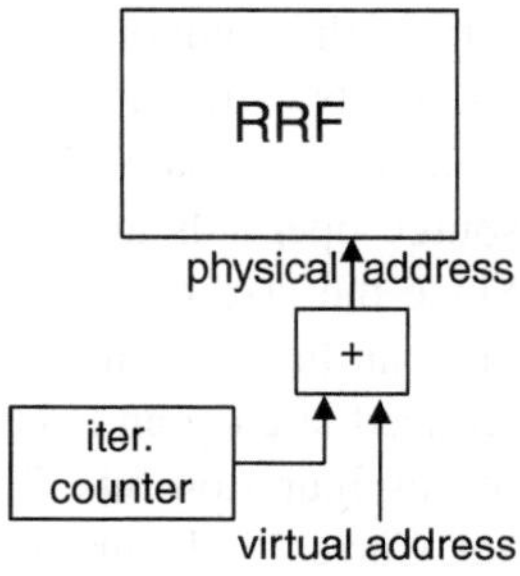

Fig. 6.4 Rotating register file

6.2.5 Routing Networks

The routing networks consist of a data network and a predicate network. The data network routes the normal data among FUs and RFs, while the predicate network directs 1-bit predicate signals. These two networks do not necessarily have the same topology and can not overlap because of different data widths.

One design principle of the ADRES architecture is compiler-friendliness and software-like design experience. Therefore, it is expected that the clock speed is predetermined and the compiler doesn't need to do timing analysis like in FPGA design. This imposes constraints on how the routing networks are constructed. In ADRES, most routing is done by direct point-to-point interconnections, consisting of wires and multiplexors. Since they are direct links, the timing can be statically analyzed at design time for an ADRES instance. When the compiler maps different kernels, the mapping will not change the timing behaviour for a given ADRES fabric. If long interconnections are needed, register can be introduced in the (data) network to limit the delay of the critical path. FUs can be viewed as part of the routing network by using copying instructions. As discussed in Section 6.2.3, FUs are enhanced to support several routing instructions for both predicate and data parts. Therefore, an indirect connection between two FUs can be implemented by using another FU as a bridge. Since FUs are enhanced with swapping logic at inputs, the load of multiplexors can be distributed between two source ports of an FU. For example, in a 8NN (nearest neighbour) topology, each FU accepts the output from 8 neighbour FUs. However, using one 4-input multiplexor is sufficient for each data input port of the FU.

6.2.6 Configuration Layer

The configuration memory provides bits to control all the components within the array. The total number of configuration bits required can be calculated from Equation 6.0. $Bits_{FU}$ refers to the bits needed to control an FU. $Bits_{port}$ stands for the bits needed for the address associated with a RF port. One extra bit is needed for WE (write enable) or RE (read enable) signals. $Bits_{mux}$ is required for selection on a multiplexor. $Bits_{bus}$ are the bits to control a bus. Equation 6.2 describes the components of $Bits_{FU}$. $Bits_{pred\_opc}$ specifies the opcode, e.g., CON1 and CON2, for the predicate datapath. $Bits_{main\_opc}$ specifies the opcode for the main FU datapath. Bit_{swap} is a one bit signal indicating whether the two data source operands are exchanged. $Bits_{const}$ are required for the constant operand. Since the size of a configuration is not as strict as a processor instruction, these bits can be saved in uncoded format in the configuration memory. This results in reduced latency and energy of the decoding step at the expense of a large number of configuration bits. The tradeoff between these two options is not investigated yet. Currently the uncoded configuration bits are stored directly in the configuration memory.

$$Bits_{total} = \sum_{\forall FUs} Bits_{FU} + \sum_{\forall RFs} \sum_{\forall ports} (Bits_{port} + 1) + \sum_{\forall muxes} Bits_{mux} + \sum_{\forall busses} Bits_{bus} \quad (6.1)$$

$$Bits_{FU} = Bits_{pred\_opc} + Bits_{main\_opc} + Bit_{swap} + Bits_{const} \quad (6.2)$$

The configuration memory can store a number of contexts. Each kernel may use one or more contexts, determined by the modulo scheduling algorithm. When a kernel is executed, these contexts are cyclically loaded until the loop ends. If there is only one context needed for a kernel loop, the context is buffered in the register so that the context doesn't need to be fetched from the configuration memory every cycle. If the configuration memory is not big enough to accommodate all kernels, the configuration contexts have to be loaded from the main instruction memory hierarchy to the configuration memory. As will be explained in Section 6.7.2 and Section 6.7.4, for a typical ADRES instance with 64 FUs a configuration context requires about 1500–2500 bits.

6.2.7 XML-Based Architecture Description Flow

Unlike other processor architecture description languages [7, 10, 24], our XML-based description focuses on only high-level features of an architecture like the amount of resources and their topology, which are mostly needed information for the compiler. We didn't provide mechanisms to define the semantics of each individual operation. Instead we assume the ADRES architecture will inherit the operation set generated by the compiler frontend. This assumption simplifies the compiler support. We also simplify the pipeline description by assuming all the FUs are fully pipelined. Therefore, only the latency of each type of operation needs to be specified. The current description language is extensible. In future, more architectural features can be added.

6.2.8 Improved Performance with the VLIW Processor

Many CGRAs consist of a reconfigurable array and a RISC processor, e.g., TinyRisc in MorphoSys [32] and ARM7 in PACT SMeXPP [22]. The execution generally follows the processor/coprocessor model. The reconfigurable array normally accelerates the regular, dataflow-like kernels, while the RISC processor performs the control-intensive or rarely executed code. The RISC processors are highly flexible but only have limited performance because of their inability to exploit parallelism. Though the non-accelerated part only represents a small portion of total execution time, it still can have significant impact on the overall system performance due to the huge performance gap between the RISC processor and the reconfigurable array. According to Amdahl's law [23], the performance gain that can be obtained by improving some portion of an application can be calculated as Equation 6.3. If the speedup of the fraction mapped to the array is very high, then the fraction executed by the RISC becomes a bottleneck.

$$Speedup_{overall} = \frac{1}{(1 - Fraction_{enhanced}) + \frac{Fraction_{enhanced}}{Speedup_{enhanced}}} \quad (6.3)$$

Figure 6.5 shows the overall system speedup in relation to the kernel fraction and the kernel speedup. If the kernel accounts for less than 90 % of execution time, the overall system speedup is pretty low no matter how high is the kernel speedup. The non-kernel part becomes a bottleneck that seriously constrains the overall system performance. If we can find a way to accelerate the non-kernel code even moderately, it will have a significant impact on the overall system performance. Figure 6.6 shows the impact of speedup of non-kernel code when the kernel speedup is fixed to 30 and the kernel fraction varies from 70 % to 100 %. Even with non-kernel speedup of only three, the overall system performance is more than doubled when the kernel fraction is smaller than 90 %. In ADRES, a VLIW processor is used instead

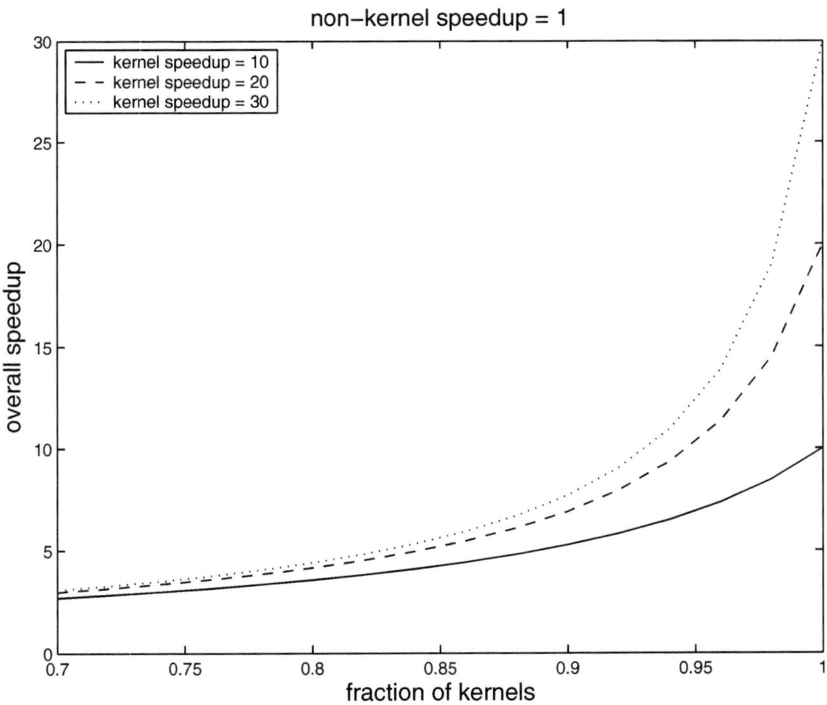

Fig. 6.5 Impact of kernel fraction and kernel speedup

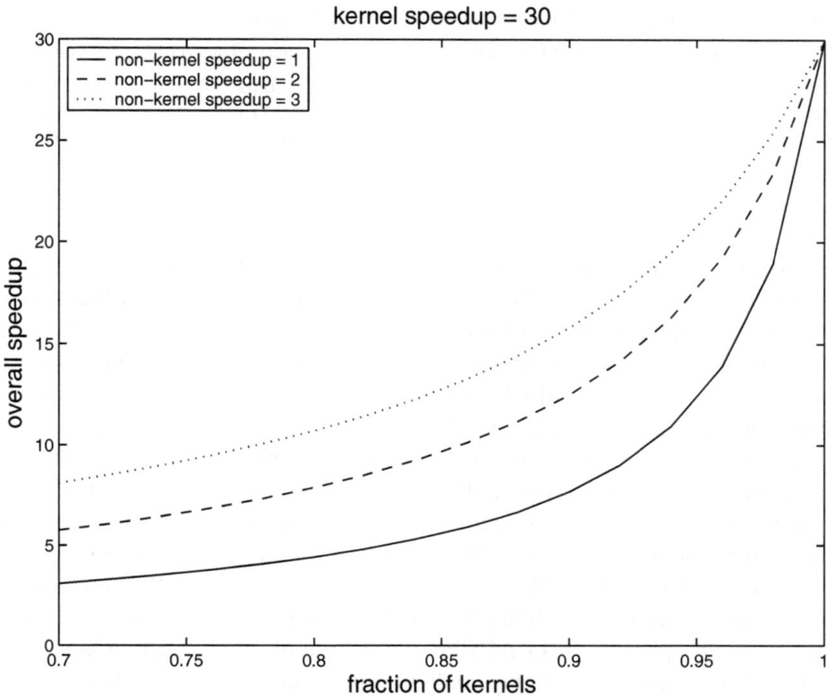

Fig. 6.6 Impact of kernel fraction and non-kernel speedup

of the RISC processor to exploit instruction-level parallelism for the non-kernel part, where 2–4 times speedup over the RISC processor is reasonable. In Chapter 6, the VLIW processor achieves approximately 2.0 instruction-per-cycle (IPC) for the non-kernel part of an MPEG-2 decoder.

6.2.9 Power Efficiency

Power is one of the most important issues in embedded system design. Here some qualitative analysis of power efficiency of the ADRES architecture is presented.

The power of a computation engine generally comprises several major components: computation power, communication power and memory power. In the ADRES architecture, the computation is performed by FUs. For a given application, the total number of instructions executed is close for both VLIW and ADRES if the instructions sets are similar. Therefore, there is no real significant difference in computation power between them. The big difference comes from the communication power. In the ADRES architecture, data is transfered between FUs through partial interconnect. Though still not as efficient as dedicated point-to-point interconnect, it is much more power-efficient than the general communication schemes like the big multi-port RF used in VLIW processor. According to [27], the power dissipation of a centralized multi-port RF grows as N^3 for N FUs. Therefore, for an 8-FU VLIW

Table 6.1 Power of loading instruction/configuration in VLIW and ADRES

arch.	FUs	inst./conf. width	entries	energy/(inst./conf.)	energy/FU
VLIW	8	256	512	0.26 nJ	0.033 nJ
ADRES1	64	2048	64	0.79 nJ	0.012 nJ
ADRES2	64	2048	32	0.69 nJ	0.011 nJ
ADRES3	64	2048	16	0.63 nJ	0.010 nJ

processor, the communication power associated with each instruction is 512 times bigger than that of a 1-FU RISC processor if nothing is done. Various RF design and partitioning techniques can be applied to reduce power of multi-port RF [27], nevertheless, the power is still quite significant. In the ADRES array, the communication power is consumed by the multiplexers, distributed RFs and driving wires to several destinations. Compared with a central VLIW RF, the power of these components is much lower. The last part is memory power, including both data and instruction memory hierarchies. For a given application, an ADRES array and a VLIW processor have similar memory access without further optimization techniques. The power of this part should be similar as well. However, in future, power can be greatly reduced if distributed memory blocks are implemented into the ADRES architecture. The instruction memory (configuration memory for CGRAs) contributes a significant part of the power [3]. In ADRES, most time of an application is spent on executing kernels from the on-chip configuration memory. It has much less entries, ranging from 16 to 64, than a typical instruction cache used in a VLIW processor, e.g., 512 entries in TI 64 $\times$ series [34]. Since the power of each access is largely determined by the depth of the memory [30], it is more power-efficient for ADRES to load configurations than for a VLIW processor. Table 6.1 compares energy of some typical scenarios. We assume the ADRES instance is 8 $\times$ 8 and needs 2048 bits for one context, while the VLIW has 8 FUs and needs 256 bits for 8 operations. A CACTI3.0 model scaled to 130nm technology is used to estimate energy [1, 30]. It shows that energy/FU for loading configuration on ADRES is only one third of that on VLIW.

The ADRES architecture does come with some power overhead. Due to aggressive hyperblock optimization, it generates more operations than a VLIW processor. It also needs more NOP operations to fill unused FUs in case that some kernels have low parallelism. Loading and distributing configuration from main memory hierarchy to the on-chip reconfiguration memory requires much power if it happens often. These problems can be addressed partly by circuit design or other techniques. For example, FU can be designed in a way that nullified operations in hyperblocks only consume little power.

6.3 DRESC Compiler Flow

The DRESC framework is shown in Fig. 6.7. A design starts from a C-language description of the application. The profiling/partitioning step identifies the

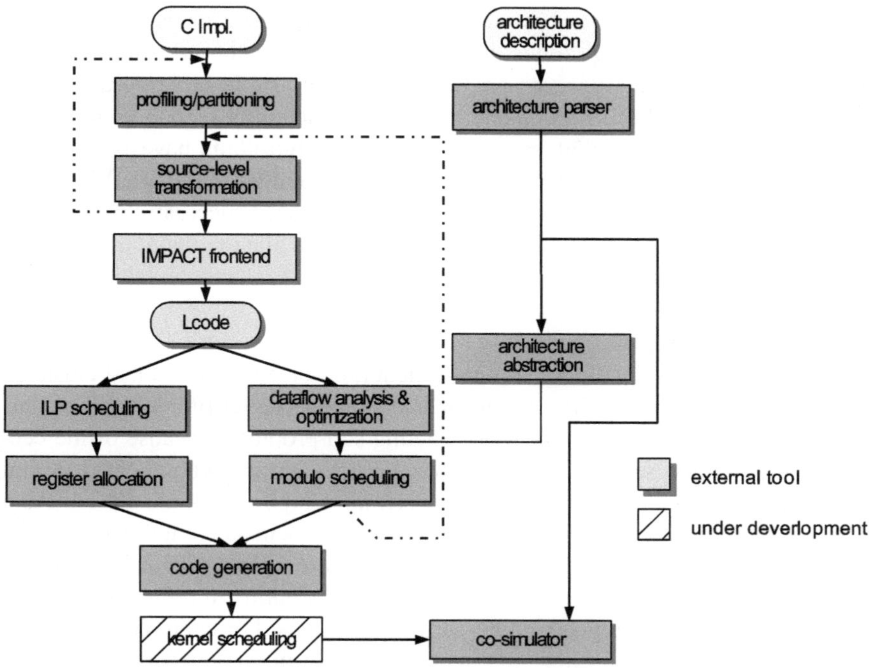

Fig. 6.7 DRESC design framework

candidate loops for mapping on the reconfigurable array based on the execution time and possible speed-up. Source-level transformations try to rewrite the kernel in order to make it pipelineable and to maximize the performance. In the next step, we use IMPACT, a VLIW compiler framework [12], to parse the C code and do some analysis and optimization. IMPACT emits an intermediate representation, called *Lcode*, which is used as the input for scheduling. On the right side of Fig. 6.7, the target architecture is described in an XML-based language. The parser and abstraction steps transform the architecture to an internal graph representation [18]. Taking the program and architecture representations as input, a novel modulo scheduling algorithm is applied to achieve high parallelism for the kernels, whereas traditional ILP scheduling techniques are applied to discover the available moderate parallelism for the non-kernel code. The communication between these two parts is automatically identified and handled by our tools. Finally, the tools generate scheduled code for both the reconfigurable array and the VLIW. The result is simulated by a co-simulator.

6.4 Modulo Scheduling

Modulo scheduling is a widely used software pipelining technique [13]. The objective of modulo scheduling is to engineer a schedule for one iteration of the loop such that this same schedule is repeated at regular intervals with respect to intra- and inter-iteration dependences and resource constraints. This interval is termed *initiation interval* (II), essentially reflecting the performance of the scheduled loop. In modern VLIW-based DSP processors, it plays a central role in exploiting parallelism [33]. In the ADRES architecture, modulo scheduling is also identified as the main technique to exploit parallelism. Various algorithms have been developed to solve this problem for both unified [16, 26] and clustered VLIW processors [2, 8, 21]. To the best of our knowledge, it has not been successfully applied to coarse-grained reconfigurable architectures (CGRA). While the main idea of modulo scheduling remains the same when applied to CGRAs, the complexity is much higher due to the more complex architecture of CGRAs.

Table 6.2 compares the complexity of modulo scheduling for CGRAs with several similar problems, including modulo scheduling for both unified and clustered VLIW processors, and FPGA placement and routing (P&R) problems. The unified VLIW doesn't have placement and routing sub-problems because of the centralized RF. Modulo scheduling for clustered VLIW is more complicated in that the scheduler has to choose on which cluster to assign operation. FPGA P&R doesn't need to decide in which cycle an operation is scheduled and doesn't have the modulo constraint. All these related problems are known as NP-hard, which means it is impossible to find a theoretically optimal solution in polynominal time. These problems are usually solved by various heuristics. For example, the VLIW scheduling is based on list scheduling, while the FPGA placement is usually solved by simulated annealing. The modulo scheduling for CGRAs combines these sub-problems and modulo constraint all together so that its complexity is even higher. Therefore, we can expect the solution to this problem is also based on heuristics and sub-optimal. In this section, a novel modulo scheduling algorithm is developed for the ADRES architecture. It solves these sub-problems in one framework and respects the modulo constraint by utilizing a novel abstract architecture representation. The algorithm is integrated into the DRESC compiler for the ADRES architecture. Nonetheless, the algorithm is applicable to generic CGRAs that have support for multi-context configuration memory and compiler-determined timing.

This section first illustrates the problem, then describes the basic scheduling algorithm in details, including the abstract architecture representation, complexity analysis and experiments on a 8×8 ADRES instance for a set of multimedia and DSP benchmarks.

Table 6.2 Modulo scheduling of CGRAs in comparison with similar problems

	Unified VLIW modulo sched.	Clustered VLIW modulo sched.	FPGAs P&R	CGRAs modulo sched.
Scheduling	Yes	Yes	No	Yes
Placement	No	Yes	Yes	Yes
Routing	No	No	Yes	Yes
Modulo constraint	Yes	Yes	No	Yes

6.4.1 Problem Illustrated

To illustrate the problem, let's consider a simple data dependence graph (DDG), representing a loop body (Fig. 6.8a) and a 2×2 array (Fig. 6.8b). The scheduled loop is depicted in Figure 1.9a, which is a space-time representation of the scheduling space. The 2 × 2 array is flattened to 1 × 4 for convenience of drawing. The dashed lines represent routing possibility between the FUs. The outputs of the FUs can only be connected to the FUs in the next cycle because the FUs are registered at the output.

From Fig. 6.9a, we see that modulo scheduling on CGRAs is a combination of 3 sub-problems: *placement, routing and scheduling*. Placement determines on which FU of a 2D (2-dimensional) array to place an operation. Scheduling, in its literal meaning, determines in which cycle to execute that operation. Routing connects the placed and scheduled operations according to their data dependences. If we view *time* as an axis, the modulo scheduling can be simplified as a placement and routing problems in a modulo-constrained 3D space. The routing resources in the 3D space are asymmetric because data can only be routed from earlier time to later time, as shown in Fig. 6.9a. Moreover, all resources are modulo-constrained because the execution of consecutive iterations which are in distinct stages is overlapped. The number of stages in one iteration is termed *stage count* (SC). In this example, II = 1 and SC = 3. The schedule on the 2 × 2 array is shown in Fig. 6.9b. FU1 to FU4 are configured to execute n2, n4, n1 and n3 respectively. In this example, there is only one configuration needed. By overlapping different iterations of a loop, we are able to exploit a higher degree of ILP. In this simple example, the instruction per cycle (IPC) is 4. As a comparison, it takes 3 cycles to execute one iteration in a non-pipelined schedule due to the data dependences, corresponding to an IPC of 1.33, no matter how many FUs are in the array.

Many factors such as resource constraints and recurrence dependences can have a big impact on the II. For example, assuming there is only one load/store FU (fu3) in Fig. 6.8b, and operation n1 and n2 are memory operations in Fig. 6.8a, the II would be at least 2 to map the same DDG (Fig. 6.10a). It requires two configurations, which are loaded cyclically during the execution of the loop. It can be easily observed that the number of configurations is equal to II.

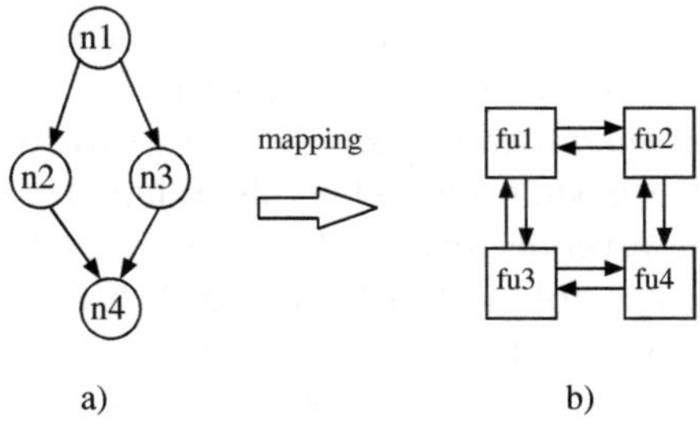

Fig. 6.8 a) A data dependence graph; b) A 2 × 2 reconfigurable matrix

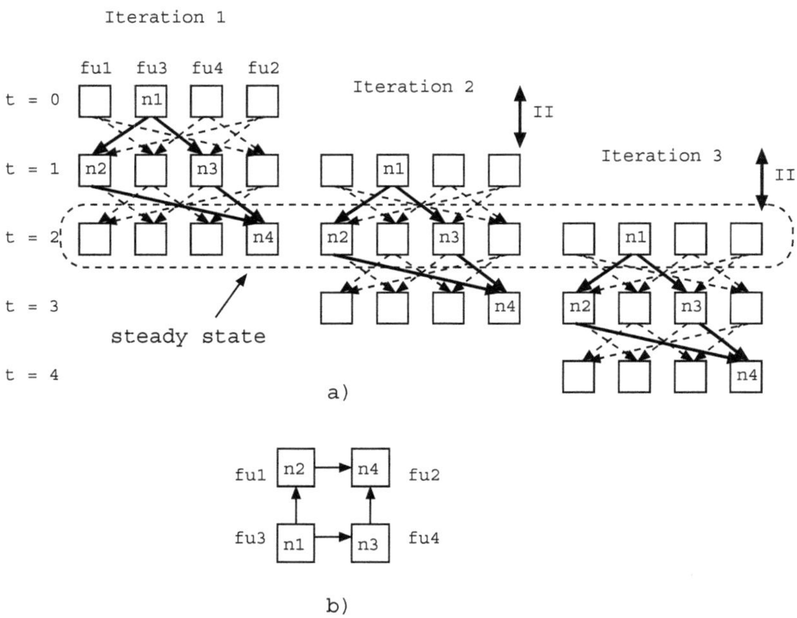

Fig. 6.9 a) Schedule with II = 1; b) One configuration

6.4.2 *Modulo Routing Resource Graph*

The modulo scheduling problem for CGRAs is based on an abstract architecture representation.

As shown in the previous section, the modulo scheduling problem for coarse-grained architectures is essentially a P&R problem in a modulo-constrained 3D space. In CGRAs, many components can be viewed as routing resources in the 3D space. For example, an FU may be used to take data from one input port and copy it to the output port. For a register file, data is written into the register file through one input port, and is read out later. This is essentially a routing capability along the time axis. From the scheduler point of view, it is important to model all these heterogeneous routing resources in a simple way to expose routing possibilities. Another problem is how to enforce the modulo constraints to make any modulo scheduling algorithm easier. Taking care of placement and routing problems in 3D space is already a difficult task, it would be much simpler for the scheduler if an architecture representation could automatically impose the modulo constraints.

To address the above problems, we propose a graph representation, namely *modulo routing resource graph* (MRRG), to model the architecture internally for the modulo scheduling algorithm. The MRRG combines features of the *modulo reservation table* (MRT) [13] for software pipelining and the *routing resource graph* [6] used in FPGA P&R. It only exposes the necessary information to the modulo scheduling algorithm. The MRRG is a directed graph $G = \{V, E\}$ which is constructed by composing sub-graphs representing the different resources of the

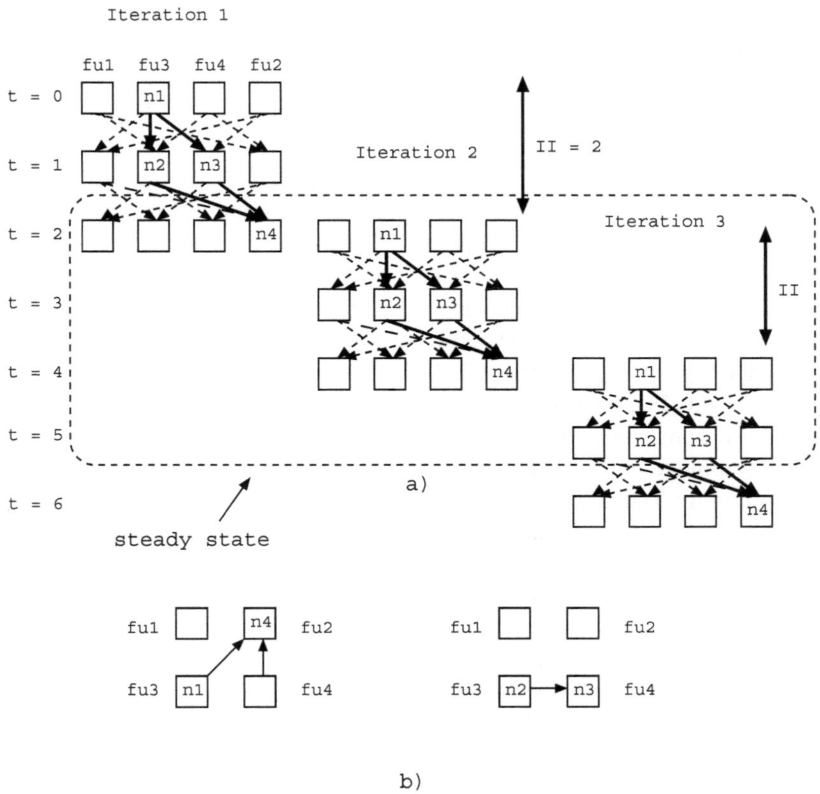

Fig. 6.10 a) Schedule with II = 2; b) Two configurations

ADRES architecture. Because the MRRG is a time-space representation of the architecture, every subgraph is replicated each cycle along the time axis. Hence each node v in the set of nodes V is a tuple (r, t) where r refers to the port or wire of the resource and t refers to the time stamp. The edge set $E = \{(v_m, v_n)|t(v_m) <= t(v_n)\}$ corresponds to switches that connect these nodes – the restriction $t(v_m) <= t(v_n)$ modeling the asymmetric nature of the MRRG. Finally, an II (initiation interval) is associated with each MRRG. The MRRG has two important properties. First, it is a modulo graph. If scheduling an operation involves the use of node (r, t_j), then all the nodes $\{(r, t_k)|t_j \text{mod} II = t_k \text{mod} II\}$ are used too. Second, it is an asymmetric graph. It is impossible to find a route from node v_i to v_j, where $t(v_i) > t(v_j)$. As we will see in section 6.4.4, this asymmetric nature imposes major constraints on the scheduling algorithm.

During scheduling we start with a minimal II and iteratively increase the II until we find a valid schedule (see Section 6.4.3). The MRRG is constructed from the architecture specification and the II under evaluation. Each component of the ADRES architecture is converted to a subgraph in the MRRG. Each node in the MRRG is associated with a *base_cost*, which is used in the scheduling algorithm (Section 1.14). We can assign different weights to the *base_cost* to help the scheduler. For example, currently, most nodes have a *base_cost* of 1. However, the *base_cost* for nodes representing the output ports of the VLIW RF are 2 because they are a bottleneck. By raising the cost of using these ports, the scheduler would be more likely to find other alternatives. The internal nodes of RFs have a *base_cost* of 0.5 because using these nodes is cheaper than using other nodes for routing.

Figure 6.11 shows how components of the ADRES architecture are converted to the MRRG. Figure 6.11a is the functional unit. Each input and output port has corresponding nodes in the MRRG graph. Virtual edges are created between src1 and dst, src2 and dst to model the fact that a FU can be used as routing resource to directly connect src1 or src2 to dst, acting just like a multiplexor or demultiplexor. In addition, two types of artificial node are created, namely *source* and *sink* to model the exchange capability enhanced with the FU (Section 6.2.1). When an operation is scheduled on this FU, the source or sink node are used as routing terminals instead of the nodes representing ports. Thus the router can freely choose which port to use. This technique improves the flexibility of the routing algorithm, and leads to higher routability. The outputs of an FU are connected to nodes representing output registers at the next cycle. Figure 6.11b shows conversion of a register file with one write port and two read ports. The idea is partly from [28]. Similar to the FU, the subgraph has nodes corresponding to each input and output port, which are replicated over each cycle. Additionally, an internal node is created which has a capacity equal to the RF size. All internal nodes along the time axis are connected one by one. The input nodes are connected to the internal node of the next cycle, whereas the output nodes are connected to the internal node of this cycle. In this way, the routing capability of the register file is effectively modeled out of its write-store-read functionality. Moreover, the register allocation problem is implicitly solved by the modulo scheduling algorithm (see section 6.4.4). Figure 6.11c shows the MRRG sub-graph of a multiplexor. It is simply replicated along the time axis. Figure 6.11d shows a bus with one cycle latency. Two nodes are created to model the latency.

By this abstraction, all routing resources, whether physical or virtual, are modeled in a universal way using nodes and edges. This unified abstract view of the architecture only exposes strictly necessary information to the scheduler and enforces the modulo constraint automatically. It greatly reduces the complexity of the scheduling algorithm. The modulo scheduling problem is therefore transformed to how to place and route a DDG representing a loop body on an MRRG representing an ADRES instance. With the MRRG abstraction, the compiler is easily retargetable to the different instances of the ADRES architecture template. The DRESC compiler reads an XML-based architecture description and transform it to an MRRG. In implementation, because the nodes and edges are simply replicated along the time axis, the MRRG can be stored in a more compact way instead of storing every node and every edge. Since the nodes and edges of an MRRG are simply replicated along the time axis, the MRRG can be implemented in a way to save memory and simplify its manipulation (Figure 6.12). All the properties of an MRRG node can be classified by three types: cycle-invariant node property, modulo node property and cycle-unique node property. For the first type, we only need to store one copy for all nodes with the same r but different $t((r, t_j)\forall j)$. For the second type of properties such as occupancy, we need to store II copies for all the node with the same r. The

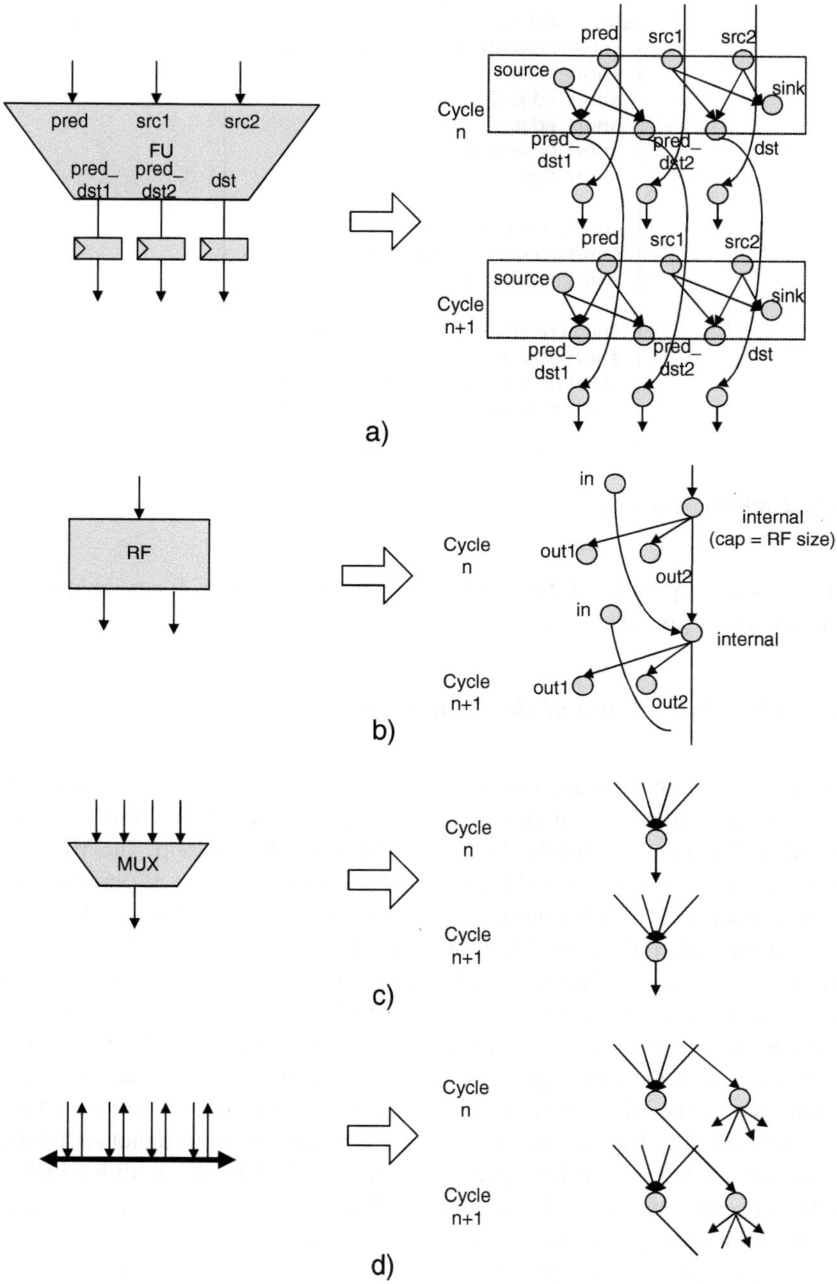

Fig. 6.11 MRRG sub-graphs: a) functional unit; b) register file c) multiplexor; d) bus with 1 cycle latency

```
struct MRRG{
  //node property shared by all cycles
  short index;
  short edges[];
  short edge_latency[];
  short base_cost;
  short cap;

  //node property related to II
  short occ[II];    //occupancy
  short cost[II];   //computerd cost

  //node property that is cycle-related
  short prev_node[max_sched_length];
  short prev_time[max_sched_length];
  short path_cost[max_sched_length];
};
```

Fig. 6.12 MRRG implementation

last one is usually required during routing to record the trace. Every unique (r, t) needs a copy to store these properties.

6.4.3 The Outer Loop of the Scheduler

The outer loop of the modulo scheduler is described in Fig. 6.13. Like other modulo scheduling algorithms, the modulo scheduler starts from a minimal initiation interval (MII), which is the theoretical lower bound of the II for a loop. The MII is calculated as: $MII = max(ResMII, RecMII)$. ResMII refers to resource minimal initiation interval, which is constrained by availability of resources needed by the loop. For example, if there are 4 memory operations in the loop but only 2 load/store units in the architecture, the II cannot be smaller than 2 because each iteration needs the load/store units for at least two cycles. RecMII refers to recurrence minimal initiation interval. A loop contains a recurrence if an operation has direct or indirect dependence upon the same operation from a previous iteration. It means there is a recurrence in the DDG of the loop. The existence of such a recurrence imposes constraints on MII. The recurrence operation in a new iteration cannot start until it gets all data it depends on from previous iterations. In DRESC, both ResMII and RecMII are calculated using the algorithms adapted from [26]. Readers are referred to it for extensive explanation on these concepts and algorithms.

If the scheduler fails to find a feasible schedule at certain II, it increases the II by 1 and tries again. With a larger II, more resources in the MRRG can be used for scheduling because of the modulo property. Moreover, a larger II results in a later deadline for the recurrence operations, which results in more freedom for scheduling.

Before calling the scheduler core, the algorithm has to recalculate some II-dependent data for every II. First all operations are ordered by the technique

```
procedure ModuloSchedule (DDG)
begin
 II := ComputeMII(DDG)

  while (!success) do
   op_list := SortOps(DDG, II);
   ComputeASAPandALAP(DDG, II);
   max_sched_length = AjustALAP(relax_factor);
   InitMRRG(II, max_sched_length);
   success := ModuloScheduleCore(op_list, II);
   II++;
  endwhile;
end;
```

Fig. 6.13 The outer loop of the scheduler

described in [16]. Using this algorithm, priority is given to operations on the critical path and an operation is scheduled as close as possible to both its predecessors and successors, which effectively reduces the routing length between operations. Next, the ASAP (as soon as possible) and ALAP (as late as possible) values of each operation have to be computed for each II. The ASAP and ALAP are computed using the algorithms described in [16]. The algorithms are shown in Equation 6.4 and 6.5. $Pred()$ and $Succ()$ are the predecessor and successor sets of an operation. λ refers to the latency of an operation. $\delta_{v,u}$ denotes the distance between operation v and u, which means the operation u of iteration I depends on the operation v of iteration $I - \delta_{v,u}$. V refers to all the operations in the loop.

$$ASAP_u = \begin{cases} ASAP_u = 0; & if Pred(u) = \emptyset \\ ASAP_u = max(ASAP_v + \lambda_v - \delta_{v,u} \times II); & \forall v \in Pred(u) \end{cases} \tag{6.4}$$

$$ALAP_u = \begin{cases} ALAP_u = max(ASAP_v); & \forall v \in V \\ ALAP_u = min(ALAP_v - \lambda_u + \delta_{u,v} \times II); & \forall v \in Succ(u) \end{cases} \tag{6.5}$$

It should be noted that in [16], ASAP and ALAP is only computed for MII. To obtain more accurate information, they are computed in DRESC for every II because they are II-dependent. The difference between ALAP and ASAP of an operation is called *mobility* or *slack*. It determines the time range within which an operation can be moved. This is very important information for the scheduler. If the range is too tight, the scheduler has difficulty to move the operation around to find a good schedule. Therefore, we use a parameter, called *relax_factor*, to adjust ALAP values (Equation 6.6). In this way, the mobility of an operation is relaxed to improve schedulability. The downside is that it also increases the schedule length, which translates to more overhead in prologue and epilogue. The *relax_factor* is one of the parameters that tune the scheduler. It usually ranges form 1.3 to 2.0. Additionally, the maximal schedule length is actually the maximal ALAP value of all operations

(Equation 6.7). All the moves of operations along the time axis are constrained to be less than *max_sched_length*.

$$ALAP_u = \begin{cases} ALAP_u = max(ASAP_v) \times relax\_factor; & \forall v \in V \\ ALAP_u = min(ALAP_v - \lambda_u + \delta_{u,v} \times II); & \forall v \in Succ(u) \end{cases} \quad (6.6)$$

$$max\_sched\_length = max(ALAP_v); \forall v \in V; \quad (6.7)$$

6.4.4 The Core Algorithm

As mentioned in the previous section, by using MRRG the three sub-problems (placement, routing and scheduling) are reduced to two sub-problems (placement and routing), and modulo constraints are enforced automatically. However, the problem is still more complex than a traditional FPGA P&R problem due to the modulo and asymmetric nature of the P&R space and the scarce routing resources available. In FPGA P&R algorithms, we can comfortably run the placement algorithm first by minimizing a good cost function that measures the quality of placement. After minimal cost is reached, the routing algorithm connects placed nodes. The coupling between these two sub-problems is very loose. In our case, we can hardly separate placement and routing as two independent problems. It is almost impossible to find a placement algorithm and a cost function which can foresee the routability during the routing phase. Therefore, in the core scheduling algorithm, we try to simultaneous solve the placement and routing problems in one framework. The core scheduling algorithm is described in Fig. 6.14.

For each II, the algorithm first generates an initial schedule which respects dependence constraints, but may overuse resources (1). For example, more than one operation may be scheduled on one FU in the same cycle, or a register file is used above its capacity. Finding an initial schedule itself is a difficult problem on a coarse-grained array. Fortunately, we can take advantage of the fact that the first row of the ADRES architecture is also a VLIW processor. Therefore, we can easily apply an existing modulo scheduling algorithm for VLIW [26] to initially schedule the loop only on the VLIW. This is even easier than the original VLIW modulo scheduling problem because we allow the scheduler to overuse resources in this phase.

From the initial P&R, the algorithm iteratively reduces resource overuse and tries to come up with a legal schedule in the inner loop (2). The scheduler iterates through all the operations in the sorted operation list in a reverse order. At every iteration, an operation is ripped up from the existing schedule (3). The scheduler tries to generate a random position that comprises of (t, FU_{no}). The t is generated within the time boundary computed using Equation 6.8 and 6.9, which are similar to those described in [16]. In these equations, t_v is the cycle when the operation v is scheduled. The other symbols are the same as those in Section 6.4.3. The physical FU position

```
procedure ModuloSchedulerCore(op_list, II)
begin
 InitTemperature();
 InitPlacementRouting();                (1)

 while (not scheduled) do
   for each op in op_list do
      RipUpOp(op);                              (3)
      boundary := ComputeBoundary(op);    (4)

      for i := 1 to random_pos_to_try do
         pos := GenRandomPos(pos);
         success := PlaceAndRoute(op, pos);    (5)

        if (success) then
          new_cost := ComputeCost(op);
          accepted := EvaluateNewPos(new_cost); (6)
         if (accepted) then
            break;
         else
            continue;
        endif;
      endfor;

      if (not accepted) then
        RestoreOp(op);
      else
        CommitOp(op);

      if (no overuse) then
        return success;
   endfor;
   if( StopCritieria() ) then
     return failed;
    UpdateTemperature();                    (7)
    UpdateOverusePenalty();
 endwhile;
end;
```

Fig. 6.14 The modulo scheduling algorithm core

FU_{no} is generated within the set of FUs which supports the operation. After a valid random position is generated, the scheduler tries to place and route the operation on it (5). All the nets connected to this operation are rerouted accordingly. The routing algorithm is the basic maze router [14]. A maze router is based on Dijkstra's algorithm [5] to find the path with the lowest total cost between a source node and a sink node on a routing resource graph. If there are one source and multiple sinks for a net, this problem can be modeled as a *rectilinear steiner tree* problem, which is NP-complete [9]. It suggests no polynomial-time algorithm can solve this problem exactly. In DRESC, a simpler algorithm is used. Whenever a sink of the net is reached by the maze router, the entire routed segment is added as the source for further routing. Hence, the multiple sinks are routed one by one. The routing order of sink terminals has a big impact on routing results using this method. Currently we only follow the order specified by the sink number in the net.

$$earliest\_time = \max(ASAP(u), \max_{v \in pred(u)} t_v + \lambda_v - \delta_{v,u} \times II) \tag{6.8}$$

$$latest\_time = \min(ALAP(u), \min_{v \in succ(u)} t_v - \lambda_u + \delta_{u,v} \times II) \tag{6.9}$$

If the operation is placed and routed successfully in the new position, a cost function is computed to evaluate the new placement and routing (6). The cost is computed by accumulating the costs of all used MRRG nodes incurred by the new placement and routing of the operation. The cost function of each MRRG node is shown in Equation 6.10. It is constructed by taking into account the penalty of overused resources. The cost comprises two parts. The first part is a basic cost (*base_cost*) associated with each MRRG node. The *occ* represents the occupancy of that node. The second part is the cost penalty associated with overuse, which is calculated by Equation 6.11. The *cap* refers to the capacity of that node. Most MRRG nodes have a capacity of 1, whereas a few types of nodes such as the internal node of a register file have a capacity larger than one. Therefore, if a node is used more than its capacity, a penalty arises. The *penalty* factor associated with overused resources is increased at the end of each iteration (7). We use a simple scheme to update the penalty factor (Equation 6.12). Through a higher and higher overuse penalty, placer and router will try to find alternatives to avoid congestion. However, the penalty is increased gradually to avoid abrupt increase of the overused cost that may trap solutions into local minima. This idea, called *congestion negotiation*, is borrowed from the *Pathfinder* algorithm [6], which is used to solve FPGA P&R problems.

$$cost = base\_cost \times occ + overuse\_penalty; \tag{6.10}$$

$$overuse\_penalty = \begin{cases} 0, & occ \leq cap \\ (occ - cap) \times penalty\_factor & occ > cap \end{cases} \tag{6.11}$$

$$penalty\_factor = penalty\_factor \times multi\_factor; \tag{6.12}$$

In order to help solutions to escape from local minima, we use a *simulated annealing* (SA) strategy to decide whether each move is accepted or not (5). In this strategy, if the new cost is lower than the old one, the new P&R of this operation will be accepted. On the other hand, even if the new cost is higher, there is still a chance to accept the move, depending on *"temperature"*. Possibility of acceptance is computed as: $p = \exp^{-\delta cost/temperature}$, in which $\delta cost$ is defined as cost difference after and before the move. At the beginning, the temperature is very high so that almost every move is accepted. The temperature is decreased at the end of each iteration (8). Therefore, the operation is increasingly difficult to be moved around. One key issue of simulated annealing is how the temperature is decreased. If the process cools too fast, the state is "frozen" without conducting sufficient search at every temperature. On the other hand, if the temperature is decreased too slowly, the process requires too much time. To achieve a good balance of quality and speed, we use an adaptive annealing technique [4] to update temperature as shown in Equation 6.13. In this scheme, the accept rate, which is defined as the percentage of accepted moves out of total tries, is used to select an annealing rate among several ones (0.5 to 0.95), obtained from experiments. When the accept rate is in the middle range, the temperature is decreased slowly to ensure quality by exploiting an extensive search. When the accept rate is in the higher or lower range, the temperature decreases more rapidly to speed up the scheduling process.

$$T = \begin{cases} T \times 0.5, & accept\_rate \geq 0.96 \\ T \times 0.9, & 0.8 \leq accept\_rate < 0.96 \\ T \times 0.98, & 0.15 \leq accept\_rate < 0.8 \\ T \times 0.95, & accept\_rate < 0.15 \end{cases} \tag{6.13}$$

In the end, if the stop criteria is met without finding a valid schedule (6), i.e., the scheduler can not reduce overused nodes after a number of iterations, the schedule algorithm starts with the next II. The stop criteria is computed as in Equation 6.14. *tota_cost* refers to the sum of costs of all MRRG nodes, which are calculated by Equation 6.10. *total_net* refers to the total number of nets in the data dependency graph representing the loop. When the temperature is so low that a new move is unlikely, the scheduling process stops.

$$temperature < 0.01 * (total\_cost/total\_net) \tag{6.14}$$

The live-in and live-out variables are transformed to three types of pseudo operations: REG_SRC, REG_SINK and REG_BIDIR. Therefore, the scheduler can easily deal with these variables just like normal operations except these pseudo operation can only be assigned to the VLIW RF. In addition, REG_SRC and REG_SINK cannot move along the time axis. They are fixed to cycle 0 and N (N = schedule length) respectively. The REG_BIDIR operation can be moved freely along the time axis by the scheduler like other normal operations.

According to [16], the maximum number of simultaneously live values at any cycle is an approximation of the number of registers required, which is called *MaxLive*. In the RF modeling discussed in section 6.4.2, the size of a RF is modeled as capacity of a node in MRRG. The scheduling algorithm is a P&R algorithm performed on MRRG. If a valid schedule is found on the architecture, at any time a node is ensured to be used below its capacity. Therefore, the register allocation problem of the ADRES architecture is implicitly solved by the modulo scheduling algorithm.

6.4.5 Experimental Results

We have tested our algorithm on a mesh-based ADRES instance *arch-8mem-meshplus-homo*. There are 64 FUs in the architecture. The first row of 8 FUs is also configured as a VLIW with a multi-port RF connected to the FUs. Only these FUs have access to the memory. Each of the other 56 FUs is coupled with an 8-entry distributed data RF and a predicate RF. Each FU is not only connected to the 4 nearest neighbor FUs, but also to the FUs with one hop in the same row and column (Fig. 6.15a, in which only 4×4 array is shown for convenience of drawing). In addition, there are row buses and column buses across the matrix. All the FUs in the same row or column are connected to the corresponding bus (Fig. 6.15b). There are also rich connections between FUs and RFs. An FU is not only connected to its local RF, but also to 4 RFs in diagonal directions (Fig. 6.15c). The architecture is homogeneous, which means each FU supports all the operations apart from load/store operations.

The testbench consists of 8 kernels, from typical multimedia and telecommunication applications. The *idct1* and *idct2* are vertical and horizontal loops of an 8×8 inverse discrete cosine transformation respectively. They are extracted from an MPEG-2 decoder program. The *get_block1*, *get_block2* and *get_block3* are 3 out of 6 loops from the *get_block* function of the AVC (Advanced Video Coding) decoder [35]. The function implements the interpolation for the frame prediction and is one of the most computationally intensive kernels in the AVC decoder. The *mimo_msse* is a module that computes MSSE (minimum mean square error) in the MIMO (multiple-in multiple-out) wireless application. The *mimo_matrix* refers to

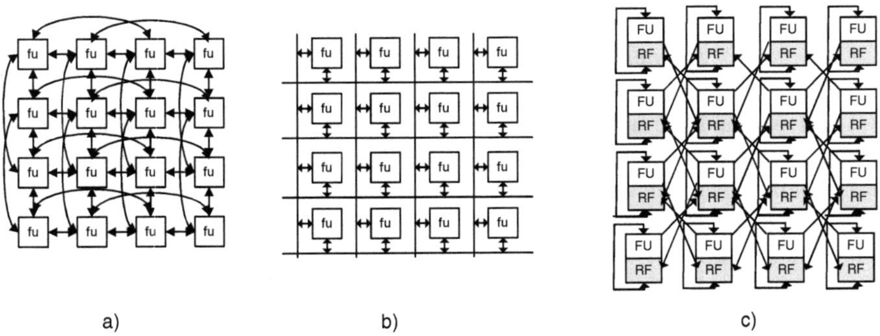

Fig. 6.15 Topology for tested architecture: a) connections between FUs; b) row and column busses; c) connections between FUs and RFs

Table 6.3 Schedule results

kernel	no. of ops	min. II	II	ins. per cycle	sched. density (excl. routing)	sched. density (incl. routing)	time (sec.)
idct1	79	2	3	26.3	41.2 %	64.6 %	99
idct2	128	2	3	42.7	66.7 %	90.1 %	340
get_block1	57	1	2	28.5	44.5 %	67.2 %	170
get_block2	86	2	3	28.7	44.8 %	74.0 %	159
get_block3	65	2	2	32.5	50.8 %	71.9 %	97
mimo_mmse	213	4	6	35.5	55.5 %	84.6 %	2845
mimo_matrix	155	3	5	31	48.4 %	76.2 %	2615
fft	79	3	4	19.8	30.9 %	75.0 %	314

a matrix calculation used in the MIMO application too. The *fft* is a radix-4 1024-point Fast Fourier Transformation program. All the kernels have been transformed at source-level to make them better suited for pipelining.

The scheduling results are shown in Table 6.3. The second column refers to the total number of operations within pipelined loops. The minimal initiation interval (MII) is the lower bound of achievable II, constrained by resources and recursive dependence, whereas the initiation interval is the value actually achieved after scheduling. The instructions per cycle (IPC) reflects how many operations are executed in one cycle on average. The sixth and the seventh columns are scheduling density excluding and including routing operations. The scheduling density is equal to $IPC/No.of FUs$. It reflects the utilization of all FUs. The last column is the CPU time to compute the schedule on a Pentium M 1.4 GHz PC.

The experiments show that the scheduler can handle a wide range of loops with different amount of operations. The achieved IPC for the tested benchmarks ranges from 19.8 to 42.7. For *idct2*, the IPC is especially high because its data dependency is mainly local. Therefore, the rich local connections available in ADRES can serve the routing task well. For other kernels, the resource utilization is usually around 50 % and the IPC is about 30. The scheduling time of kernels is quite long compared with normal software compilation techniques, especially for large kernels like *mimo_msse*. It takes the scheduler several minutes or more to find a valid schedule.

6.5 Co-Simulator Generation

The co-simulator of the ADRES architecture belongs to a category of so-called compiled simulators, in which a compiled application on the target architecture is translated back to the C code that simulates the execution of the assembly/machine code. As shown in Fig. 6.7, the compiled code and the XML architecture description are used as inputs to generate C code for the simulator. On one hand, the compiled code contains much scheduling information. On the other hand, the simulator generator checks the architecture description to derive other information, e.g., how the input ports of an FU are connected to other components through multiplexors. Combining this information, the simulator generator is able to emit C code to emulate the execution of compiled code on the target ADRES architecture. Simulation code for the VLIW part can be similarly generated. A scheduled operation on the VLIW contains information of FU, cycle and latency. A control block in the scheduled VLIW code is translated to a piece of C code that simulates the execution of these operations. At the end of each control block, the total number of cycles and other information are collected. After the simulation code for the VLIW and the array is generated and compiled by the host compiler, we can simply run the simulator to verify results and collect various statistics.

The biggest advantage of a compiled simulator is its performance [20]. It usually has much higher simulation speed than an interpretive simulator. This is specially useful for simulating big applications or for simulating a long period. However, there are some limitations with our compiled simulator. First, it doesn't give users enough interaction during simulation. It is very hard to support debugging features as *step*, *breakpoint*, *continue* commands or processor status monitoring. Second, the compiled simulator lacks flexibility to extend the simulator to support more architecture. Finally, the size of the generated simulator grows rapidly with large applications and more complex architecture. We are therefore in the process of developing a full interpretive simulator for the ADRES architecture to overcome these shortcomings and to complement the existing simulator.

6.6 Mapping an Entire Application: A Case Study

In this section, we presents how a real-life multimedia application, MPEG (Motion Picture Expert Group)-2 decoder, is mapped onto the ADRES architecture by utilizing the DRESC design flow.

MPEG-2 is a video compression standard widely used in digital video broadcasting, DVD video, etc. The block diagram of the MPEG-2 decoder is shown in Fig. 6.16. It consists of several major functional blocks: Huffman/VLD (Variable Length Decoding), dequantization, motion compensation, IDCT and summing blocks. As other multimedia applications, the MPEG-2 decoder requires very high computation power. Most execution time is spent on a few functions or kernels, which usually have high inherent parallelism. It makes the MPEG-2 decoder a good candidate to be mapped onto the ADRES architecture.

6.6.1 Profiling and Partitioning

We use a C implementation from the MPEG Software Simulation Group (MSSG) [19] as a starting point. The code is highly optimized for processors and included as a part of MediaBench [15]. The program comprises 21 files and around 10,000 lines of C code.

The MPEG-2 program is profiled by compiling and executing it on the VLIW part of the ADRES array. This can give us more accurate and relevant information than profiling on other processors. The time breakdown of all the functions is shown in Fig. 6.17 using a bitstream *mobl_015.m2v* as input (Table 6.5). In the

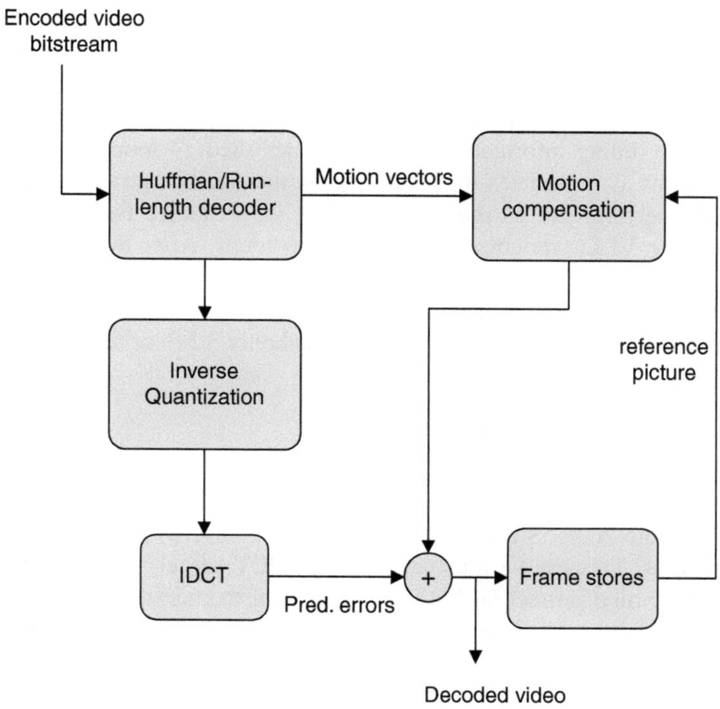

Fig. 6.16 Block diagram of the MPEG-2 decoder

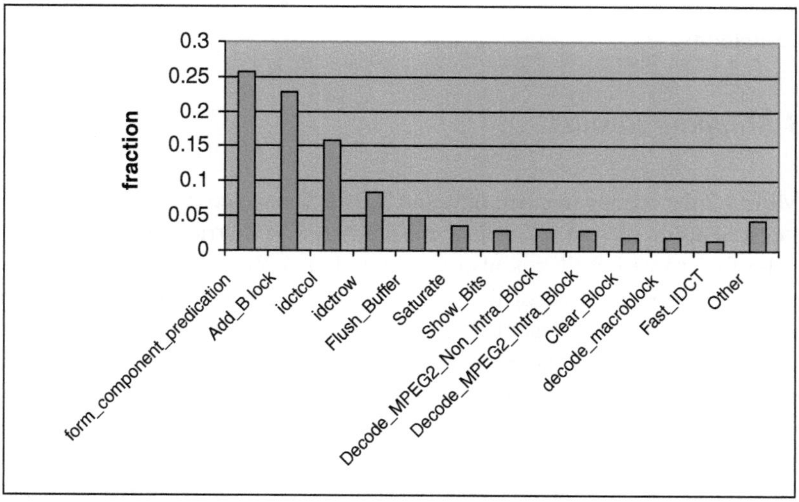

Fig. 6.17 Profiling results of the MPEG-2 decoder

figure, the top 12 functions account for the most execution time of the MPEG-2 decoder, whereas all other functions only account for 4.2 % of total execution time. Therefore, we only need to focus on a few important functions to exploit loop-level parallelism on the ADRES architecture.

From the profiling information, we easily identified 14 loops from the original applications as candidates for pipelining on the ADRES array. In addition, to improve the performance as much as possible, we extracted two dequantization loops from the VLD (variable length decoding) loops. After the transformation, two pipelineable loops are created for dequantization though at the expense of some extra operations. All the 16 candidate loops are listed in Table 6.4. These loops account for 85.4 % of the total execution time and only 3.3 % of the total code size.

6.6.2 Source-Level Transformations

Out of the 16 loops we identified, only a few of these loops can be immediately mapped onto the ADRES array. For the others we have to perform source-level transformations. This can be illustrated on the IDCT kernel. First, the inner loop body is in the form of a function that we have to inline to make the loop pipelineable. Second, all the shortcut computation is removed to make the code more regular. Finally, the *idct* is applied to a macroblock, which usually has 6 8 × 8 blocks in an MPEG-2 video stream. This increases the iteration count of each *idct*, greatly reducing the pipelining overhead (prologue and epilogue).

The above example shows that the required transformations are very diverse in a real-life application. It requires experience of the designer to figure out the appropriate ones. Fortunately, since we only need to focus on a few loops instead of the entire application, design efforts are limited to C-to-C rewriting for these loops while bearing the target reconfigurable array in mind. In this particular case, it took us less than one week to rewrite and verify the C code to reflect all the required transformations.

6.6.3 Mapping Results

The MPEG-2 decoder is mapped onto the ADRES instance *arch-meshplus-8mem-hete2* described in Section 6.7. It is a mesh-based 8 × 8 array with heterogeneous FUs. The first row of 8 FUs is also configured as a VLIW with a multi-port RF connected to the FUs. Each of the other 56 FUs is coupled with an 8-entry distributed data RF and a predicate RF. Each FU is not only connected to the 4 nearest neighbor FUs, but also to the FUs with one hop in the same row and column. In addition, there are row and column buses and some other local connections. 16 FUs out of 64 FUs are dedicated multipliers. The entire design took about one person-week to finish starting from the software implementation. The main efforts are spent on identifying pipelineable loops and source-level transformations. The modulo scheduling results for all the kernels are listed in Table 6.4. The second column of the table is the number of operations in the loop body. Initiation interval

Table 6.4 Scheduling results for kernels

kernel	no. of ops	II	IPC	stages	sched. time (secs)
clear_block	8	1	8	3	5.4
form_comp_pred1	41	2	20.5	7	146
form_comp_pred2	13	1	13	6	12.2
form_comp_pred3	57	2	28.5	11	227
form_comp_pred4	33	2	16.5	6	73
form_comp_pred5	54	2	27	11	209
form_comp_pred6	30	2	15	5	51
form_comp_pred7	67	3	22.3	6	211
form_comp_pred8	43	2	21.5	8	137
saturate	78	3	26	13	769
idct1	79	3	26.3	6	325
idct2	128	3	42.7	14	415
add_block1	48	2	24	9	136
add_block2	44	2	22	5	45
non_intra_dequant	20	1	20	14	47
intra_dequant	18	1	18	12	53

(II) means a new iteration can start at every II cycle. Instruction-per-cycle (IPC) reflects the parallelism. Stages refer to total pipeline stages which have an impact on prologue/epilogue overhead. Scheduling time is the CPU time to compute the schedule on a Pentium M 1.4GHz/Linux PC.

After all the loops are successfully mapped, we obtained the configuration contexts required for each loop. Basically, the II is equal to the number of configuration contexts required. The characteristics of the MPEG-2 decoder can help to reduce the configuration RAM requirements. In the MPEG-2 video, there are three types of frames, I-, P- and B-Frame. Different frames call different kernel sequences. I-Frames use only 6 loops, while P- and B-Frames call 10 and 14 loops respectively. A frame lasts normally more than 20ms, while loading a configuration context takes only μs. Hence, the kernels needed by a frame can be loaded before the frame starts without incurring much overhead. In MPEG-2, the maximal number of contexts required is constrained by the B-Frame, which uses 29 contexts. In case the configuration RAM is not big enough to accommodate all contexts, we should apply kernel scheduling techniques, still ongoing research, to minimize reconfiguration overhead.

6.6.4 Simulation Results and Comparisons with VLIW Processor

To test the mapped MPEG-2 decoder, we use several different video bitstreams, whose characteristics are listed in Table 6.5.

Since we view the coarse-grained reconfigurable architecture as a promising alternative competing with other established programmable architectures, we compared our results to those on the VLIW, which is widely used in DSP and multimedia applications and has mature compiler support. The IMPACT framework is used again as both compiler and simulator to obtain results for the VLIW,

Table 6.5 Characteristics of tested video bitstreams

bitstream	resolution	bit-rate	total frame
mobl_15.m2v	352×240	1.5Mb/s	450
tens_40.m2v	704×480	4.0Mb/s	450
cact_60.m2v	704×480	6.0Mb/s	450
flwr_80.m2v	704×480	8.0Mb/s	450

where aggressive optimizations are enabled. The tested VLIW has the same configuration as the first row of the tested ADRES architecture. Both the simulation results of the mapped MPEG-2 decoder and the comparisons are shown in Table 6.6. The total number of instructions on ADRES exclude those generated for routing purposes, which are simple copy instructions. The decoding frame rate is calculated by assuming that both the ADRES architecture and the VLIW run at 100 MHz.

From the results, we can see that the ADRES architecture can achieve about 12 times speed-up for the kernels, while the overall application speed-up is up to 5. There are some variations between different video streams because the time distribution of the kernels are slightly different. Overall ADRES executes more operations due to the transformation and optimization techniques. For example, loop coalescing added a few operations to calculate indices in the loop. The hyperblock formation tries to execute two branches at the same time, though operations in one branch will be nullified. At only 100 MHz, the ADRES architecture can decode 104 frames of lower quality video (352 × 240). The higher quality video (704 × 480) can also be decoded almost at real-time.

It should be noted that currently the simulator cannot simulate the memory hierarchy. Therefore, the real performance would be somewhat lower due to the cache misses.

Table 6.6 Simulation results and comparisons with VLIW processor

	mobl_15.m2v		tens_40.m2v	
	VLIW (IMPACT)	ADRES	VLIW (IMPACT)	ADRES
total ops ($\times 10^9$)	3.37	3.44	13.2	13.5
total cycles ($\times 10^8$)	20.00	4.31	78.1	15.8
frames/sec	22.5	104.4	5.8	28.5
speed-up/kernels		12.3		12.1
speed-up/overall		4.64		4.94
IPC(excl. kernels)		2.03		1.87
	cact_60.m2v		flwr_80.m2v	
	VLIW (IMPACT)	ADRES	VLIW (IMPACT)	ADRES
total ops ($\times 10^{10}$)	1.45	1.53	1.44	1.49
total cycles ($\times 10^9$)	8.52	1.87	8.55	2.12
frames/sec	5.3	24.3	5.3	21.2
speed-up/kernels		11.75		12.0
speed-up/overall		4.56		4.02
IPC(excl. kernels)		1.81		1.80

6.7 Architecture Exploration Experiments

Since ADRES is a template for a CGRA, we can easily derive an architecture instance using the XML-based architecture description language and the DRESC tools are automatically applicable to the instance. In this section, we try to explore the effects of varying some important architectural aspects: heterogeneous resources, interconnection patterns, memory ports and distributed register files. Though this approach cannot produce an optimal architecture instance, it can point to better architecture design and is an easier way to explore the vast design space of the ADRES architecture template.

In the following exploration experiments, we use the same benchmark set described in Section 6.4.5. In every experiment, we vary only one architectural parameter and fix others. To reduce the impact of the inherent randomness of our scheduling heuristics, each kernel is scheduled for 5 times with different random seeds and the best result is selected. The schedulability of an architecture instance is measured as both the achieved II and the total overused nodes in the MRRG graph when II is 1 less than the achieved II. II is used as the primary metric. However, two architectures may achieve the same II for a given benchmark but with a big difference of easiness in finding the schedule. Hence, the total amount of overused nodes in the MRRG graph is used as the secondary metric to more accurately measure the schedulability of an architecture. Since overused node is always 0 when a valid schedule is found, it is counted at II-1 where a valid schedule cannot be found.

6.7.1 Experiments with Heterogeneous Resources

In a heterogeneous architecture, there are different types of FUs in the array, e.g., ALU and MUL. The number of each type of FU and how they are distributed in the array are two important parameters of heterogeneous architectures. Since the possible combinations of these parameters are virtually unlimited, here we are only able to experiment with several cases. We assume there are only two different FUs: ALU and MUL. All the instances are 8×8 arrays (Fig. 6.18). Their interconnection topology and amount of memory ports are the same as those described in Section 6.4.5. The instances of *arch-8mem-meshplus-hete1*, *arch-8mem-meshplus-hete2* and *arch-8mem-meshplus-hete3* all have 48 ALUs and 16MULs but distributed in different ways, while the instance *arch-8mem-meshplus-hete4* only has 8 MULs together with 56 ALUs.

The experimental results show there are no significant difference between these heterogeneous instances. In most benchmarks, the same IIs are achieved for all the instances. The schedulability of *arch-8mem-meshplus-hete1*, *arch-8mem-meshplus-hete2* and *arch-8mem-meshplus-hete4* are pretty similar, while the *arch-8mem-meshplus-hete3* is slightly worse than the other three instances. Interestingly, even though *arch-8mem-meshplus-hete4* has only half the multipliers compared with

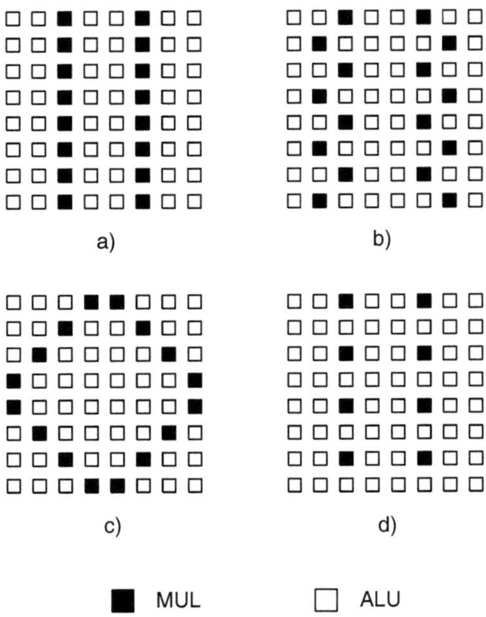

Fig. 6.18 Heterogeneous ADRES instances: a) *arch-8mem-meshplus-hete1*; b) *arch-8mem-meshplus-hete2*; c) *arch-8mem-meshplus-hete3*; d) *arch-8mem-meshplus-hete4*

other instances, it still performs quite well. The reason is probably because there are only a small portion of multiplication operations in typical kernels, and hence 8 multipliers are enough. *arch-8mem-meshplus-hete3* is slightly worse because its multipliers are not as evenly distributed as other instances.

6.7.2 Experiments with Interconnection Topologies

The interconnection topology is one of the most important parameters of the ADRES architecture template. It largely determines the schedulability of a given

Table 6.7 Scheduling results of different heterogenous instances

	hete1		hete2		hete3		hete4	
	II	overuse (II - 1)	II	overuse (II - 1)	II	overuse (II - 1)	II	overuse (II - 1)
idct1	3	10	3	10	3	16	3	10
idct2	3	69	3	68	4	1	3	72
get_block1	3	1	3	1	3	1	3	1
get_block2	3	8	3	5	3	10	3	5
get_block3	2	63	2	65	2	67	2	65
mimo_mmse	6	25	6	23	6	25	6	23
mimo_matrix	5	19	5	16	5	13	5	17
fft	4	4	4	3	4	4	4	4

ADRES instance. Generally, kernels are easier scheduled with richer interconnections. However, richer interconnections come with costs such as wider multiplexors, more wires, more configuration bits, which translate to large silicon area and higher power consumption. Even with the same amount of interconnection resources, we can expect there are differences among topologies. Choosing a good topology is an essential step of the architecture exploration. Since the ADRES architecture not only supports regular topology, but can also handle irregular ones, the possible topologies are virtually unlimited. Here we only tested several regular topologies (Fig. 6.19). The first one is the basic mesh topology with the 4 nearest-neighbour connections between FUs. The second one is the enhanced mesh topology *meshplus*, which is widely used in previous experiments. An FU is not only connected to 4 nearest neighbours, but also with FUs within one hop. The last one is a MorphoSys-like topology [32]. An 8×8 array is divided into 4 4×4 tiles. Within each tile, an FU is connected to all the FUs in the same row and the same column. Additional interconnections are needed in the boundaries of tiles. Fig. 6.19 only shows interconnection in a 4×4 array for convenience of drawing. All these topologies are enhanced with busses and diagonal connections to the register files (see Figure 6.18b).

Three ADRES instances are specified based on these three topologies. They are *arch-8mem-mesh-hete2*, *arch-8mem-meshplus-hete2* and *arch-8mem-morphosys-hete2*. Apart from the topology, other aspects of these architectures are identical. They all have 8 memory ports and heterogeneous FUs, which consists of 48 ALUs and 16 MULs and are distributed in the same way as in Figure 6.18b. Different topologies result in different multiplexor area and total configuration bits. Table 6.8 lists their characteristics. The multiplexor area is estimated using the Module Compiler synthesis tool, while the configuration bits are calculated according to Equation. Among three topologies, *arch-8mem-mesh-hete2* requires the least area and configuration bits since its topology is the simplest one. *arch-8mem-morphosys-hete2* has the highest cost in both multiplexor area and configuration bits. Other metrics such as power and delay are not studied yet though we believe they are also greatly impacted by the topology.

Table 6.9 shows the scheduling results for these different topologies. Again we use II and overused MRRG nodes at II-1 to measure the schedulability of the given

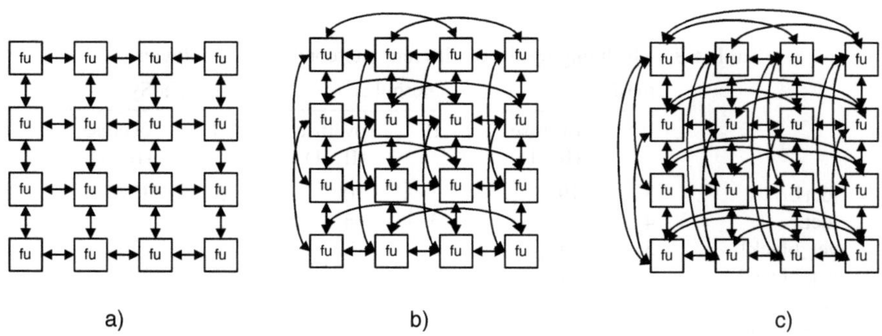

Fig. 6.19 Several interconnection topologies: a) *mesh*; b) *meshplus*; c) *MorphoSys*

Table 6.8 Characteristics of architectures with different topologies

architecture	Total config. bits	mux area (mm^2)
arch-8mem-mesh-hete2	2339	0.277
arch-8mem-meshplus-hete2	2415	0.370
arch-8mem-morphosys-hete2	2562	0.426

architecture instances. As expected, *arch-8mem-mesh-hete2* has the worst schedulability among the three topologies: achieved IIs are bigger for 4 kernels. *arch-8mem-meshplus-hete2* and *arch-8mem-morphosys-hete2* have pretty similar performance. Both architectures find schedules with the same II for all the kernels. The overused nodes in II - 1 are also very close. Since *arch-8mem-morphosys-hete2* requires considerably more silicon area and configuration bits, the *meshplus* topology is better than the MorphoSys topology.

6.7.3 Experiments with Memory Ports

Memory bandwidth is one of the greatest constraints for the ADRES architecture. In previous experiments, we assume that all the architecture instances have 8 memory ports in the first row. Nonetheless, it is not cheap to implement such memory subsystem. In this section, we experiment with ADRES instances with different number of memory ports. 4 ADRES instances are constructed. They are *arch-8mem-meshplus-hete2*, *arch-6mem-meshplus-hete2*, *arch-4mem-meshplus-hete2* and *arch-2mem-meshplus-hete2*, which have 8, 6, 4, 2 memory ports respectively. Each memory port corresponds to an FU that can execute load/store operations. The memory ports are located as shown in Fig. 6.20. Other architectural aspects are identical for these instances (see previous sections).

The scheduling results are shown in Table 6.10. Here we only compare the achieved IIs for each instance because they are enough for this particular comparison. As expected, *arch-8mem-meshplus-hete2* has the best results with the highest memory bandwidth. *arch-6mem-meshplus-hete2* is only slightly worse by finding the same IIs for 7 out of 8 kernels. When the number of memory ports decreases

Table 6.9 Scheduling results of different interconnection topologies

	mesh		meshplus		MorphoSys	
	II	overuse (II - 1)	II	overuse (II - 1)	II	overuse (II - 1)
idct1	3	19	3	10	3	17
idct2	4	4	3	68	3	71
get_block1	3	4	3	1	3	1
get_block2	3	17	3	5	3	7
get_block3	3	2	2	65	2	61
mimo_mmse	7	1	6	23	6	13
mimo_matrix	6	3	5	16	5	23
fft	4	9	4	3	4	1

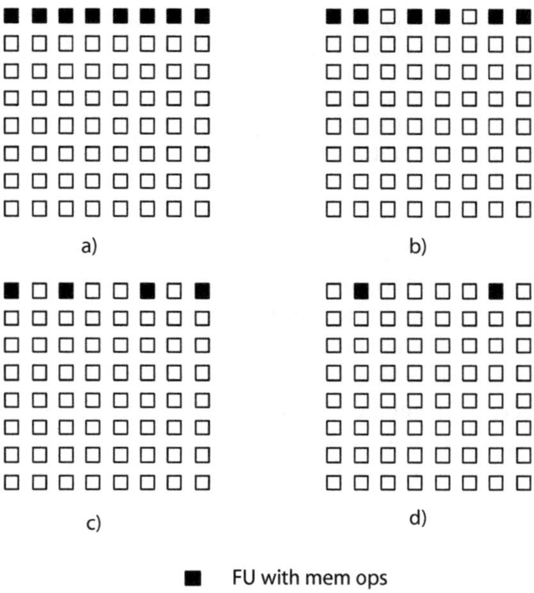

Fig. 6.20 The position of memory ports: a) arch-8mem-meshplus-hete2; b) arch-6mem-meshplus-hete2; c) arch-4mem-meshplus-hete2; d) arch-2mem-meshplus-hete2

further, the performance of scheduled kernels degrades significantly. For kernels such as *fft*, which contain many memory accesses in the loop body, II is almost tripled, which translates to only 1/3 performance of the *8mem* and *6mem* cases. For kernels such as *mimo_mmse* and *mimo_matrix*, the performance doesn't drop so much because they are not memory-intensive loops. Generally, with only 4 or 2 memory ports, a big ADRES array containing 8×8 cannot be efficiently utilized and the parallelism cannot be fully exploited.

6.7.4 Experiments with Distributed Register Files

In the ADRES architecture, register files (RF) are used to store intermediate data. In the abstract architecture representation, they can be viewed as a kind of routing

Table 6.10 Scheduling results of different memory ports

	II			
	8mem	6mem	4mem	2mem
idct1	3	3	4	8
idct2	3	3	4	8
get_block1	3	3	3	4
get_block2	3	3	3	4
get_block3	2	3	4	7
mimo_mmse	6	6	7	8
mimo_matrix	5	5	6	8
fft	4	4	6	11

resource that direct data along the time axis (see Section 6.4.2). Apart from the multi-port VLIW RFs shared between the array and the VLIW, there might be distributed RFs in the array. They usually account for a major portion of total costs because RFs themselves occupy considerable area and more RFs require more configuration bits to provide the address for each port. Wider configuration memory not only increases area but also consumes more energy. Additionally, more RFs also introduce more multiplexors and wires. How to reduce the amount of RF without hurting the performance is an important issue for the architecture exploration.

In previous experiments, fully distributed RFs are used, i.e., each FU is accompanied with a data and a predicate RF. By examining the scheduled kernels, we discovered that these RFs are not efficiently used. Typically only 1/5 of RF ports are accessed at any cycle. This indicates an opportunity for reducing the number of the distributed RFs. Here we compared three ADRES instances. *arch-8mem-meshplus-hete2* is widely used in previous experiments, and has 56 data and 56 predicate RFs respectively (Fig. 6.21a, only 4×4 array is shown). Each FU except those in the first row is coupled with a data and a predicate RF. An FU is not only connected to

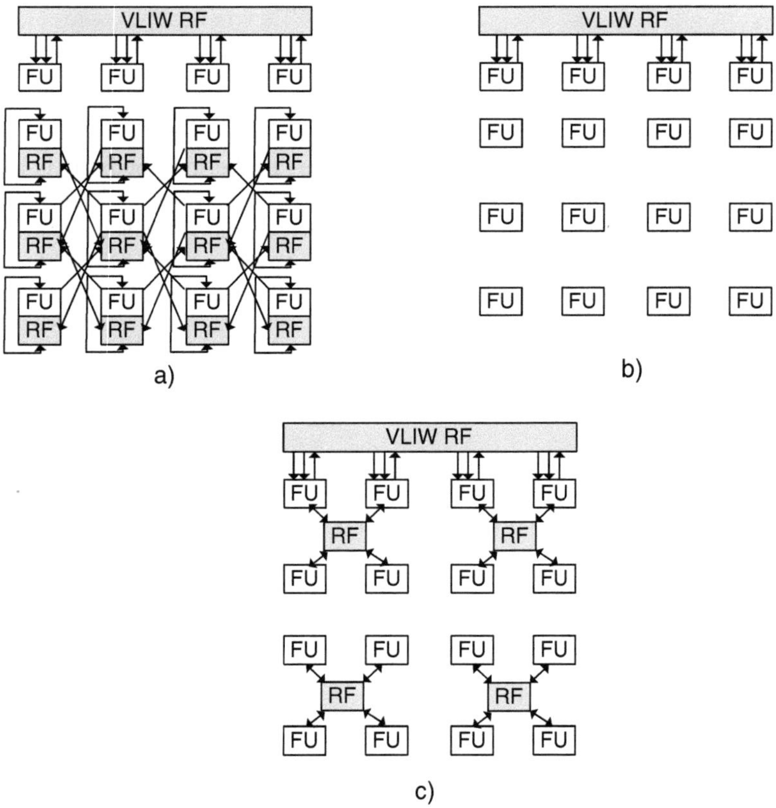

Fig. 6.21 Positions of distributed RFs and their connections to FUs: a) *arch-8mem-meshplus-hete2*; b) *arch-8mem-meshplus2-hete2*; c) *arch-8mem-meshplus3-hete2*

Table 6.11 Characteristics of architectures with different distributed RFs

architecture	total config. bits	2R1W 32×8 RF	1R1W 1×8 RF	mux area (mm^2)
arch-8mem-meshplus-hete2	2415	56	56	0.370
arch-8mem-meshplus2-hete2	1337	0	0	0.231
arch-8mem-meshplus3-hete2	1701	16	16	0.313

its local RF but also 4 RFs in diagonal directions. *arch-8mem-meshplus2-hete2* does not have any distributed RF (Fig. 6.21b). *arch-8mem-meshplus3-hete2* possesses 16 data RFs and 16 predicate RFs, which are connected to 4 neighbour FUs in a way shown in Fig. 6.21c.

Table 6.11 lists characteristics of these ADRES instances. The second column is the total configuration bits required for each instance. With the most RFs, *arch-8mem-meshplus-hete2* requires the most configuration bits. *arch-8mem-meshplus2-hete2* needs slightly more than half of the configuration bits of *arch-8mem-meshplus-hete2*. *arch-8mem-meshplus3-hete2* uses a little more than *arch-8mem-meshplus2-hete2* but still significantly less than *arch-8mem-meshplus-hete2*. The third and fourth columns list the number of distributed data and predicate RFs in each instance. The last column shows the silicon area required for multiplexors, which are different for each instance because of different amount and connection of RFs. It should be noted that the total area comparisons among these instances should not only include multiplexor area, but also RFs area, configuration memory area. For the last two, we are not able to provide data due to lack of a reliable area model.

Table 6.12 presents the scheduling results on these ADRES instances with different RFs. As expected, *arch-8mem-meshplus-hete2* performs best because it features much more RFs than the other two. *arch-8mem-meshplus2-hete2* has the worst results because it does not have distributed storage capability at all. For the biggest kernel, *mimo_mmse*, the scheduler even cannot find a valid schedule no matter what the II is. It indicates that for some kernels the RFs may be the indispensable parts. The schedulability of *arch-8mem-meshplus3-hete2* is in between of the other two instances. Interestingly, with much less resources than *arch-8mem-meshplus-hete*, its performance is not considerably lower. Only 3 out of 7 kernels have lower

Table 6.12 Scheduling results of instances with different RFs

	meshplus		meshplus2		meshplus3	
	II	overuse (II - 1)	II	overuse (II - 1)	II	overuse (II - 1)
idct1	3	10	3	32	3	11
idct2	3	68	5	1	4	2
get_block1	3	1	3	5	3	2
get_block2	3	5	3	46	3	13
get_block3	2	65	3	5	2	67
mimo_mmse	6	23	failed		7	27
mimo_matrix	5	16	7	15	6	1
fft	4	3	4	18	4	3

performance. Some of these kernels such as *idct2* and *mimo_matrix* are actually very close to finding a valid schedule with lower IIs. We believe its schedulability can be improved with some enhancements of the architecture, e.g., more local interconnections. Meanwhile, its costs can still remain much lower than those of *arch-8mem-meshplus-hete2*. It will be one of our future research topics.

6.8 Conclusions and Further Research

Coarse-grained reconfigurable architectures (CGRAs) have been emerging as a potential architecture for future embedded systems in recent years. However, there are still many outstanding issues that prevent existing CGRAs from going mainstream. One main problem is lack of a good design methodology in order to compete with other high-performance programmable architectures such as VLIW-based DSPs. The methodology should be able to exploit high performance and efficiency on CGRAs while still provides a software-like design experience.

To address this issue, this work presents a solution package that combines both a novel CGRA template, ADRES, and a C compiler framework, DRESC. Unlike other CGRAs, the ADRES architecture tightly couples a VLIW processor and a reconfigurable array by providing two virtual views on the same physical resources. Kernels (loops) of an application are mapped onto the array in a highly parallel way, whereas the remaining code is mapped onto the VLIW processor by exploiting modest instruction-level parallelism. This unique feature brings many advantages, including improved performance, ease-of-programming, lower communication costs and resource sharing.

The ADRES architecture is a very flexible template. An ADRES instance can be specified by an XML-based architecture description language. Architectural aspects such as number of resources, interconnection topology and operation set supported by each FU can be easily specified by the description language. The DRESC compiler is automatically retargetable within the ADRES template. This opens the door to research such as architecture exploration for domain-specific ADRES architectures.

The DRESC framework is centered around a novel modulo scheduling algorithm. It is capable of mapping a loop in a pipelined way onto a reconfigurable array. The algorithm can map arbitrary size loops onto a fixed size CGRA by making use of multiple reconfiguration contexts at run-time. The modulo scheduling problem for CGRAs is a combination of placement, scheduling and routing in a modulo constrained space. The complexity of the problem is much higher than the related domains. The proposed algorithm can solve these subproblems simultaneously in reasonable time. To the best of our knowledge, this is the first algorithm to solve this kind of problem.

Experiments have been done by mapping both small kernels and a complete application, MPEG-2 decoder, onto the ADRES architecture. The experiments show 30.9 %–66.7 % resource utilization and high parallelism with 19.8–42.7 instruction-per-cycle (IPC) for the given kernel set. For the MPEG-2 decoder, the 8×8 ADRES instance achieves up to 12 times speed-up for kernels and 5 times for the overall performance over a VLIW processor of 8 FUs. With the C-based design methodology, the mapping effort is comparable to software development. Our experience reveals that the MPEG-2 decoder can be mapped within 1 week starting from a software implementation. Main efforts are spent on identifying and transforming loops.

The architecture exploration is investigated in this work. The impact of different FU distributions, topologies, memory ports and distributed register files are evaluated. First, we discovered that the distribution patterns of heterogeneous FUs does not make a significant difference as long as the distribution is more or less even. Second, the interconnection topology has an important impact on cost and performance. The enhanced mesh topology, *meshplus*, has a good balance of performance and costs. Third, the memory bandwidth is one of the most significant constraints for the ADRES architecture. With less available memory ports, the performance degrades considerably for most kernels. Finally, distributed register files account for a major portion of total costs in the ADRES architecture. The distributed register files can be largely reduced without sacrificing much performance.

Many opportunities exist to perform further research around this work. Research to overcome the current limitations has the highest priority:

- Limited memory bandwidth is a pressing issue for the ADRES architecture. As we see in Section 6.7.3, the performance degrades significantly when the amount of memory ports decreases. As designing a memory subsystem with multiple ports is not a cheap option, we should study other ways to increase memory bandwidth. Two possible solutions are single-instruction-multiple-data (SIMD) access and distributed memory blocks. SIMD access packs multiple variables of small data types to a bigger one. Hence, one load/store can effectively replace several ones. With distributed memory blocks, multiple data structures can be assigned to different blocks and accessed in parallel so that the effective memory bandwidth is increased multi-fold. Both approaches can be combined. Nonetheless, both approaches have serious constraints on possible data layout. They require considerable architectural changes and compiler improvements over the existing approach.
- The current compiled simulator has many limitations compared to an interpretive simulator. It is hard to include simulation of memory hierarchy and other architectural features so that the simulation results may not be very accurate. It is also very difficult for an inexperienced user to debug an application mapped onto ADRES due to lack of interactivity. Therefore, an interpretive simulator is highly desirable though its simulator speed is much lower. The simulator is expected to be retargetable and can be generated automatically from the architecture description.

In longer term, we expect the following topics are interesting to further extend the capability of the ADRES architecture:

- Currently, the ADRES is essentially a single-processor architecture though it can exploit high degree of parallelism from the pipelineable loops. To further improve the performance and reduce power consumption, it would be interesting to exploit task-level parallelism (TLP) on top of instruction-level parallelism (ILP) and loop-level parallelism (LLP) that have already been exploited by ADRES/DRESC. Multi-task or multi-thread programming has been studied for decades. Many architecture, language and compilation techniques have been developed in past. The ADRES architecture can leverage these ideas and techniques to build a multiple-processor based coarse-grained reconfigurable architecture. Different levels can utilize different programming models. For example, multi-thread programming can be used at task-level, while our modulo scheduling technique can still be used at loop-level within a task.

- The ADRES architecture relies on predicate and if-conversion to handle control-flow. Basically, multiple branches of a control-flow are executed simultaneously. Only one of them produces valid results in the end while other branches are nullified. In our experiments with AVC encoder/decoder, we found that modern multimedia applications feature rich control-flow even inside loops. All the AVC encoder/decoder loops are almost full of *if-else-if* constructs. In many cases, the control flows are too wide and deep to efficiently apply if-conversion. Hence, we should find a way to improve the ADRES architecture so that it not only handles dataflow and limited control-flow but also more complex control-flow efficiently.

References

1. V. Agarwal, S. W. Keckler, and D. Burger. The effect of technology scaling on microarchitectural structures. Technical report, University of Texas at Austin: Tech Report TR2000-02, 2000.
2. C. Akturan and M. F. Jacome. CALiBeR: A software pipelining algorithm for clustered embedded VLIW processors. In *Proc. of International Conference on Compute-Aided Design (ICCAD)*, pages 112–118, 2001.
3. L. Benini, D. Bruni, M. Chinosi, R. Zafalon, C. Silvano, and V. Zaccaria. A power modeling and estimation framework for VLIW-based embedded systems. *ST Journal of System Research*, 3(1):110–118, Apr. 2002.
4. V. Betz, J. Rose, and A. Marguardt. *Architecture and CAD for Deep-Submicron FPGAs*. Kluwer Academic Publishers, 1999.
5. E. Dijkstra. A note on two problems in connexion with graphs. *Numer. Math.*, 1:269–271, 1959.
6. C. Ebeling, L. McMurchie, S. Hauck, and S. Burns. Placement and routing tools for the Triptych FPGA. *IEEE Trans. on VLSI*, 3:473–482, Dec. 1995.
7. A. Fauth, J. Van Praet, and M. Freericks. Describing instruction set processors using nML. In *Proc. of Design Automation Conference (DAC)*, pages 503–507, 1995.
8. M. M. Fernandes, J. Llosa, and N. P. Topham. Distributed modulo scheduling. In *Proc. of International Symposium on High Performance Computer Architecture (HPCA)*, pages 130–134, 1999.
9. M. Garey and D. Johnson. The rectilinear steiner tree problem is npcomplete. *SIAM Journal Appl. Math.*, 32:826–834, 1977.
10. G. Hadjiyiannis, S. Hanono, and S. Devadas. ISDL: An instruction set description language for retargetability. In *Proc. of Design Automation Conference (DAC)*, pages 299–302, 1997.
11. R. Hartenstein, M. Hertz, Th. Hoffmann, and U. Nageldinger. Mapping applications onto recon?gurable kressarrays. In *Proc. of Field Programmable Logic and Applications (FPL)*, pages 385–390, Aug. 1999.

12. The IMPACT group. http://www.crhc.uiuc.edu/impact.
13. M. S. Lam. Software pipelining: An effecive scheduling technique for VLIW machines. In *Proc. of ACM SIGPLAN '88 Conference on Programming Language Design and Implementation*, pages 318–327, 1988.
14. C. Y. Lee. An algorithm for path connections and its applications. *IRE Trans. Electron. Comput.*, EC=10:346–365, 1961.
15. C. Lee, M. Potkonjak, and W. H. Mangione-Smith. Mediabench: A tool for evaluating and synthesizing multimedia and communicatons systems. In *Proc. of International Symposium on Microarchitecture*, pages 330–335, 1997.
16. J. Llosa, E. Ayguade, A. Gonzalez, M. Valero, and J. Eckhardt. Lifetime-sensitive modulo scheduling in a production environment. *IEEE Transactions on Computers*, 50(3):234–249, 2001.
17. R. Maestre, F. J. Kurdahi, M. Fernández, R. Hermida, N. Bagherzadeh, and H. Singh. A framework for recon?gurable computing: Task scheduling and context management. *IEEE Trans. on VLSI Systems*, 9(6):858–873, Dec. 2001.
18. B. Mei, S. Vernalde, D. Verkest, H. De Man, and R. Lauwereins. Exploiting loop-level parallelism for coarse-grained recon?gurable architecture using modulo scheduling. In *Proc. of Design, Automation and Test in Europe (DATE)*, pages 296–301, 2003.
19. MPEG software simulation group. http://www.mpeg.org/mpeg/mssg.
20. A. Nohl, G. Braun, O. Schliebusch, R. Leupers, H. Meyr, and A. Hoffmann. A universal technique for fast and ?exible instruction-set architecture simulation. In *Proc. of ACM/IEEE Design Automation Conference (DAC)*, pages 62–67, 2002.
21. E. Nystrom and A. E. Eichenberger. Effective cluster assignment for modulo scheduling. In *Proc. of International Symposium on Microarchitecture (MICRO)*, pages 103–114, 1998.
22. *PACT XPP Technologies*, 2005. http://www.pactcorp.com.
23. D. A. Patterson and J. L. Hennessy. *Computer Architecture: A Quantitative Approach*. Morgan Kaufmann Publishers, Inc., 1996.
24. S. Pees, A. Hoffmann, V. Zivojnovic, and H. Meyr. LISA -machine description language for cycle-accurate models of programmable DSP architectures. In *Proc. of Design Automation Conference*, pages 933–938, 1999.
25. B. R. Rau, M. Lee, P. P. Tirumalai, and M. S. Schlansker. Register allocation for software pipelined loops. In *Proc. of ACM SIGPLAN Conf. Programming Language Design and Implementation*, pages 283–299, 1992.
26. B. Ramakrishna Rau. Iterative modulo scheduling. Technical report, Hewlett-Packard Lab: HPL-94-115, 1995.
27. S. Rixner, W. J. Dally, B. Khailany, P. Mattson, and U. J. Kapasi an J. D. Owens. Register organization for media processing. In *Proc. of International Symposium on High Performance Computer Architecture*, pages 375–386, 2000.
28. S. Roos. Scheduling for ReMove and other partially connected architectures. Technical report, Laboratory of Computer Enginnering, Delft University of Technology, 2001.
29. H. Schmit, D. Whelihan, A. Tsai, M. Moe, B. Levine, and R. Reed Taylor. PipeRench: A virtualized programmable datapath in 0.18 micron technology. In *Proc. of IEEE Custom Integrated Circuits Conference*, 2002.
30. P. Shivakumar and N. P. Jouppi. CACTI3.0: A integrated cache timing, power, and area model. Technical report, COMPAQ Western Research Laboratory, Aug. 2001.
31. Siliconhive, 2004. http://www.silicon-hive.com.
32. H. Singh, M-H Lee, G. Lu, F. J. Kurdahi, N. Bagherzadeh, and E. M. Chaves Filho. Morphosys: An integrated recon?gurable system for data-parallel and computation-intensive applications. *IEEE Trans. on Computers*, 49(5):465–481, May 2000.
33. TI Inc. *TMS320C6000 Programmer's Guide*, 2002. http://www.ti.com/.
34. TI Inc. *TMS320C64x DSP Two-Level Internal Memory Reference Guide*, 2002. http://www.ti.com/.
35. T. Wiegand, G. J. Sullivan, G. Bjontegaard, and A. Luthra. Overview of the H.264/AVC video coding standard. *IEEE Trans. on Circuits and Systems for Video Technology*, 13(7):560–576, July 2003.

Chapter 7
A Taxonomy of Field-Programmable Custom Computing Machines

An Architectural Approach

Mihai Sima, Stamatis Vassiliadis and Sorin Cotofana

Abstract The ability for providing a hardware platform which can be customized on a per-application basis under software control has established the *Reconfigurable Computing* ($\mathcal{RC}$) as a new computing paradigm. A machine employing the $\mathcal{RC}$ paradigm is referred to as a *Field-Programmable Custom Computing Machine* (FCCM). So far the FCCMs have been classified according to implementation criteria. In this presentation we propose a new classification of the FCCMs according to architectural criteria. Specifically, to analyze the phenomena inside FCCMs, we introduce a new formalism based on microcode, in which any reconfigurable operation is executed as a microprogrammed sequence with two basic stages: SET, and EXECUTE. We also introduce the concepts of spatial and temporal constructions to describe the structure of set of the reconfigurable units. The SET/EXECUTE formalism and the spatial/temporal constructions are the two most important criteria we use in building the taxonomy of field-programmable custom computing machines.

Key words: Reconfigurable computing · Field-programmable gate arrays · Microcode

7.1 Introduction

One of the fundamental trade-offs in the design of a computing system involves the balance between flexibility and performance. A general purpose computing platform can be used in large number of applications with an acceptable performance. For a specific application, however, the general-purpose platform performs less when compared to an *Application-Specific Integrated Circuit* (ASIC) that achieves a high performance for such application at the expense of flexibility. The ability for providing a hardware platform which can be metamorphosed under software control has established a new computing paradigm referred to as *Reconfigurable Computing* ($\mathcal{RC}$) [Gray and Kean, 1989; Mangione-Smith and Hutchings, 1997; Villasenor and Mangione-Smith, 1997; Mangione-Smith et al., 1997; Hauck, 1998c; Hauck, 1998a;

Kastrup et al., 1999b], as an emerging technology for more than ten years. According to this paradigm, the main idea in improving the performance of a computing machine is to define customized computing resources on a per-application basis, and to dynamically configure them onto a *Field-Programmable Array* (FPA) [Brown and Rose, 1996]. Since a virtually infinite hardware can be emulated in this way, the execution of the critical parts of the algorithm being implemented is accelerated. As a general view, a computing machine working under the new $\mathcal{RC}$ paradigm typically includes a *General-Purpose Processor* (GPP) augmented with an FPA. The basic idea is to exploit both the GPP flexibility to achieve medium performance for a large class of applications, and FPA capability to implement application-specific computations. Such a hybrid is referred to as a *Field-Programmable Custom Computing Machine* (CCM) [Buell and Pocek, 1995], [Hartenstein et al., 1996]. The synergism of GPP and FPA can achieve orders of magnitude improvements in performance over a GPP alone, while preserving the flexibility of the programmed machines over ASICs in implementing a large number of applications. However, the FCCM performance in terms of speed and power may still be of orders of magnitude lower than the performance of an ASIC.

By successive instantiation of computing resources on the reconfigurable platform, a FCCM can rely on more hardware than actually has. Moreover, the computing resources which are to be instantiated can be beforehand adapted to application. The idea of adaptating the architecture to application is not new. Before the $\mathcal{RC}$ paradigm emerged, the adaptation had been performed by rewriting the machine microcode. However, the efficiency of such procedure was limited by the inflexibility of the hardware platform. Augmenting a processor with reconfigurable hardware opens up a new way for architecture customization, in which the microcode and underlying hardware are both entirely flexible. As opposed to a program for a classical processor which includes only a software image of the algorithm for the static hardware platform, an *FCCM program* includes a hardware image of the computing units, and also a software image that will run on these units.

The paper constitutes a survey of FCCMs that have been proposed so far. We will organize the survey under the form of a taxonomy of such machines. Former attempts in FCCM classification used implementation criteria [Guccione and Gonzales, 1995], [Radunovi¢ and Milutinovi¢, 1998], [Kastrup et al., 1999b], [Wittig and Chow, 1996], [Jacob and Chow, 1999]. As the user observes only the architecture of a computing machine [Blaauw and Brooks, Jr., 1997], the previous classifications do not seize well the impact of the new $\mathcal{RC}$ paradigm to the user. In this presentation, we propose to classify the FCCMs according to architectural criteria. In order to analyze the phenomena inside FCCMs, we introduce a new formalism based on microcode, in which any operation executed on field-programmable hardware is carried out as a microprogrammed sequence with two basic stages: `SET` a new configuration on the FPGA, and `EXECUTE` the newly configured operation. We also introduce the concepts of *spatial* and *temporal constructions* to describe the structure of the reconfigurable microcoded units. The `SET/EXECUTE` formalism and the spatial/temporal constructions are the two most important criteria we will use in building the taxonomy of field-programmable custom computing machines.

This presentation has the following format. For background purpose, the most important concepts related to microcode such as control store, control store address register, computing facilities, vertical and horizontal microinstructions are outlined in Section 7.2. In Section 7.3, the basics of reconfigurable devices are reviewed, and the major FPGA architectural classes are discussed: globally reconfigurable (single-context, and multiple-context), and partially reconfigurable. Section 7.4 introduces the new SET/EXECUTE formalism as well as the spatial/temporal constructions, emphasizing that these are the basic criteria in analyzing the FCCM architectures from the microcode point of view. In Section 7.5 we build the FCCM taxonomy. Section 7.6 completes the paper with some conclusions and closing remarks.

7.2 Microcode. What it is and How it Appears

The memory devices have played the role of the *technology driver* of the computer industry for many decades. As one of the major source of bottlenecks, the characteristics and capacity of the memory devices have had a large impact on the performance of different computer generations. Usually, a memory device with slow access speed entails a poor performance of the machine that uses it. This was the situation when computers were introduced and constituted a strong motivation for developing schemes exhibiting a reduced number of memory access cycles. A solution to improve the performance of systems with slow-speed memory is to encode as much information as possible into the *operation code* (opcode) of an instruction, and assign multiple short-lasting decode and execute cycles per long-lasting fetch cycle. In this way, the number of fetch cycles, and, consequently, the overhead associated with main memory access were reduced [Johnson, 1991]. The engines using this technique are generally referred to as *Complex Instruction Set Computers* (CISC) [Johnson, 1991], [Patterson and Hennessy, 1996]. Figure 7.1 depicts the execution strategy of a CISC instruction.

Generally speaking, a CISC instruction specifies a complex operation which is, in turn, composed by a sequence of primitive operations or micro-operations. There have been two strategies of implementing such a complex operation: hardwired and microprogrammed. In the *hardwired implementation*, an automaton is designed for each complex operation. Such an automaton sequentially generates the control signals for the hardware computing units. In the *microprogrammed implementation*, the information needed to generate the control signals is stored under the form of a *microprogram* [Agrawala and Rauscher, 1976], [Rauscher and Agrawala, 1978], in a faster but expensive and low capacity memory, called *Control Store* (CS), or *Control Memory*. In this case a complex operation is implemented as a sequence of *microinstructions*. Like any software unit, a microprogrammed control

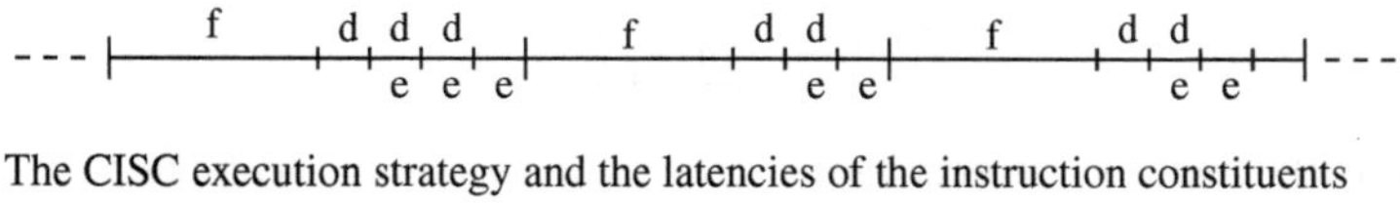

Fig. 7.1 The CISC execution strategy and the latencies of the instruction constituents

unit overcomes the inflexibility of a hardwired control unit; the penalty is a lower operation speed.

Advancements in technology reduced the performance gap between main memory and control store [Athanas and Silverman, 1993]. As the instruction fetch latency became comparable with both instruction decode and execute latencies, the complex encoding strategy was not needed any longer. This resulted in the development of *Reduced Instruction Set Computers* (RISC) [Patterson and Sequin, 1981], [Hennessy, 1989]. In a RISC machine, the instructions are of reduced complexity, much like of earlier microinstructions. Now, one decode and one execute cycles are assigned for each and every fetch cycle, as depicted in Fig. 7.2. Actually, such an architecture can be thought as exposing the microcode (we will refer to both microinstructions and microprogram as *microcode*; the meaning of the microcode will become obvious from the context) to the user, i.e., the instruction set itself is composed of microinstructions. Consequently, the control store is a part of the main memory and the microprogram resides in the program segment. It is worth mentioning here that having an instruction set of reduced complexity may also increase the compiler efficiency in optimizing the program code (it is well known that compilers cannot choose optimal encoding; as such, for a CISC 70% of instructions occupies 99% of code, and 50% of instructions occupies 95% of code) [Johnson, 1991].

The section further contains three subsections. In the first subsection we will briefly expound the basic structure of a microprogrammed computer. Then we will present issues related with the architecture of the microcode. In the third subsection we will discuss several issues related to adapting the architecture to application. Finally, we will provide for several explanations connected with microcode terminology.

7.2.1 The Basic Microprogrammed Computer

The concept of microprogramming was originally developed by Maurice Wilkes in 1951 [Wilkes, 1951]. Figure 7.3 depicts the organization of a microprogrammed computer as it is described by Agrawala and Rauscher [Agrawala and Rauscher, 1976], and Rauscher and Adams [Rauscher and Adams, 1980]. It must be noted that a microprogram in control store is associated with each incoming instruction. This microprogram is to be executed on the *Microprogrammed Loop* under the control of the *Sequencer*, as follows:

1. The sequencer maps the incoming instruction code into a control store address, and stores this address into the *Control Store Address Register* (CSAR).
2. The microinstruction addressed by CSAR is read from control store into the *MicroInstruction Register* (MIR).

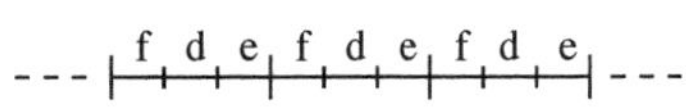

Fig. 7.2 The RISC execution strategy and the latencies of the instruction constituents

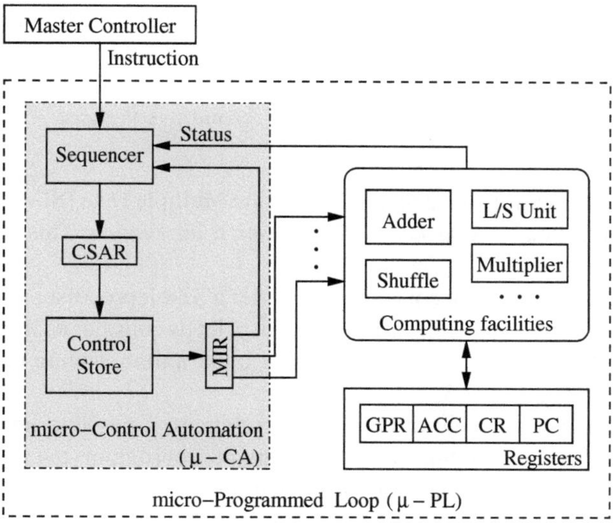

Fig. 7.3 The organization of a microprogrammed computer

3. The microoperations specified by the microinstruction in MIR are decoded, and the control signals are subsequently generated.
4. The computing resources perform the computation according to such control signals.
5. The sequencer uses the status information generated by the computing facilities as well as some information originating from MIR to prepare the address of the next microinstruction. This address is then stored into CSAR.
6. If an *end-of-operation* microinstruction is detected, the control is granted to the Master Controller which will issue a new instruction. The new incoming instruction initiates a new cycle of the microprogrammed loop.

These are just a few introductory considerations regarding microcode. In the next subsection we will emphasize only those architectural issues that are of interest for our taxonomy. For more details regarding microcode we note that there is a rich literature in this field. For example, the interested reader can consult [Andrews, 1980; Cline, 1981; Vassiliadis et al., 2003].

7.2.2 The Architecture of the Microcode

The microinstructions may be classified by the number of controlled computing resources. Given a hardware implementation which provides a number of computing resources, the amount of explicitly controlled resources during the same time unit (cycle) determines the verticality or horizontality of the microcode as follows:

- A microinstruction which controls multiple resources per cycle is *horizontal* (a horizontal microcoded machine is also referred to as a *Very Long Instruction*

Word (VLIW) machine [Ramakrishna Rau and Fisher, 1993]). In the extreme case, all the resources of the data path are controlled, as it is depicted in Fig. 7.4–a.
- A microinstruction which controls a single resource is *vertical*. This situation is pictured in Fig. 7.4–b.

We would like to mention that a Single Instruction Multiple Data (SIMD) computing resource is controlled by a *packed horizontal microinstruction*. This case is shown in Fig. 7.4–c.

In order to clarify the concepts, let us consider a 32-bit processor whose instruction set includes a bit-wise AND instruction. Also, let us consider an implementation that performs the AND operation serially, one bit at a time, having the computing resources depicted in Fig. 7.5.

Based on this arrangement of computing resources, a possible implementation of the bit-wise AND instruction is the following microprogram (the numbers in the left margin are not part of the actual microprogram. They are shown here so that the explanation can refer to individual lines of the microprogram conveniently):

```
01          | LOAD     A                     |
02          | LOAD     B                     |
03 loop:    | SHIFT    A                     |
04          | SHIFT    B                     |
05          | AND      A_bit,B_bit,C_bit     |
06          | SHIFT    C                     |
07          | BACK     loop, 31_TIMES        |
08          | STORE    C                     |
09          | ENDOP                          |
```

As it can be easily observed, only one facility is being controlled during each time unit; therefore, the constituent microinstructions are *vertical*. For the same arrangement of computing resources, the next microprogram implements the bit-wise AND instruction with a smaller number of microinstructions:

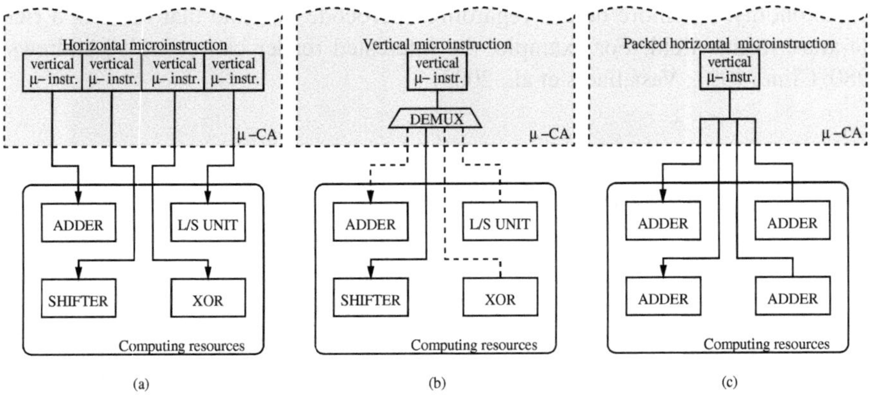

Fig. 7.4 Resource controlling strategies: (a) – horizontal microinstruction; (b) – vertical microinstruction; (c) – packed horizontal microinstruction

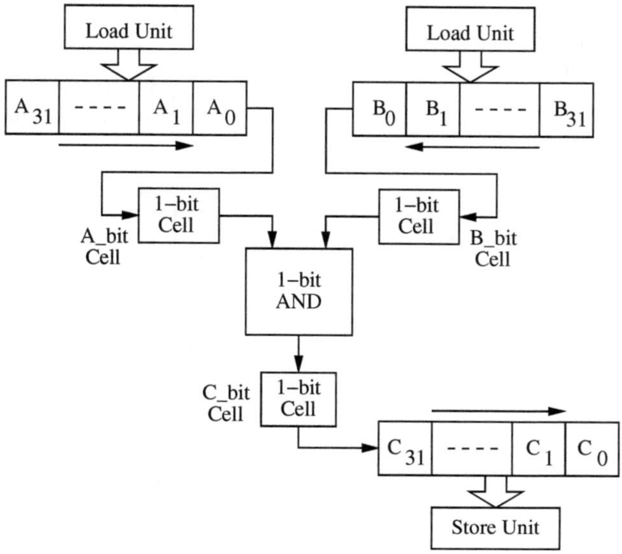

Fig. 7.5 Serial implementation of the 32-bit bitwise AND operator

```
        | LOAD    A                    | LOAD    B               |
loop:   | SHIFT   A                    | SHIFT   B               |
        | AND     A_bit,B_bit,C_bit    | NOP                     |
        | SHIFT   C                    | BACK    loop, 31_TIMES  |
        | STORE   C                    | ENDOP                   |
```

In this case, two computing resources are controlled per time unit (cycle); therefore, the microinstructions are *horizontal.*

In order to avoid any confusions regarding the classification of microinstructions in vertical and horizontal, we will present two typical mistakes the novices do. The first one is related to classifying the microinstructions in vertical and horizontal based on the encoding level of the microinstruction word bits. The fact that the bit fields are not encoded, each and every bit driving a distinct unit as shown in Fig. 7.6–(a), or the bit fields are encoded as shown in Fig. 7.6–(b), has nothing to do with the verticality or horizontality of the microcode. That is, the encoding level does not define the verticality or horizontality of the microcode; it just has implications related to implementation in the sense that the bit-width of the control store is smaller at the expense of a more complex microinstruction decoding.

A second typical mistake is related to the correlation between bit fields. The bits may not be correlated (as pictured in Fig. 7.7–(a)) or may be correlated (as shown in Fig. 7.7–(b)); however, this has nothing to do with the verticality or horizontality of the microcode. The degree of correlation between bit fields is related to the microcode implementation, and *not* with the verticality or horizontality of the microcode. When the bit fields are correlated, the value in one field determines the meaning of the other fields in the microinstruction. The typical example is a

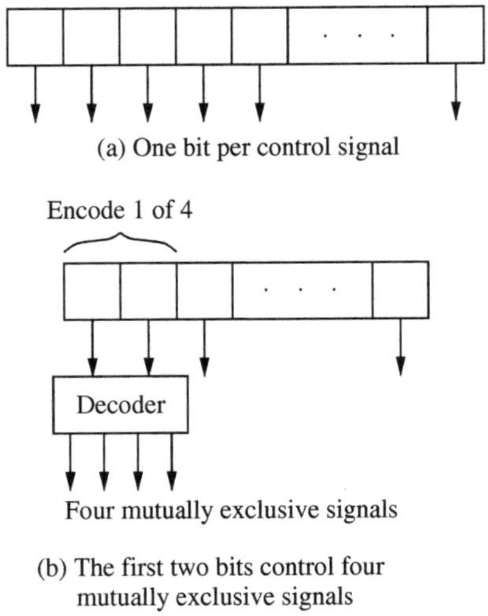

Fig. 7.6 Bit-field encoding strategies: (a) without bit encoding; (b) with bit encoding

microinstruction that contains an *opcode* field, for which a micro-opcode decoding cycle is needed.

Both vertical and horizontal architectures have benefits and shortcomings as well. The vertical microcode exhibits low latency in execution, but requires only a narrow-width control store. The horizontal microcode is faster since it can execute multiple micro-operations at a time, but it requires a wide-width control store. Moreover, some slots of a horizontal microinstruction may contain *No Operation* (NOP), and that translates into a suboptimal utilization of the control store space.

7.2.3 Adapting the Architecture to Application

Concerning the computing efficiency, the architecture of a *general-purpose computer* intercepts (meets) the computational characteristics of a large class of applications. In this way, an acceptable performance can be provided for all of these

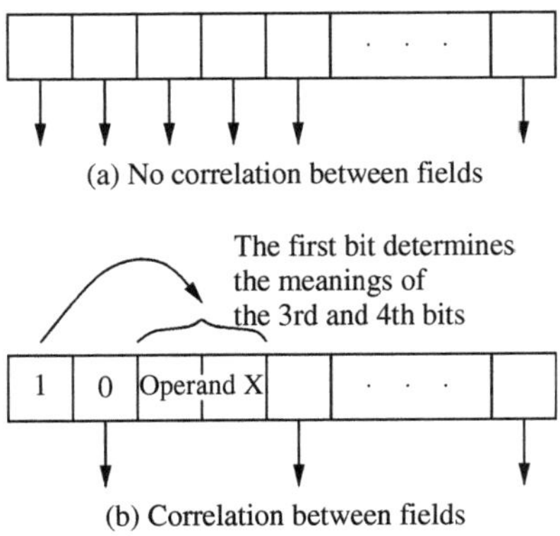

Fig. 7.7 Bit-field encoding strategies: (a) without correlation between fields; (b) with correlation between fields

applications. If *application specific computing* is addressed, then the architecture has to be tuned towards the application. Let us assume we have a *Computing Machine* (CM) and its instruction set. An *implementation* of the CM can be formalized by the doublet:

$$\text{CM} = \{\mu P\ ,\ \mathcal{R}\} \tag{7.1}$$

where μP is the microprogram including all the microroutines that implement the instruction set, and $\mathcal{R}$ is the set of N *computing (micro-)resources*, each of them being controlled by a microinstruction:

$$\mathcal{R} = \{r_1\ ,\ r_2\ ,\ \ldots\ ,\ r_N\} \tag{7.2}$$

Let us assume the computing resources are hardwired. If the microcode is exposed to the user, i.e., the instruction set is composed of microinstructions, there is no way to adapt the architecture to application but by custom-redesigning the computing facilities set, $\mathcal{R}$. When the microcode is not exposed to the user, i.e., a microroutine is associated with each instruction, then the architecture can be adapted by rewriting the microprogram μP. Changing the microprogram in the control store requires a *Writable Control Store* (WCS). The microprogram is generated at compile time and subsequently dynamically loaded into control store [Agrawala and Rauscher, 1976; Rauscher and Adams, 1980]. When only a limited amount of writable control storage is available, some form of overlaying is needed. In this situation, different pages of control store are swapped during computation [Liu and Mowle, 1978; Baldwin et al., 1991; Johnson et al., 1998; Teich et al., 1999].

With hardwired computing resources, the architecture adaptation by rewriting the microprogram has a limited efficiency. A new instruction is created by threading the operations of hardwired units, and not by custom redesigning the computing resources. However, when the resources themselves are microcoded, the formalism recursively propagates to lower levels. Therefore, the implementation of each resource can be viewed as a doublet composed of a *nanoprogram* (nP) and a *nanoresource set* ($n\mathcal{R}$):

$$r_i = \{nP\ ,\ n\mathcal{R}\}\ , \qquad i = 1, 2, \ldots, N \tag{7.3}$$

Now it is the rewriting of the nanocode which is limited by the fixed set of nanoresources.

The presence of the reconfigurable hardware opens up new ways to adapt the architecture to application. Assuming the computing resources are implemented on a field-programmable array, adapting the resources is now entire flexible and can be performed on-line. Thus, the resource set $\mathcal{R}$ metamorphoses into a new one, $\mathcal{R}^*$:

$$\mathcal{R} \longrightarrow \mathcal{R}^* = \{r_1^*\ ,\ r_2^*\ ,\ \ldots\ ,\ r_M^*\}, \tag{7.4}$$

and so does the set of associated microinstructions. Writing new microprograms with application-tuned microinstructions is clearly more effective than with fixed microinstructions.

It is worth mentioning now that the microcode implementation concerns a *recursive formalism*. The micro and nano prefixes should be used against an *implementation reference level* (If it will not be specified explicitly, the *implementation reference level* will be considered as being the level defined by the instruction set. For example, although the microcode is exposed to the user in the RISC machines, the RISC operations are specified by *instructions*, rather than by microinstructions). Once such a level is set, the operations performed at this level are specified by *instructions*, and are under the explicit control of the *user*. As such, the operations below this level are specified by *microinstructions*, those on the subsequent level are specified by *nanoinstructions*, and so on.

At the end of this section, we would like to clarify some aspects related with terminology. When the user is given the facility to write its own microprograms and load these microprograms into the control store, then the machine is called *microprogrammable*. Otherwise, the machine is only *microprogrammed*. Machines in which the control store can be loaded under program control are called *dynamically microprogrammable*. Loading at either application launch-time or run-time may be carried out; in the later case, the control store may have a resident and page-able storage space. The hardwired machines are called *non-microprogrammed*.

7.3 Reconfigurable Arrays: Terminology and Concepts

In the pre-FPGA era, a full-custom unit was designed for each particular task. Such circuit exhibits the highest performance for the smallest die size at the expenses of large design and manufacturing time. For this reason, it is not suitable for a number of tasks, such as prototyping or low-volume manufacturing scenarios. A trade-off between advantages and shortcomings of full-custom circuits has been provided by the new class of *semi-custom* circuits. For such circuits, the user defines the connections linking pre-manufactured processing elements [Smith, 1997; Brown et al., 1992]. The semi-custom devices can be programmed by ordering to a vendor at fabrication time only, and as such they are referred to as *mask-programmable* devices.

The next step in the evolution of programmable devices is represented by *Field-Programmable Devices* [Brown et al., 1992; Trimberger, 1994; Chan and Mourad, 1994; Jenkins, 1994; Brown, 1994; Oldfield and Dorf, 1995; Salcic and Smailagic, 1997]. In a very general view, a field-programmable device is composed of *Raw Hardware* (processing elements and interconnecting resources) and *Configuration Memory*. The function performed by the raw hardware is defined by the information stored into configuration memory. As opposed to mask-programmable devices, the field-programmable devices can be configured *in the field* by the end user either by placing them into a special programming unit or even directly *in-system*.

Based on architecture, the field-programmable devices can be classified in two classes: *Programmable Logic Devices* (PLD) and *Field-Programmable Logic Arrays*

(FPGA) [Brown and Rose, 1996]. A programmable-logic device contains essentially two levels of logic, in which an AND-plane has its outputs connected to the inputs of an OR-plane; thus, the logic functions are implemented as sum-of-products. The interconnection structure is quite simple, and the delay is highly predictable for a programmable-logic device. On the other side, an FPGA includes a set of *Processing Elements* (also commonly referred to as *Active Logic*), which defines the logic primitives, an *Interconnection Network* linking the processing elements together, and I/O Blocks. For both PLD and FPGA, a *Configuration Memory* is available on chip.

A generic FPGA processing element is pictured in Fig. 7.8. It is highly configurable and typically composed of a combinational logic element, a storage element (a flip-flop or a register), and an output selector. the combinational logic element (which, in fact, implements the logic function) may take different forms, ranging from a 2-input gate [Carter, 1987; Camarota et al., 1992; Camarota et al., 1993], to a fully-programmable *Look-Up Table* (LUT) or even pairs of Look-Up Tables, each LUT having 3, 4, or more inputs, and one or more outputs [Xilinx, 1999; Atmel, 1999g; Atmel, 1999i; Altera, 1999d; Altera, 1999c; Furtek et al., 2000b; Amerson et al., 1996; Miyazaki et al., 1995; Bertin et al., 1989b]. The output of the storage element serves as input to both the combinational logic element and the output selector. The output selector allows for the storage element to be bypassed if a combinational output is needed [Carter, 1987; Bertin et al., 1989b].

The *interconnection network* also takes different forms. It can range from a *mesh* network connecting the processing elements according to an orthogonal [Camarota et al., 1992; Camarota et al., 1993] and/or diagonal [Furtek et al., 2000a] nearest neighbour pattern, to the very area consumming *full-crossbar* providing 100% connectivity of the processing elements. In order to reduce the silicon area, the processing elements are commonly partitioned in clusters [Marquardt et al., 1999; Betz and Rose, 1998]. In this case, the interconnection network contains *local buses* connecting the processing elements in the same cluster, and *express buses* providing intercluster connectivity. The typical architectures of FPGA's raw hardware are presented in Fig. 7.9. Different circuits are mapped with different efficiency on different architectures. Since mapping efficiency issues are not of interest for our taxonomy we do not go into further details. However, the interested reader can consult a rich bibliography [Brown et al., 1992; Brown and Rose, 1996; Xilinx, 1999; Atmel, 1999g; Atmel, 1999i; Altera, 1999d; Altera, 1999c].

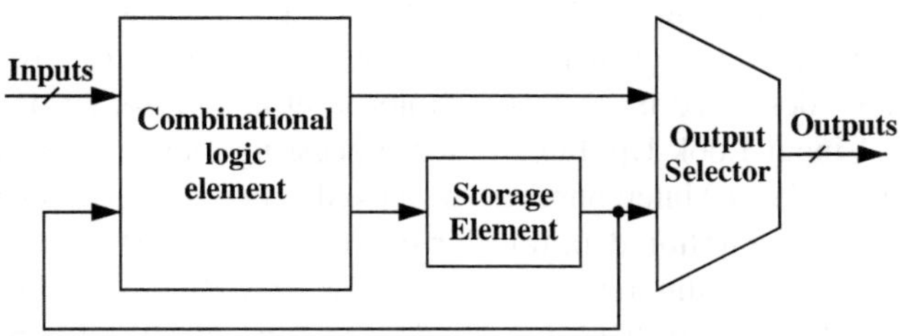

Fig. 7.8 Generic architecture of an FPGA Processing Element

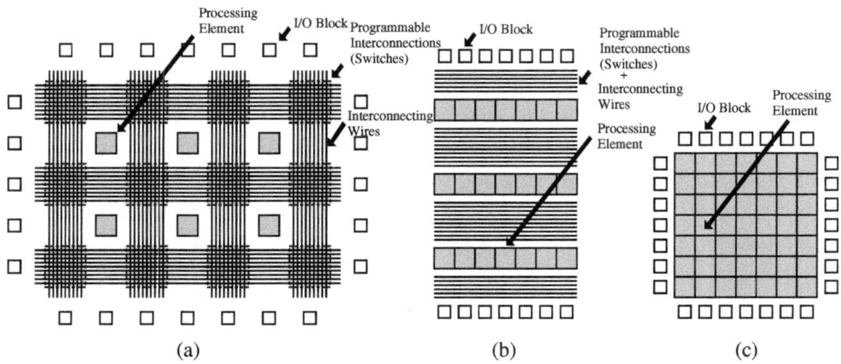

Fig. 7.9 Raw hardware structures: (a) – 2-D oriented; (b) – 1-D oriented; (c) – Mesh

Regarding the configurability and reconfigurability, we would like to mention that several *programming technologies* are available. As outlined by Compton and Hauck [Compton and Hauck, 2002], some FPGAs can be configured only once by burning fuses [Actel, 1996a; Actel, 1996b; Actel, 1998a; Actel, 1999a; Actel, 1999b]. Other FPGAs can be reconfigured an indefinite number of times, since their configuration memory is RAM based. Typically, the configuration memory drives both the look-up tables and switches in RAM-based FPGAs [Carter et al., 1986; Trimberger, 1994]. By programming and reprogramming the RAM cells, the FPGA is configured and reconfigured esentially as any other standard memory device is. There are also EEPROM-based FPGAs, e.g., MAX families from Altera [Altera, 1999e; Altera, 1999f], but their reprogramability is very limited. For this reasons, from the point of view of reconfigurable computing the non-RAM based FPGAs can be considered as being ASIC replacements; therefore, they will not be considered any longer in this presentation.

7.3.1 FPGA Reconfiguration Patterns

In order to benefit from virtual hardware, programs need to reconfigure the FPGA on-the-fly. That is, different configurations are swapped in and out the reconfigurable hardware during program execution. To achieve run-time reconfiguration, a different configuration bit stream has to be specified per *configurable unit* (processing element or interconnection switch), and computing cycle (Fig. 7.10). Following the methodology described by Bolotski [Bolotski et al., 1994], let us consider an array of 100 4-input Look-Up Table-based processing elements (such as Xilinx XC4003E [Xilinx, 1999]) interconnected by a fixed network. Assuming a cycle time of 10^{-7} sec., a configuration data transfer rate of $100 \cdot 2^4/10^{-7} = 16$ Gb/sec is needed to change the configuration every cycle at 10 MHz. Such a huge reconfiguration data bandwidth is difficult to provide, and consequently it translates into a large *reconfiguration time*.

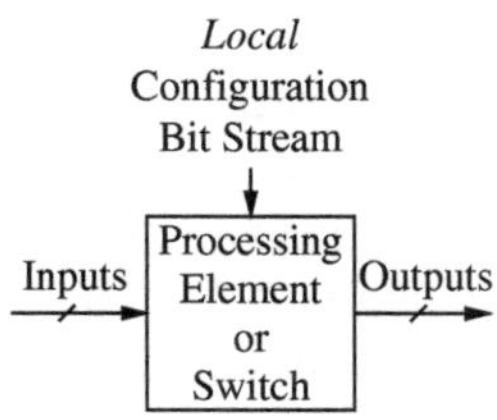

Fig. 7.10 FPGA configuration bit-stream

According to Wirthlin [Wirthlin, 1997], the following five techniques can be employed for reducing the *reconfiguration time*:

1. Increasing the reconfiguration bandwidth at the expense of additional resources for configuration data distribution.
2. Reducing configuration data by compressing the configuration bit stream, or increasing the granularity of the processing element.
3. Providing means FPGA partial reconfiguration.
4. Overlapping the reconfiguration and execution stages.
5. Caching the configuration data.

The first two techniques are rather general and can be implemented in any FPGA. The last three techniques, however, are orthogonal and lead to different reconfiguration classes: *Single-Context*, *Multiple-Context*, and *Partial Reconfigurable*. Each of these classes will be discussed below.

7.3.1.1 Single-contexts FPGAs

Only a single configuration can be stored on chip at a time. That is, for each configurable resource (processing element or interconnect switch), there is a single set of bits which define its behaviour. Single-context devices are best represented by commercial FPGAs [Xilinx, 1999; Altera, 1999d; Altera, 1999c; Atmel, 1999g]. Being used primarily as multimode hardware for hardwired circuit emulation when the same operation per element is required, these devices make use of no techniques for reconfiguration time reduction. A *single-context* device is usually a serially programmed chip that requires a complete reconfiguration even for changing 1 bit of its configuration information. Their reconfiguration time typically ranges between tens and hundreds of miliseconds [Xilinx, 1999]. The execution and configuration of a single-context device are mutually exclusive. The *atomic unit of reconfiguration* is the whole chip in a single-context device, i.e., partial modification of a configuration that is already loaded is not possible. Therefore, a full-lenght configuration bit stream must be loaded once the reconfiguration process is initiated.

7.3.1.2 Multiple-context FPGAs

At the expense of larger configuration memory, a way to reduce the reconfiguration time is to prestore multiple configuration images or *contexts* on the chip, and activate these contexts as needed. By broadcasting a *context identifier* on a global selection bus every cycle, the contexts are made active one at a time, as depicted in Fig. 7.11. In this way, a very fast global reconfiguration of both processing elements and interconnection switches [DeHon, 1994; Trimberger et al., 1997] or only of the interconnection switches [Bhat and Chaudhary, 1997] is provided. However, loading new context configuration is still limited by the low off-chip reconfiguration bandwidth. Since each layer of the configuration memory can be independently written, the circuit defined by the active configuration layer may continue its execution, while the non-active configuration layers are being updated.

Two scenarios are possible with multiple-context reconfigurable arrays:

1. Preloading all the contexts and activating them one by one as needed. In this scenario, the amount of emulated virtual hardware is limited to N times the physical hardware in FPGA, where N is the number of contexts.
2. Operating on one context while other contexts are being prefetched [Tau et al., 1995a; Tau et al., 1995b]. In this way, the low efficiency reconfiguration beyond N contexts is overcomed and the amount of efficiently emulated virtual hardware is not limited by the number of contexts any more.

Several multi-context devices have been proposed so far: *Dynamically Programmable Gate Array* [DeHon, 1996a; DeHon, 1996b; DeHon, 1996c; DeHon et al., 1998], *Time-Multiplexed FPGA* [Ong, 1995; Trimberger et al., 1997; Trimberger et al., 1998; Trimberger et al., 1999], and *Context Switching Reconfigurable Computing* [Scalera and Vázquez, 1998]. The differences between these architectures are rather small. Since a detailed analysis of multi-context devices is beyond the paper scope, we will not go into further details.

7.3.1.3 Partially Reconfigurable FPGAs

In such a device, it is possible to reconfigure only a selected subset of the raw hardware, while the portions of the array which are not being configured may continue execution. Consequently, it is possible to overlap the computation and

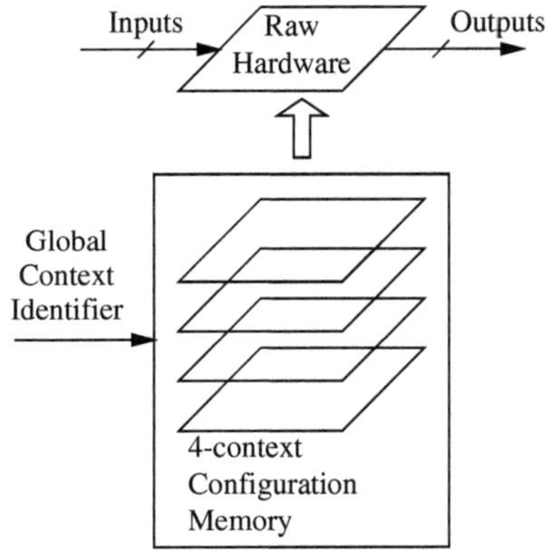

Fig. 7.11 Broadcasting a context identifier

reconfiguration. The smallest part which can be reconfigured at a time is called a *reconfiguration atom.* In a partially reconfigurable device, the underlying configuration layer operates much like a block-based memory device in which each block can be individually accessed. Typically, the reconfiguration process does not call for restrictions upon the portions that are not being reconfigured; these portions may continue execution. It is worth mentioning that while there are multiple device-resident configuration images for each bit processing element in a multiple-context device, and the reconfiguration atom is the whole device, there is only one device-resident configuration image in a partial-reconfigurable device, the reconfiguration atom being configuration-memory block.

As shown in Fig. 7.12, one can emulate the partial reconfiguration with a global reconfigurable device. Unfortunately, such emulation can be performed only in a spatial inefficient way, as a part of the area should stay remain unchanged. Likewise, one can emulate the global reconfiguration of a partially reconfigurable device but only in a temporal inefficient way. Since the address of each configuration memory location must be supplied along with the configuration information itself, the total amount of information needed to globally reconfigure a partial reconfigurable device is larger than that of a global reconfigurable device.

The highest flexibility in partial reconfiguration is achieved by a random reconfiguration pattern. In this case the configuration atom is usually as small as a functional block that can be independently reconfigured. Despite of these characteristics, the reconfiguration speed is limited by the narrow interface to the configuration memory that is typically a 32-bit bus. Among devices with partial reconfiguration capabilities we mention Xilinx XC6200 [Xilinx, 1996c], Atmel AT6000 [Atmel, 1999i], Lucent ORCA 2 and 3 Series [Lucent, 1999a], and the [Lucent, 1999b] National Semiconductor CLAy family [Rupp, 1995; Garverick et al., 1994].

It is also worth mentioning the configuration compression algorithm provided by Atmel [Mason et al., 1999; Atmel, 1999j; Atmel, 1999h] for AT6000 FPGA family [Atmel, 1999i]. As pictured in Fig. 7.13, the configuration bit stream produced by the compression algorithm programs only the elements whose configuration differ from the target configuration. Portions of the device not being modified remain operational during reconfiguration.

A distinct class of partial reconfiguration mechanisms called *pipeline reconfiguration* or *striping* has been proposed by Schmit [Schmit, 1997], in which the array is reconfigured at a granularity that corresponds to a pipeline stage of the application

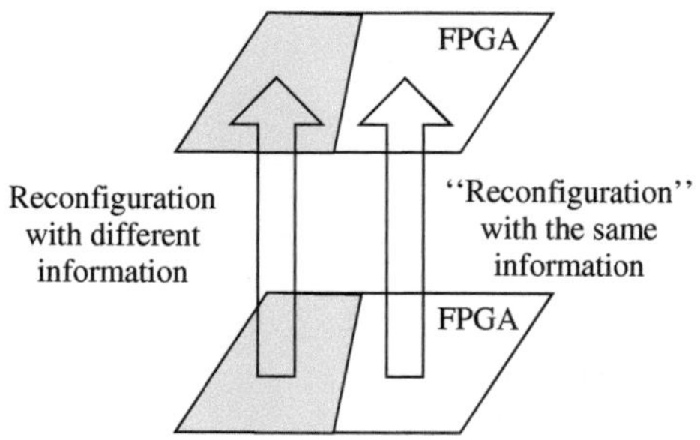

Fig. 7.12 Partial reconfiguration using a globally reconfigurable device

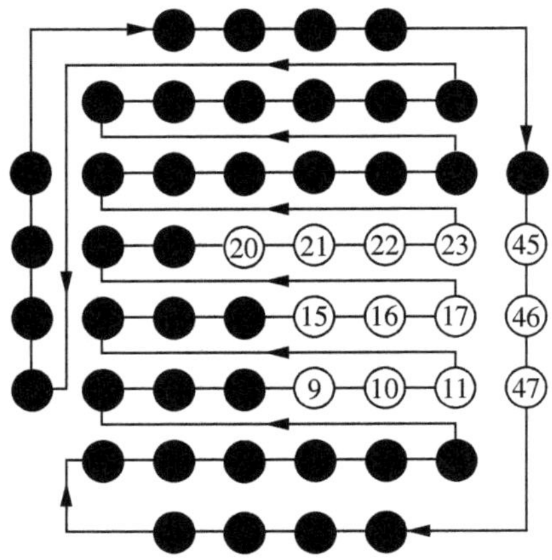

Fig. 7.13 Partial reconfiguration for Atmel AT6000 Series

being implemented. Consequently, the atomic unit of reconfiguration is a pipeline stage, and the partial reconfiguration occurs in increment of application pipeline stages. The generic architecture of a striped FPGA is shown in Fig. 7.14. The on-chip configuration cache memory allows the virtual pipeline stages to be loaded into the reconfigurable fabric at a very high speed.

7.3.2 FPA granularity

The *granularity of an FPA* can be defined as the size and complexity of the processing tile. A *small or fine granularity* refers to an FPGA having processing tiles of low complexity. Likewise, a *coarse granularity* refers to an FPA with processing elements of high complexity. In this paragraph we will present the most common devices in each granularity class.

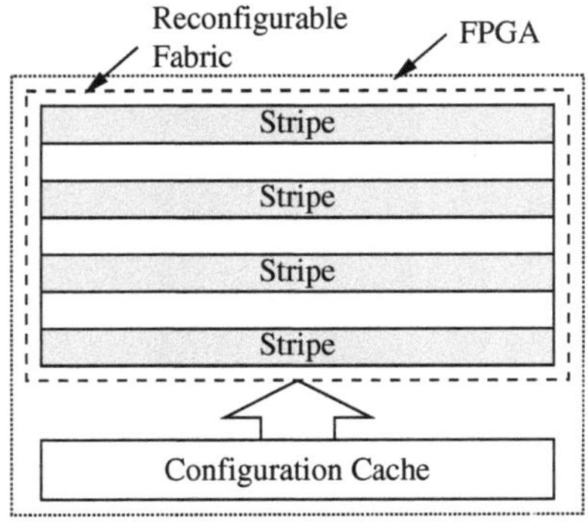

Fig. 7.14 Striped reconfiguration

7.3.2.1 Fine-grain FPAs

Are best represented by the commercial FPGAs. The computing tile is a combinational element that ranges from fixed-function 2-input gates [Carter, 1987; Camarota et al., 1992; Camarota et al., 1993], to fully-programmable *Look-Up Tables* or pairs of Look-Up Tables, with 3, 4, or more inputs, and one or more outputs [Xilinx, 1999; Atmel, 1999g; Atmel, 1999i; Altera, 1999d; Altera, 1999c; Furtek et al., 2000b; Amerson et al., 1996; Miyazaki et al., 1995; Bertin et al., 1989b].

Garp fine-grain reconfigurable array [Hauser and Wawrzynek, 1997; Hauser, 1997] is a matrix of small computing blocks interconnected by a network of wires. One block per row is a *control block*, while all the others are *logic blocks*. There are 24 blocks in a row. The number of rows is implementation-specific, but a typical value will be at least 32. Four *memory buses* run vertically through the rows for moving information into and out of the array. Each logic block can perform a simple logical or arithmetic operation on 2-bit operands. Wider-bit computations are achieved by aggregating logic blocks along a row into larger computing units. An example of several layouts of multi-bit functions is presented in Fig. 7.15.

7.3.2.2 Major Coarse-Grain FPAs

A typical member of this class is the *Reconfigurable Pipelined Datapath* (RaPiD). It is an array of coarse-grain functional units aimed to implement linear pipelines, much like those encountered in DSP applications [Ebeling et al., 1996; Cronquist et al., 1999]. The RaPiD's basic cell is depicted in Fig. 7.16. It comprises an integer multiplier, two integer ALUs, six general-purpose registers and three small local memories. The registers, ALU, RAM, multiplier and buses all operate on 16-bit signed or unsigned data. The multiplier produces 32-bit result, which can be further shifted to maintain the appropriate fixed-point representation.

The functional units are interconnected in mostly a nearest neighbor fashion through a set of segmented buses that run over the length of the datapath. A pipeline of functional units can be constructed on the RaPiD datapath. Therefore, a functional unit can get its input operands from, and send its outputs to its adjacent functional

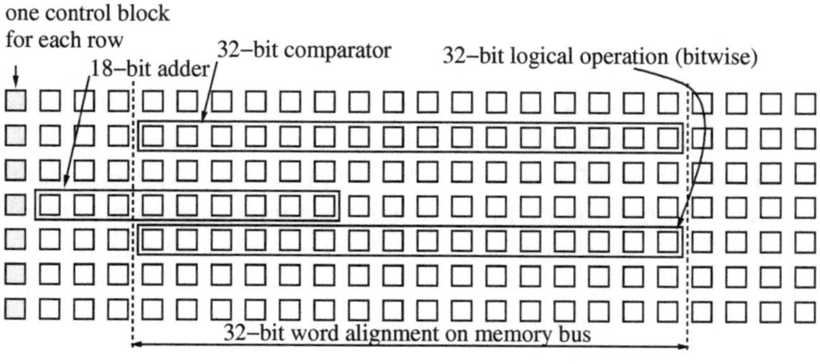

Fig. 7.15 Garp fine-grain reconfigurable array

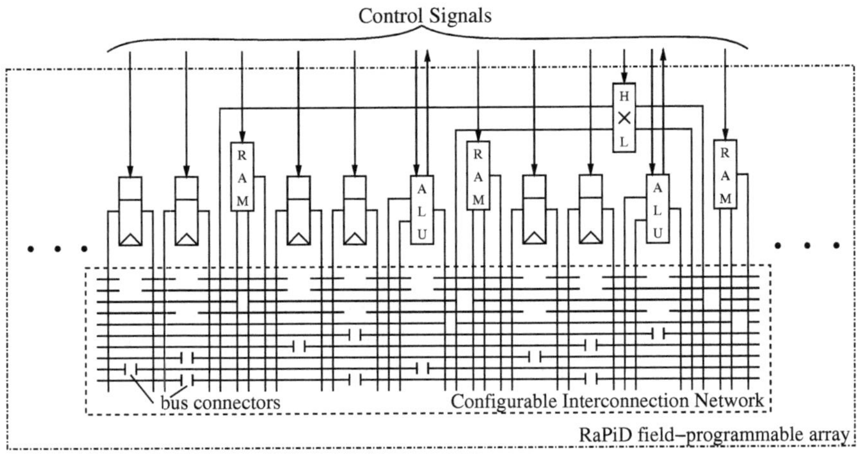

Fig. 7.16 The RaPiD's basic cell

units only. Each functional unit output is registered; however, the output register can be bypassed via configuration. The short parallel lines on the buses represent *bus connectors*, which can be configured to connect adjacent bus segments by either a buffer or a register.

The datapath is controlled using a combination of static and dynamic control signals. The static control signals are defined by RaPiD configuration, and determine the configuration of the interconnection network, and the insertion of or bypassing the pipeline registers. The RaPiD configuration is stored in a configuration memory as in ordinary FPGAs. The dynamic control signals are managed outside of the array and schedule the datapath operations over time.

A second typical member of the coarse-grain class is the *Reconfigurable Multimedia Array Coprocessor* (REMARC) [Miyamori and Olukotun, 1998b], [Miyamori and Olukotun, 1999]. It consists of 64 coarse-grain programmable logic blocks called Nano Processors, which are arranged as 8 × 8 array and a Global Control Unit (Fig. 7.17). Briefly, the nano processor has a 32-entry instruction RAM (nano instruction RAM), an ALU, a 16-entry data RAM, and thirteen 16-bit data registers. The ALU operates on 16-bit data. Each nano processor can comunicate to the four adjacent nano processors through dedicated connections and to the processors in the same row and the same column through a Horizontal Bus and a Vertical Bus, respectively. The global control unit issues an operation code ("nano Program Counter" – nano PC) to the nano processors. All nano processors execute the instructions indexed by the nano PC. The additional function of the global control unit is to provide the interface between REMARC and a host processor. The very small number of entries in nano instruction RAM and the flexible interconnection network make us consider REMARC as being a 32-context field-programmable device.

The *Wormhole* [Bittner, Jr., 1997; Bittner, Jr. and Athanas, 1997b; Athanas and Bittner, Jr., 1998] is a field-programmable array which consists of a number of Reconfigurable Processing Units (RPU) interconnected through a mesh network. Multiple independent streams are injected into the fabric. Each stream contains a

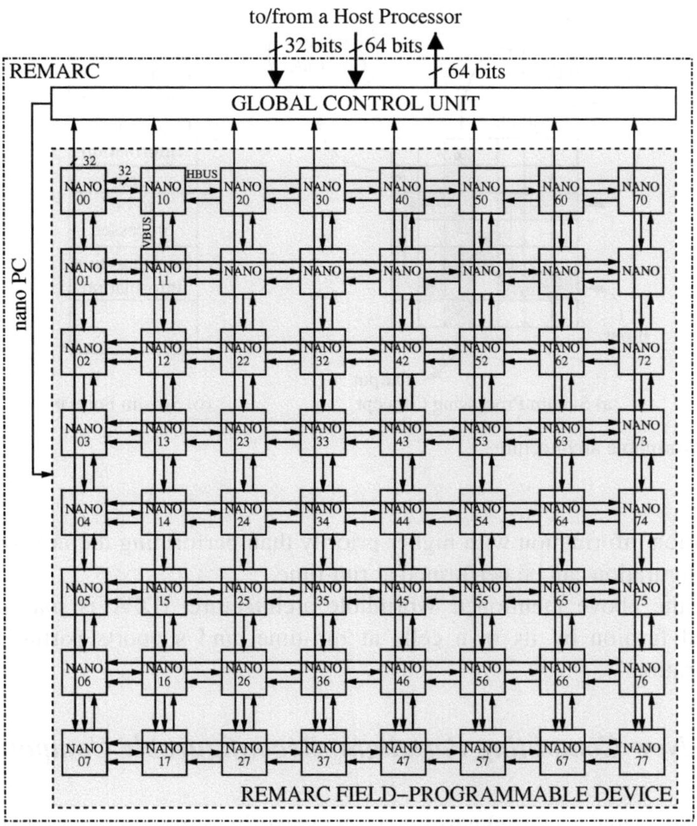

Fig. 7.17 Reconfigurable multimedia array coprocessor (REMARC)

header that includes information needed to route the stream through the fabric, and configure all RFUs along the path. The stream also contains data to be processed. In this way, the streams are self-steering, and can simultaneously configure the fabric and initiate the computation. Therefore, the reconfiguration and computation processes take place along one or more traces in the 2-D space. This process is depicted in Fig. 7.18.

The *reconfigurable Data Path Architecture* (rDPA) is also a self-steering autonomous reconfigurable architecture. As pictured in Fig. 7.19, it consists of a mesh of identical Data Path Units (DPU) [Hartenstein et al., 1994b; Hartenstein et al., 1994a]. The data-flow direction through the mesh is only from west and/or north to east and/or south and is also data-driven. A word entering rDPA contains a configuration bit which is used to distinguish the configuration information from data. Therefore, a word can specify either a SET or an EXECUTE instruction, the arguments of the instructions being the configuration information or data to be processed, respectively. One serial link at the array boundary is sufficient to completely configure the array, but multiple ports can be used to save time. As all DPU can be accessed individually, the reconfiguration is partial. Because every DPU forwards a

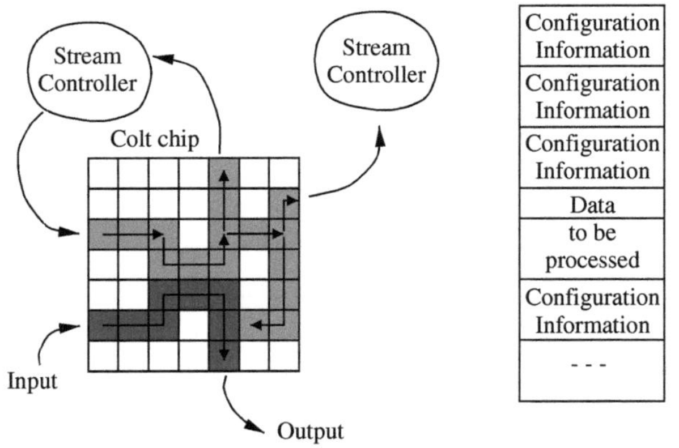

(a) Stream Processing Concept (b) Stream Format

Fig. 7.18 Wormhole architecture

configuration information with higher priority than performing the next operation, the reconfiguration can be performed at run-time.

As in the above mentioned Wormhole architecture, rDPA is managed in a distributed fashion by its own cells at run-time, and supports partial run-time reconfiguration.

7.3.3 A New Computing Paradigm: Reconfigurable Computing

Initially considered as a weakness due to the volatility of programming data, the *in-system* reprogramming capabilities of RAM-based FPGAs led to the Reconfigurable Computing ($\mathcal{RC}$) paradigm [Gray and Kean, 1989; Mangione-Smith and Hutchings, 1997; Villasenor and Mangione-Smith, 1997; Mangione-Smith et al., 1997; Hauck, 1998c; Hauck, 1998a; Kastrup et al., 1999b]. As we already mentioned, the $\mathcal{RC}$ paradigm assumes that the FPGA reconfiguration is performed under software control. This allows application-geared computing units to be instantiated on-the-fly,

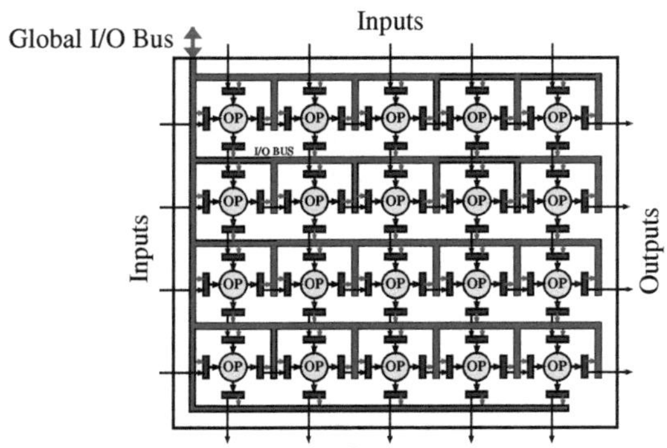

Fig. 7.19 Reconfigurable datapath architecture (rDPA)

and requires the $\mathcal{RC}$ program code to include also the description of the architecture on which it will run.

With the $\mathcal{RC}$ paradigm the user can navigate on *two dimensions* to implement customized computing units: a *spatial dimension* and a *temporal dimension.* In the *spatial dimension*, tasks which are computationally demanding can be efficiently executed by a properly designed hardware. In the *temporal dimension*, a sequential reconfiguration strategy is used in order to overcome insufficient configurable hardware. In this way, by swapping the configurations in and out of the FPGA upon demand and in real-time, only the necessary hardware is instantiated at any given time. Consequently, with a limited hardware resource, virtually infinite hardware is emulated.

The spatial and temporal dimensions generate a universe. We will refer to it as the *space-time universe.* A Mask-Programmed Gate Array (MPGA), an antifused- or PROM-based array, and an ASIC exhibit only *spatiality*, since it is not possible to reprogram them. Also, a microprocessor-based FPGA emulator, like the one proposed by Trimberger [Trimberger, 1995], exhibits at most *temporality*, since it cannot accomodate spatial computing patterns. *It is the freedom to navigate along both spatial and temporal dimensions which qualifies the computing paradigm as reconfigurable computing*. Due to these considerations, non-RAM-based FPGAs cannot be used with the $\mathcal{RC}$ paradigm. We will refer to RAM-based FPGAs as FPGAs hereafter, and will provide additional explanations only when necessary to avoid confusions.

Adapting the architecture of an FPGA-based computing machine is a new concept that appears to have common characteristics with the old microcode approach. It is the architecture of the microcode that constitutes the main element we will use to propose a formalism for analyzing the FCCMs. This is the topic of the next section.

7.4 From FPGA to Microcode Mapping

As we already mentioned, the microcode-based formalism originates in the observation that every instruction of a FCCM can be mapped into a microprogram, in which both FPGA configuration memory and instantiated computing units are regarded as controlled units. In particular, we will focus on the structure of these units and will introduce the concepts of temporal and spatial constructions. Moreover, two computational topologies (sea and pipeline of computing resources) are subsequently discussed. It is our oppinion that all the FCCM embodiments which have been proposed by the FCCM comunity can be classified by means of the microcode architecture, spatial/temporal constructions, and computational topologies.

7.4.1 The Microcoded Formalism

The main advantage of an FPGA-based computing engine is that new customized computing units can be instantiated. Since the information stored in FPGA's configuration memory determines the functionality of the raw hardware, the dynamic implementation of an instruction on FPGAs can be formalized by means of a microcoded structure. Assuming the FPGA configuration memory is written under the control of *Load Unit*, then the micro-programmed controller, the FPGA, and the configuration memory load unit may have a Δ arrangement (the particular method by which the microcode is not exposed to the upper (architectural) level is employed by way of illustration and not as a limitation of the formalism. The principles and features of the formalism apply also when the microcode is exposed to the upper (architectural) level), in which the micro-programmed controller is positioned on the top vertex, while the FPGA and the load unit are positioned on the bottom line (Fig. 7.20). The circuits instantiated on the raw hardware along with load unit(s) are regarded as controlled resources. Each such resource is given a special class of microinstructions: `SET` for configuration memory load unit, and `EXECUTE` for the circuits instantiated on raw hardware. The `SET` microinstruction initiates the reconfiguration of the raw hardware, and the `EXECUTE` microinstruction launches the operations performed by the newly instantiated circuits.

In this way, the execution of an FPGA-mapped instruction is performed as a microprogrammed sequence with two basic stages: `SET` the FPGA configuration stage, and `EXECUTE` the reconfigurable operation stage. By contrast, only `EXECUTE FIX` (micro)instructions can be associated with fixed computing facilities (such facilities cannot be reconfigured). If a multiple-context FPGA is used, the resource for activating an idle context is controlled by an `ACTIVATE` context microinstruction; this can actually be regarded a flavor of the `SET` configuration microinstruction. We will refer to load unit(s) and context activate units as *Configuring Resources* hereafter. Therefore, a field-programmable custom computing machine includes two classes of resources: *computing* and *configuring* resources.

Given this resource classification, the statement regarding the verticality or horizontality of the microcode as defined in Section 7.2 needs to be adjusted. For an FCCM hardware implementation which provides a number of *computing and configuring facilities*, the amount of explicitly controlled *computing or configuring facilities* during the same time unit (cycle) determines the verticality or horizontality of the microcode. Therefore, any of the `SET` configuration, `EXECUTE` reconfigurable

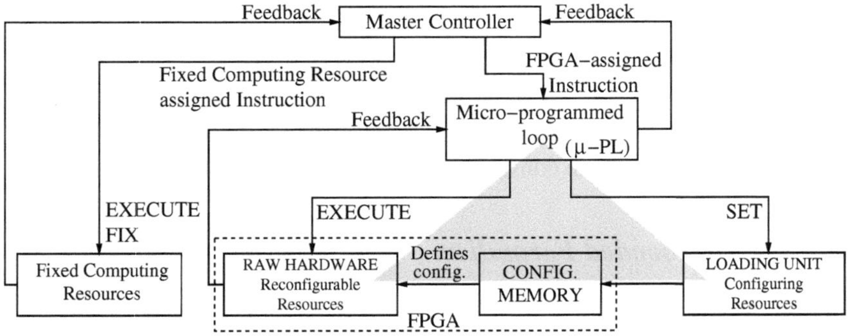

Fig. 7.20 The microcode concept applied to FCCMs – the Δ arrangement.

operation, and EXECUTE FIX (micro)instructions can be either vertical or horizontal, and may participate in a horizontal (micro)instruction. We would like to emphasize that it is the SET/EXECUTE formalism we will use in building the taxonomy of custom computing machines.

Referring to the microcode concept (Section 7.2), let us set the *implementation reference level* as being the level of instructions in Fig. 7.20. Assuming that the microcode is not exposed to the user, then an explicit SET instruction is not available. In this case, the user does not have an explicit control on the resource instantiation process; the system performs automatically the management of the active configuration. In this case, the user "sees" only the FPGA-mapped instruction that can be regarded as an EXECUTE configuration microinstruction reflected at the instruction level. Otherwise, when the reconfigurable microcode is exposed to the user, an explicit SET instruction is available, and the management of the active configuration becomes the responsibility of the user. Concerning the EXECUTE FIX (micro)instruction, we would like to mention that from our formalism standpoint such an instruction is always exposed to the user.

To clarify the microcode-based formalism, let us revisit the bitwise 32-bit AND operation example. Here, we have to point out that the following example has only a didactical value, as implementing a bitwise 32-bit AND operation rises no technical problems.

7.4.2 An Example

For the following discussion, we assume that the circuitry implementing the bitwise 32-bit AND operation is to be configured on a global reconfigurable single-context FPGA. If sufficient reconfigurable raw hardware is available, a *parallel implementation* of the bitwise AND operator comprising 32 1-bit AND blocks is feasible, as depicted in Fig. 7.21. Consequently, the bitwise AND operation can be performed in a parallel fashion, bit-by-bit of the same rank. A possible microprogram associated with circuit *configuration* and *execution* of the associated instruction is:

```
| SET      parallel_AND   |
| LOAD     A              |
| LOAD     B              |
| AND      A,B,C          |  ; C = A & B
| STORE    C              |
| ENDOP                   |
```

If sufficient reconfigurable hardware is not available to accomodate a parallel implementation of the AND operator, a *serial implementation* including an 1-bit AND unit, three memory cells, three shifting registers, two load and one store units is deployed on FPGA (Fig. 7.5). The microprogram associated with circuit *configuration* and *execution* of the 32-bit AND instruction is:

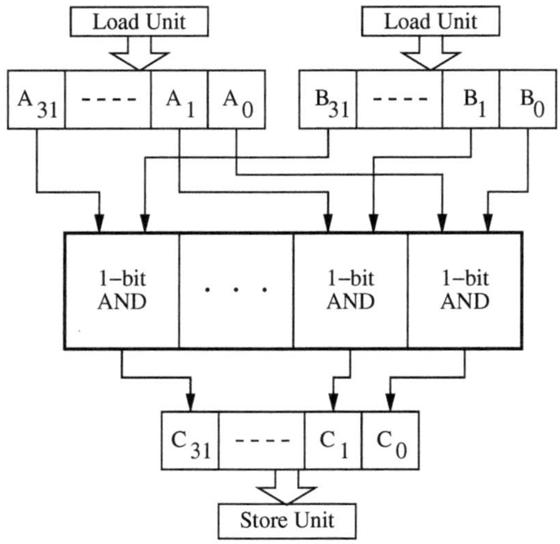

Fig. 7.21 Parallel implementation of the 32-bit bitwise AND

```
01          | SET      serial_AND            | ; global reconfiguration
02          | LOAD     A                     |
03          | LOAD     B                     |
04 loop:    | SHIFT    A                     |
05          | SHIFT    B                     |
06          | AND      A_bit,B_bit,C_bit     | ; C_bit = A_bit · B_bit
07          | SHIFT    C                     |
08          | BACK     loop, 31_TIMES        |
09          | STORE    C                     |
10          | ENDOP                          |
```

Assuming that only a very limited amount of raw hardware is available, swapping different parts of the AND circuit in and out of the reconfigurable platform is mandatory. Consequently, only a few different parts of the circuit (or even a single one) are configured on the array at a time. We will call that a *swappable implementation*. In this situation, the program is:

```
01          | SET      shifter               | ; global reconfiguration
            |          load unit             |
            |          store unit            |
            |          1-bit memory cell     |
02 loop:    | LOAD     A                     |
03          | SHIFT    A                     | ; A ≫ 1
04          | STORE    A                     |
05          | STORE    A_bit                 |
06          | LOAD     B                     |
07          | SHIFT    B                     | ; B ≫ 1
10          | STORE    B                     |
11          | STORE    B_bit                 |
12          | SET      1-bit_AND             | ; global reconfiguration
            |          load unit             | ; also removes the
            |          store unit            | ; old configuration
            |          three 1-bit memory cells
```

```
13          | LOAD      A_bit                | 
14          | LOAD      B_bit                | 
15          | AND       A_bit,B_bit,C_bit    | ; C_bit = A_bit · B_bit
16          | STORE     C_bit                | 
17          | SET       shifter              | ; global reconfiguration
                        load unit            | ; also removes the
                        store unit           | ; old configuration
                        1-bit memory cell    | 
18          | LOAD      C                    | 
19          | LOAD      C_bit                | 
20          | SHIFT     C                    | ; C ≫ 1
21          | STORE     C                    | 
22          | BACK      loop,31_TIMES        | 
23          | ENDOP                          | 
```

We would like to observe that in all three mentioned implementations, only one controlled resource (computing resource or configuring resource) is explicitly controlled at a time. Therefore, the microcode is vertical.

With partially-reconfigurable array, an efficient management of the circuits mapped on the array may be possible (recall that the partial reconfiguration can be emulated by global reconfiguration with the caveat of a possible disturbance of the "non-reconfiguring" part). In this situation, the microprogram is:

```
01          | SET       shifter              | ; partial reconfiguration
02          | SET       load unit            | ; partial reconfiguration
03          | SET       store unit           | ; partial reconfiguration
04          | SET       A_bit cell           | ; partial reconfiguration
05          | SET       B_bit cell           | ; partial reconfiguration
06          | SET       C_bit cell           | ; partial reconfiguration
07 loop:    | LOAD      A                    | 
08          | EXECUTE   A                    | ; A ≫ 1
09          | STORE     A                    | 
10          | LOAD      B                    | 
11          | EXECUTE   B                    | ; B ≫ 1
12          | STORE     B                    | 
13          | SET       1-bit_AND            | ; partial reconfiguration
            |                                | ; also removes the shifter
12          | EXECUTE   A_bit,B_bit,C_bit    | ; C_bit = A_bit · B_bit
13          | SET       shifter              | ; partial reconfiguration
            |                                | ; also removes 1-bit AND
14          | LOAD      C                    | 
15          | EXECUTE   C                    | ; C ≫ 1
16          | STORE     C                    | 
17          | BACK      loop,31_TIMES        | 
18          | ENDOP                          | 
```

The limit situation in partial reconfigurability is reached when the reconfiguration atom is smaller that the computing resource being instantiated. In this case, a partial reconfiguration strategy may require multiple SET instructions for each EXECUTE instruction. For example, several stages may be needed to configure the 32-bit bitwise AND unit: SET $\mathtt{atom}_1$, SET $\mathtt{atom}_2, \ldots,$ SET $\mathtt{atom}_\mathtt{m}$.

Finally, we would like to note that in the most general case of a partial reconfigurable FPGA, a reconfigurable operation is mapped to a microprogram as follows: $\texttt{Instruction} \longrightarrow$ SET $\texttt{hardware}_{11}$, SET $\texttt{hardware}_{12}$, ..., SET $\texttt{hardware}_{1m}$, EXECUTE $\texttt{hardware}'_{11}$, EXECUTE $\texttt{hardware}'_{12}$, ..., EXECUTE $\texttt{hardware}'_{1p}$, SET $\texttt{hardware}_{21}$, SET $\texttt{hardware}_{22}$, ..., SET $\texttt{hardware}_{2n}$, EXECUTE $\texttt{hardware}'_{21}$, EXECUTE $\texttt{hardware}'_{22}$, ..., EXECUTE $\texttt{hardware}'_{2q}$, ..., where $m, n \geq 0$, and $p, q \geq 1$. An example of such a complex mapping can be seen in the last microprogram, where the lines 01, 02, 03, and 04 act as SET microinstructions, while the lines 06, and 09 act as EXECUTE microinstructions. Also, we note that there may be no SET microinstructions if the circuit is properly configured, but there should be at least one EXECUTE microinstruction.

7.4.3 The Configuring Resources Set

An FCCM hardware implementation includes one or more configuring resources, each of them being controlled by a microinstruction. The reconfiguration of a single-context FPGA is managed by a single configuring resource; thus, a single SET microinstruction needs to be provided. A single configuring resource is also possible for multiple-context FPGAs (Fig. 7.22–a), but the SET will be an instruction with parameter. The extreme situation when the reconfiguration of each context is managed by a dedicated configuring resource is shown in Fig. 7.22–b. In this case, a SET microinstruction per configuring resource is provided. The multiplexer, which is also a configuring resource, is used for activating an idle context; it is given an ACTIVATE microinstruction.

The same analysis is performed for partially-reconfigurable FPGAs. In Fig. 7.23–a, the configuration memory of such an FPGA is written under the control of a single configuring resource; in this case, only one SET microinstruction is provided. The extreme situation when each reconfigurable atom is under the control of a dedicated configuring resource is pictured in Fig. 7.23–b; here, a SET microinstruction per reconfigurable atom is provided.

The number of the configuring resources and the way they are controlled (e.g., sequentially by a vertical microinstruction or in parallel by horizontal microinstruction) is user visible and has implications in the way the reconfigurable program written. For this reason, this number and the structure of the configuring resources set will constitute a classification criterion.

7.4.4 Spatial and Temporal Reconfigurable Constructions

The structure of the computing units to be mapped on the raw hardware is also subject to generate a clasification criterion. To refer to a set of computing units that are deployed on the raw hardware at the same time, we will use the term of *Spatial Construction*. Likewise, we will use the term of *Temporal Construction* to refer to a number of computing resources which are still resident on the configuration memory

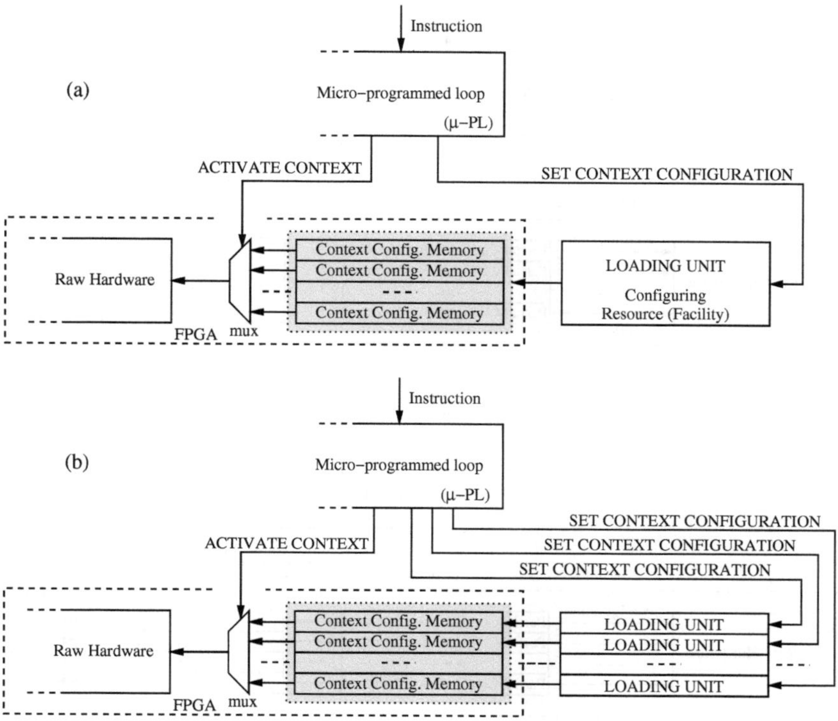

Fig. 7.22 (a) – A single configuring resource for a multiple-context FPGA (excluding the resource for activating an idle context); (b) – Multiple configuring resources for a multiple-context FPGA (excluding the resource for activating an idle context).

of a multiple-context FPGA, only one being active on raw hardware at a time. Given an FPGA-based hardware implementation which provides a number of *computing* resources (facilities), the amount of FPGA-resident resources during a time unit (cycle) determines the spatiality and/or temporality of the computing resources set. It is worth noting that this terminology is inspired from the work done by Brebner [Brebner, 1998] and Hudson *et al.* [Hudson et al., 1998].

While a temporal construction can be built only on multiple-context FPGAs, a spatial construction can be built either on a global or a partial reconfigurable FPGA. Different constructions can be generated by navigating in this space-time universe. In this way, the system architecture can be changed along the temporal dimension, spatial dimension, or even both, allowing the optimal computing architecture for the algorithm being implemented to be deployed.

Assume that the number of computing resources which are configured on the raw hardware at a given time is R, and the number of FPGA contexts is C. Then, a temporal construction will have $C > 1$, and a spatial construction will have $R > 1$. Temporality and spatiality are orthogonal features. For example, there are N contexts and M resources per context in Fig. 7.24.

For the origin of the space ($R = C = 1$) we will use the term of *Primordial Construction* (PrC). A primordial constructions will refer to a single computing resource

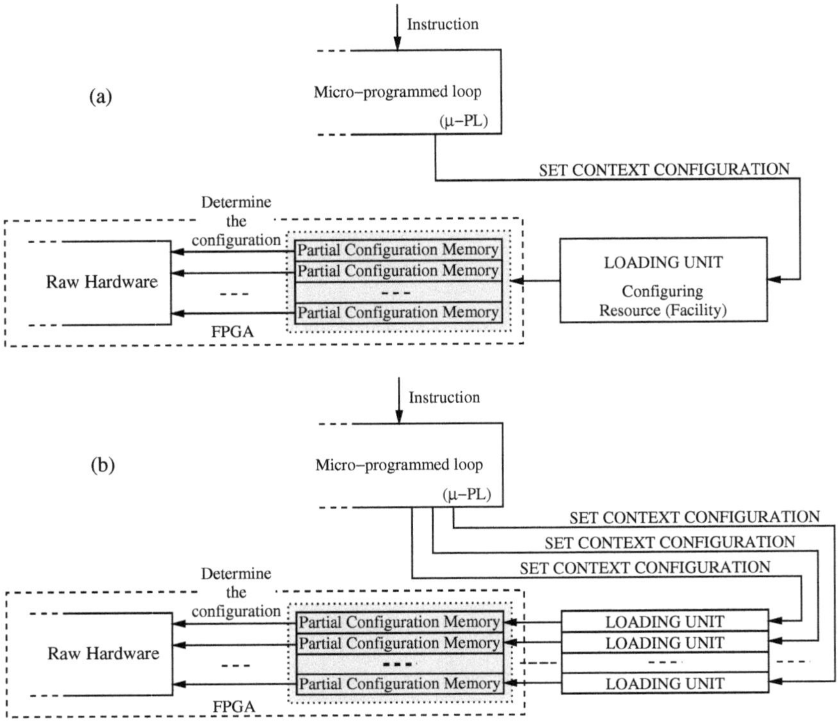

Fig. 7.23 (a) – A single configuring resource for a partial reconfigurable FPGA; (b) – Multiple configuring resources for a partial reconfigurable FPGA

instantiated on a single-context (globally-reconfigurable) FPGA, as depicted in Fig. 7.25. The operations initiated by SET configuration and EXECUTE the instantiated operation microinstructions are time-exclusive. That is, EXECUTE can be launched only after the SET has completed, and vice-versa. Therefore, the microcode can only be vertical in a primordial construction.

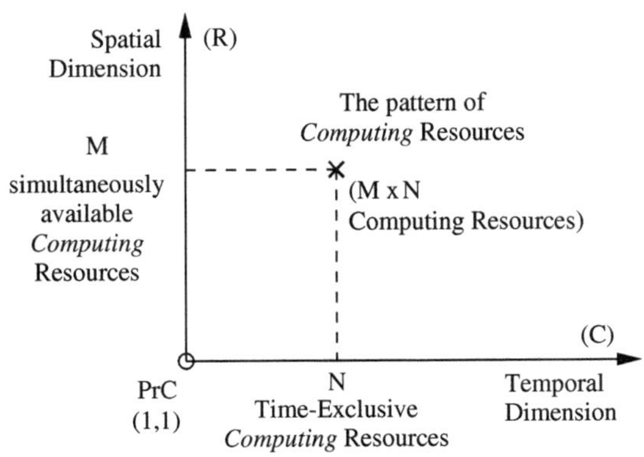

Fig. 7.24 Temporal and spatial dimensions as orthogonal features

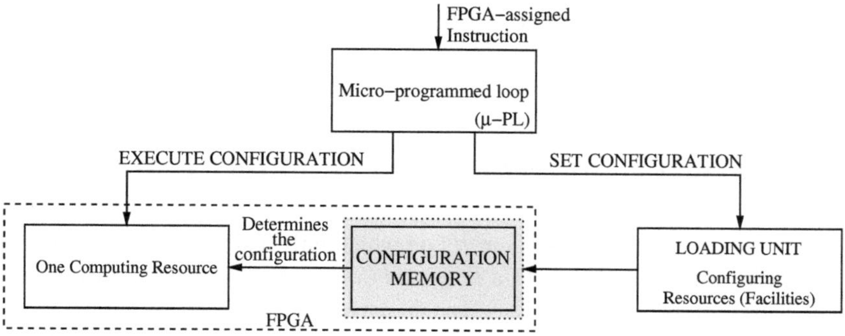

Fig. 7.25 Primordial construction built with single-context global reconfigurable FPGA

Finally, we would like to note that a primordial construction can also be built on a partial reconfigurable FPGA, as it is depicted in Fig. 7.26. However, this technique is generally not used, since changing the only computing resource in the raw hardware typically implies a global reconfiguration of the FPGA. As we already mentioned in Section 7.3, emulating the global reconfiguration with partial reconfiguration is performed in an inefficient way.

In the next two subsections we will analyze the temporal and spatial constructions in relationship with the microcode architecture. Our main goal is to build the

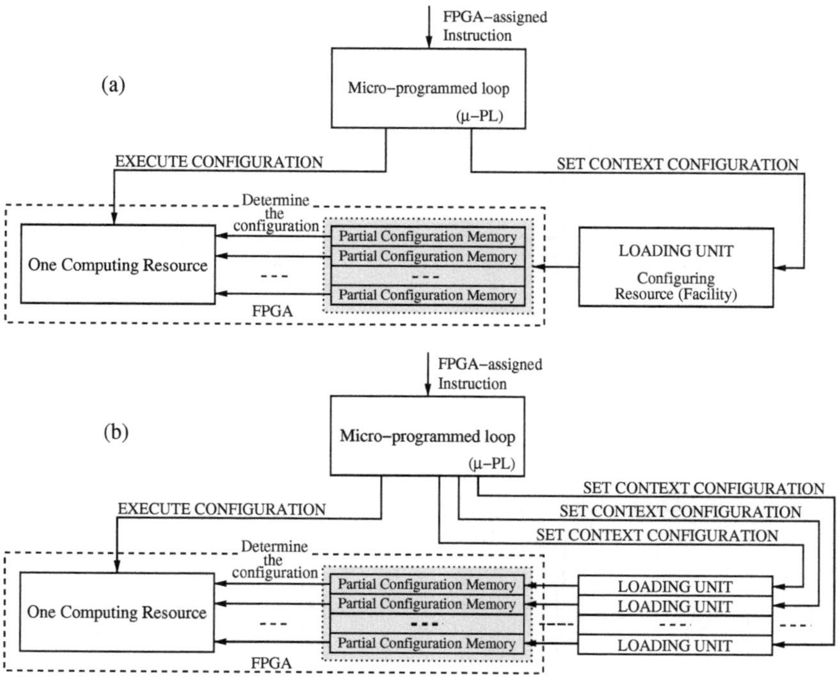

Fig. 7.26 Primordial construction built with partially reconfigurable FPGA: (a) – a single configuring resource is assumed; (b) – multiple configuring resources are assumed

framework around the verticality and horizontality of the microcode as the main classification criterion of the taxonomy.

7.4.5 Temporal Constructions built on FPGA Data Path

With a multiple-context FPGA ($C = N$) on which only a single computing facility per context is instantiated ($R = 1$), a *Temporal Construction* of computing resources is built. This is shown in Fig. 7.27.

A temporal construction in connection with the microcode architecture leads to two major possibilities (Table 7.1):

- *Vertical microCode + Temporal Construction* (VμC-TC), in which the SET and ACTIVATE microinstructions associated with configuring facilities, and EXECUTE microinstruction associated with the computing resources are launched at different times. Obviously, the following tuples of microinstructions are time exclusive: ACTIVATE with EXECUTE, EXECUTE with SET the active context, and ACTIVATE with two SET corresponding to the contexts being swapped.
- *Horizontal microCode + Temporal Construction* (HμC-TC), in which the SET, ACTIVATE and EXECUTE may be scheduled to be launched concurrently. The same restrictions mentioned above apply here, too.

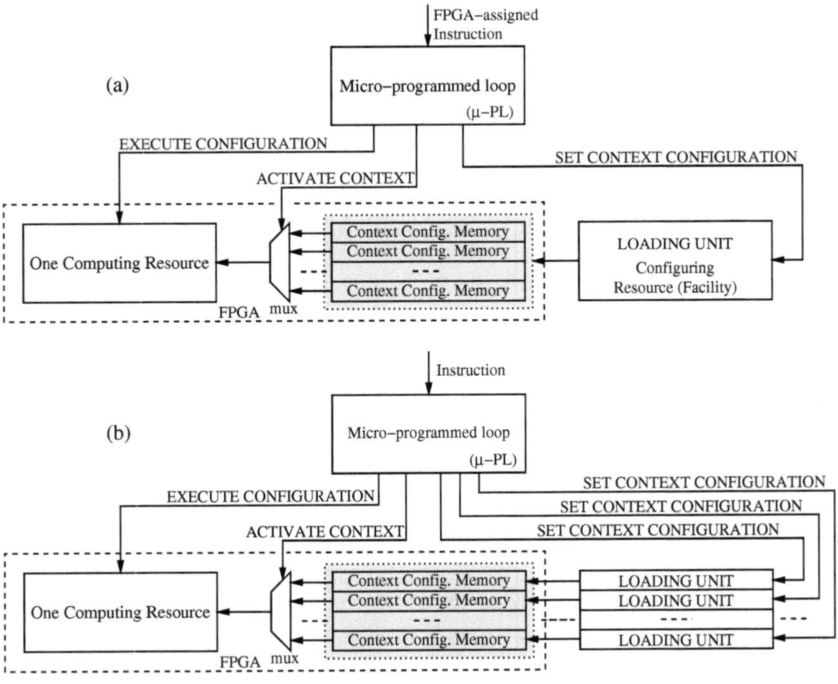

Fig. 7.27 Temporal construction built with multiple-context FPGA. Excluding the mux, (a) – a single configuring resource is assumed; (b) – multiple configuring resources are assumed

Table 7.1 Temporal constructions versus microcode

Acronym	Construction	Number of computing resources on raw hardware (R)	FPGA's number of contexts (C)	Number of configuring resources excluding MUX	Architecture of the microcode
PrC	Primordial	1	1	1	Vertical
VμC-TC	Temporal Construction	1	N	1	Vertical
				N	Vertical
HμC-TC	Temporal Construction	1	N	1	Horizontal (no. of slots ≤ 2)
				N	Horizontal (no. of slots $\leq N$)

7.4.6 Spatial Constructions built on FPGA Data Path

In this subsection, *spatiality* is the main concern, while *temporality* is a non-relevant issue. Consequently, R>1 and C=1 (i.e., the FPGA is single-context). Also, since R>1, multiple computing resources are now configured on the raw hardware and can simultaneously be controlled per time cycle. In this way, a spatial construction is indeed able to exploit the application parallelism.

The *Primordial Construction* (R=1) is also the starting point for analyzing the constructions built on spatial dimension. Assuming a single-context FPGA (C=1), a spatial construction of R=M computing facilities can be configured on the raw hardware. Spatial constructions can be built either on partial reconfigurable FPGA (Fig. 7.28) or global reconfigurable FPGA (Fig. 7.29), the differences having implications in performance:

- Reconfiguring all the computing resources (i.e., the entire context) is more efficient with a globally reconfigurable FPGA than with a partially reconfigurable FPGA where an address along with configuration information must be supplied for each atom being reconfigured. Moreover, assuming a global reconfigurable FPGA, the configuration and execution sequences are time exclusive. For example, the horizontal microinstructions may look like | `SET` | `NOP` | `NOP` | for configuration setting, and | `EXECUTE` | `EXECUTE` | `EXECUTE` | for parallel execution.
- Reconfiguring a single computing resource out of a set mapped on the raw hardware takes a shorter time with a partially reconfigurable FPGA than with a globally reconfigurable FPGA. Moreover, the partial reconfiguration allows for an efficient management of spatial constructions. Assuming two configuring resources, the configuration may be overlapped with execution: | `SET` | `EXECUTE` | `EXECUTE` | `SET` | `EXECUTE` |.

A spatial construction in connection with the architecture of the microcode leads to two major possibilities (Table 7.2):

- *Vertical microCode + Spatial Construction* (VμC-SC), in which the `SET`, and `EXECUTE` microinstructions are launched at different times.

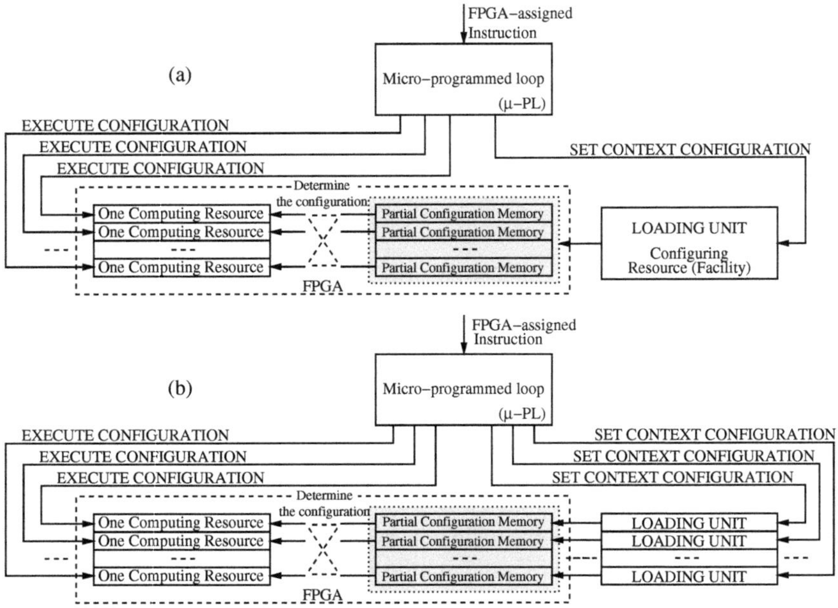

Fig. 7.28 Spatial construction built with partial reconfigurable FPGA: (a) – a single configuring resource is assumed; (b) – multiple configuring resources are assumed

- *Horizontal microCode + Spatial Construction* (HμC-SC), in which the concurrency of `SET`, and `EXECUTE` microinstructions should be discussed against the reconfiguration pattern, i.e., global or partial.

It is worth mentioning that hardwired computing resources [Dales, 1999] can be associated with spatial dimension, as they may be thought of as being continuously "configured" on the "array". For this reason, we will consider a VLIW processor having a single reconfigurable unit mapped on raw hardware at a time as exhibiting a spatial construction.

At the end of the temporal and spatial constructions presentation, several considerations concerning efficiency are worth to be provided. When the microcode

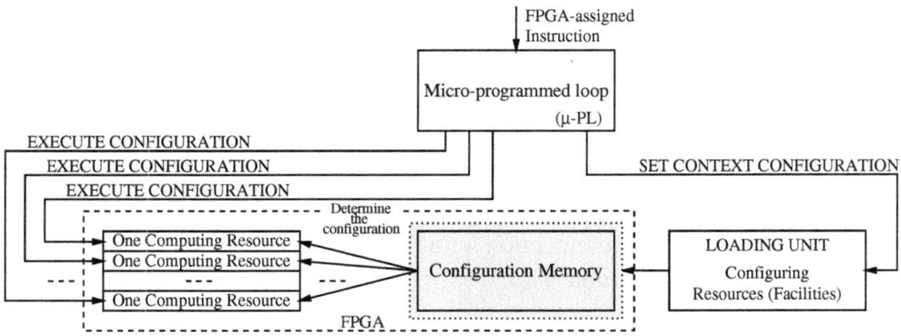

Fig. 7.29 Spatial construction built with global reconfigurable FPGA

Table 7.2 Spatial constructions versus microcode

Acronym	Construction (R)	Number of computing resources on raw hardware (C)	FPGA's number of contexts MUX	Number of configuring resources excluding	Architecture of the microcode
PrC	Primordial construction	1	1	2	Vertical
VμC-SC	Spatial Construction	M	1	1	Vertical
				N	Vertical
HμC-SC	Spatial Construction	M	1	1	Horizontal (no. of slots $\leq M$)
				N	Horizontal (no. of slots $\leq M + N - 1$)

is vertical, the constructions behave like caches. A cache of computing facilities can be constructed either on temporal or spatial dimensions. The temporal cache is more efficient from the point of view of occupied silicon area, as duplicating the configuration memory is cheaper than duplicating the raw hardware [DeHon, 1996d]. On the other hand, the spatial cache has a lower response time, because the `ACTIVATE` microinstruction used for activating a context of a multiple-context FPGA is not required. It is yet to be determined which of the spatial or temporal caching is more efficient from the point of view of both occupied silicon area and access time to a computing resource.

Regarding the complexity of temporal versus spatial constructions, we notice that *time* has only one dimension. Therefore, a temporal construction can be only of a relatively low complexity. But *space* is multi-dimensional, and the constructions built in *space* are more complex than those built in *time*. In addition to the number of facilities mapped on raw hardware, their relative position is also important. One can have *1-D*, *2-D*, or *3-D spatial caches* with different topologies, while only a *temporal 1-D cache* is possible.

7.4.7 Reconfigurable Computing Topologies

According to Brebner [Brebner, 1996; Brebner, 1997], the reconfigurable computing topologies are classified as *Sea of Accelerators* and *Parallel Harness*. In the former topology, the reconfigurable units are given individual access to operands in register file or memory (Fig. 7.30–a). For the later topology, the reconfigurable units read the operands from and provide the results to their neighbours in a pipeline fashion (Fig. 7.30–b). In order to have a consistent terminology, we will refer to the sea of accelerators as a *Sea of Computing Resources* (SoCR) topology, and to the parallel harness as a *Pipeline of Computing Resources* (PoCR) topology.

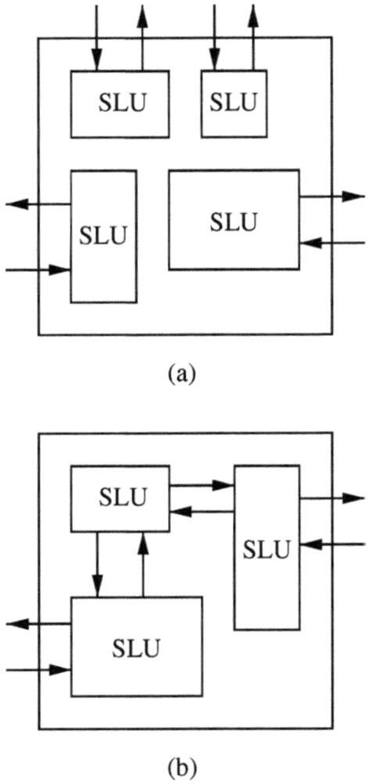

Fig. 7.30 Computing topologies: (a) – Sea of Computing Resources; (b) – Pipeline of Computing Resources

The SoCR and PoCR concepts are obvious in the framework of spatial constructions. For temporal constructions, however, several comments have to be provided. As long as each and every context uses input and output operands in register file or memory, the topology resembles the SoCR model (although the reconfigurable units are not simultaneously configured on the raw hardware). But when a context uses operands that remain stored on the reconfigurable array over a context switch as presented in Section 7.3, then the PoCR topology is in charge.

7.5 The Proposed Taxonomy of FCCMs

Based on microcode formalism, temporal and spatial constructions, and computational topologies described in the previous section, we now proceed to build the taxonomy of FCCMs. According to the first classification criterion, i.e., the **architecture of the microcode**, the FCCMs will be divided into two classes: *vertical microcoded FCCMs* and *horizontal microcoded FCCMs*. As part of the microcode architecture, the **availability of an explicit** SET instruction will be reported, too. The second classification criterion is the **structure of the set of the computing resources** that are to be configured on the raw hardware. According to this criterion, an FCCM can support either a *primordial*, *temporal*, or *spatial construction*. According to the **computing topology** criterion, both the temporal and spatial constructions can be organized either as a SoCR or PoCR. The **number of configuring resources** is the last classification criteria.

This section is organized as folows. First, a survey of the previous work concerning FCCMs classification is presented. Then, different reconfigurable computing engines will be analized and classified using the mentioned criteria. Finally the taxonomy is summarized under the form of a table.

7.5.1 Previous Work in CCM Classification

In [Guccione, 1995; Guccione and Gonzales, 1995] two parameters for classifying FCCMs are used: *Reconfigurable Processing Unit (RPU) size* (small or large) and *availability of RPU-dedicated local memory*. Consequently, FCCMs are divided into four classes:

1. Custom Instruction Set Architectures that have a small RPU and no dedicated RPU memory.
2. Application Specific Architectures that have a large RPU and no dedicated RPU memory.
3. Reconfigurable Logic Coprocessors that have a small RPU and dedicated RPU memory.
4. Reconfigurable Supercomputers that have a large RPU and dedicated RPU local memory.

Since what exactly means small and what exactly means large is subject to the complexity of the algorithms being implemented, the differences between classes are rather fuzzy. Also, providing dedicated RPU memory is an issue which belongs to the *implementation level* [Blaauw and Brooks, Jr., 1997]. Consequently, the implications to the *architectural level*, if any, are not clear.

The Processing Element (PE) *granularity*, *RPU integration level* with a host processor, and the *reconfigurability of the external interconnection network* are used as classification criteria in [Radunovi¢ and Milutinovi¢, 1998]. According to the first criterion, the FCCMs are classified as *fine-*, *medium-*, and *coarse-grain* PE based systems. The second criterion divides the machines into *dynamic* systems that are not controlled by any external device, *closely-coupled static* systems in which the RPUs are coupled on the processor's datapath, and *loosely-coupled static* systems that have RPUs attached to the host as a coprocessor. According to the last criterion, the FCCMs can have a *reconfigurable* or *fixed* interconnection network. This classification is based on the architecture of the raw hardware itself and FCCM implementation issues rather than FCCM architectural criteria.

For FCCM classification, the *loosely coupling* versus *tightly coupling* criterion is also used by other members of the FCCM community [Kastrup et al., 1999b; Wittig and Chow, 1996; Wittig, 1995; Jacob and Chow, 1999; Jacob, 1998; Compton and Hauck, 2002; Miyazaki, 1998; DeHon, 1994; DeHon, 1995; DeHon et al., 2000]. In the *loosely coupling* embodiment, the RPU is used as an *application-specific attached processor*, being connected via a bus to, and operating asynchronously with the host processor. The host will supply the reconfigurable unit with a small number of parameters, and then the reconfigurable unit will begin execution. While the reconfigurable unit performs its task, the host is free to perform other tasks. Usually, the reconfigurable unit has direct acces to the memory (DMA) and runs asynchronously with the host. At the end of the computation, a resynchronization procedure is needed. In the *tightly coupling* embodiment, the RPU is used as a *functional unit* or a *coprocessor* which augments the instruction set of the host processor. This model eliminates the problem of synchronization and reduces the communication latency between host and RPU.

We want to emphasize that all the above classifications are build using CCM *implementation* criteria. As the user observes only the architecture of a computing machine, classifying the CCMs according to architectural criteria is more appropriate. This is why we propose to classify the FCCMs according to the following criteria:

1. The verticality/horizontality of the microcode.
2. The explicit availability of a SET instruction.
3. The structure of the set of the computing resources.
4. The computing topology.
5. The number of configuring resources.

After all the background issues were presented, we can now proceed to build our taxonomy. Each FCCM will be briefly analyzed and its main architectural characteristics will be highlighted. At the end, the taxonomy will be summarized under the form of a table.

7.5.2 Vertical (Micro)coded CCMs

The FCCMs in this class are vertical (micro)coded engines, i.e., a single computing or configuring resource is explicitly controlled during a time cycle. Based on the structure of the computing units set, the vertical (micro)coded FCCMs class splits into (micro)coded FCCMs with primordial, temporal, and spatial constructions, each of which being discussed subsequently.

7.5.2.1 Vertical Microcoded FCCMs with Primordial Constructions

In such FCCMs, only a single computing resource benefits from reconfigurable hardware support at a time. The following FCCMs belong to this subclass:

- Programmable Reduced Instruction Set Computer proposed by Razdan with Harvard University, and its improvement proposed by Scott Hauck with University of Washington at Seattle.
- Garp designed by Hauser and Wawrzynek with University of California at Berkeley.
- MIPS + REconfigurable Multimedia ARray Coprocessor proposed by Miyamori with Toshiba and Olukotun with Stanford University.

The *Programmable Reduced Instruction Set Computer (PRISC)* [Razdan et al., 1994], [Razdan and Smith, 1994] maintains the instruction format of the standard MIPS processor. As depicted in Fig. 7.31, a Programmable Functional Unit (PFU) is attached directly to the register file of the RISC processor and shares the register file ports with other hardwired functional units. Application-specific instructions can be mapped onto the PFU. To control the PFU, the fixed-format of 32-bit R-type RISC instructions (two input and one output registers) [Patterson and Hennessy, 1996] is used (Fig. 7.32).

When the value of the *opcode* field in the R-type format is equal with `expfu` (*execute PFU*), then a PFU instruction is being called. The 11-bit value of the *Logical PFU function* (LPnum) indicates the required configuration (i.e., the function to be executed). The PFU is implemented on a global-reconfigurable single-context FPGA. The active configuration identifier is stored in the dedicated register Pnum. If the LPnum value $\neq$ Pnum value, a reconfiguration is needed. An exception is then raised, the processor is stalled, and a long latency reconfiguration process is initiated. As the reconfiguration process is not launched explicitly under the control of the user, a dedicated instruction for reconfiguring purpose, i.e., a `SET`, is not available.

As many as $2^{11} = 2048$ different PFU configurations can be encoded by the LPnum field. However, only a single computing unit can be controlled per cycle by a PFU instruction. Thus, PRISC is a vertical machine. Since the PRISC's control path can manage only a single reconfigurable computing unit, it is of no use to implement multiple computing units per FPGA configuration; therefore, the construction is primordial.

To reduce the overhead connected with FPGA reconfiguration, Hauck proposed a slightly modification of PRISC architecture [Hauck, 1998b], in which a `SET` instruction is explicitly provided to the user. By inserting the `SET` instruction long time before it is actually required, a *configuration prefetching* is initiated. However, the `SET` instruction becomes a `NOP` if the required circuit is already configured on the array, or is in the process of being configured. At this point the host procesor is free to perform other computations, overlapping the PFU reconfiguration with other useful work. When an `EXECUTE` instruction requires the configuration specified by the last `SET` instruction, then the PFU either will not stall the processor if the prefetching procedure has completed, or will stall the processor for a shorter period

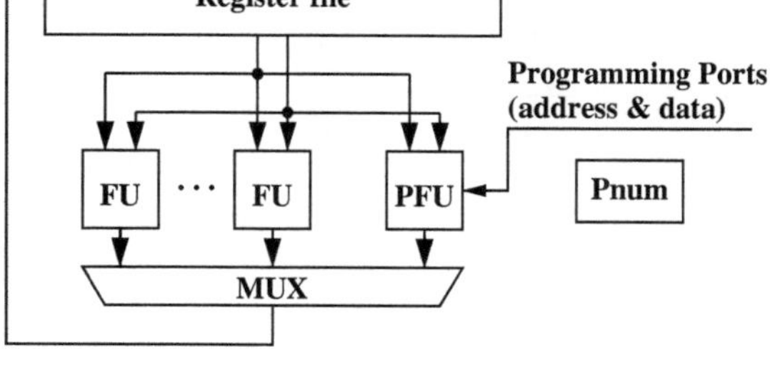

Fig. 7.31 PRISC architecture

PRISC instruction format	expfu	rs	rt	rd	LPnum
Number of bits	6	5	5	5	11

Fig. 7.32 The PRISC PFU instruction format

till the remaining reconfiguration is completed. Obviously, if the EXECUTE instruction requires a different configuration than that specified by the last SET instruction, the host processor will be stalled for the entire reconfiguration delay.

Garp designed by Hauser and Wawrzynek [Hauser and Wawrzynek, 1997; Hauser, 1997; Callahan et al., 2000] is another example of an FCCM derived from MIPS. Its basic organization is presented in Fig. 7.33. The standard MIPS instruction set is augmented with a number of instructions for loading a new configuration, initiating the execution of the newly configured computing facilities, moving data between the array and the processor's own registers, saving/retriving the array states, branching on conditions provided by the array, etc.

The coprocessor is aimed to run autonomously with the host processor. Array execution and synchronization with the host processor is governed by a countdown counter that is referred to as *the array clock counter*. As long as the clock counter is nonzero, it is decremented by 1 on every array clock cycle. When the array clock counter reaches zero, the array is stopped. For synchronization purposes, the execution of the most processor instructions that interact with the array is stalled until the clock counter reaches a zero value. For example, a new configuration can be loaded into the array only when the clock counter is zero. After a configuration has been loaded, the host processor launches the array into execution by setting the array clock counter to a nonzero value.

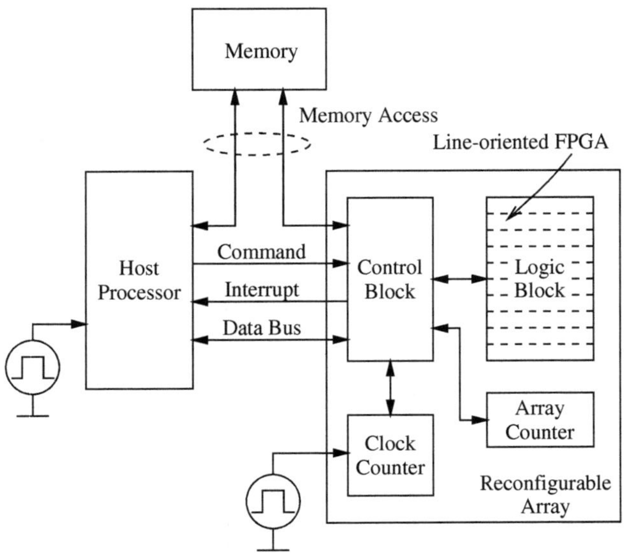

Fig. 7.33 Basic architecture of Garp

From the above mentioned considerations, we can conclude that the microcode is vertical. A SET instruction is provided to the user, although it is not explicitly named so. Also, since Garp needs only a nonzero value for the clock counter to start the computation, the instruction which loads the clock counter plays the role of an EXECUTE instruction.

The *Garp*'s FPGA is organized on rows. Only a single computing unit can be configured at a time. The smallest configuration is one row, and every configuration must fill exactly some number of contiguous rows. When a configuration uses less than the entire array, the unused rows are automatically made inactive. Therefore, the construction is primordial.

The *REconfigurable Multimedia ARray Coprocessor* (REMARC) proposed by Miyamori and Olukotun [Miyamori and Olukotun, 1998a] augments the instruction set of a MIPS core. The organization of the MIPS+REMARC system is depicted in Fig. 7.34. The reconfigurable coprocessor consists of a Global Control Unit, Data Registers, and a Reconfigurable Array. As the coprocessor does not have a direct access to the main memory, the host procesor has to write the input data to the coprocessor data registers, initiate the execution, and finally read the results from the coprocessor data registers.

The extension of the MIPS instruction set consists of instructions for downloading the configuration from memory and store it into the REMARC configuration memory, launching the execution of a reconfigurable coprocessor instruction, and transfering data between memory or host registers and coprocessor data or control registers. Therefore, we can conclude that a single configuring resource along with an explicit SET instruction are provided. Only primordial constructions having the latency of one machine cycle can be configured on the coprocessor. The microcode is also vertical.

7.5.2.2 Vertical Microcoded FCCMs with Temporal Constructions

For the machines in this subclass multiple computing resources are resident on the FPGA chip. Only a single resource is active at a time, while the others are all idle. The following FCCMs are analized:

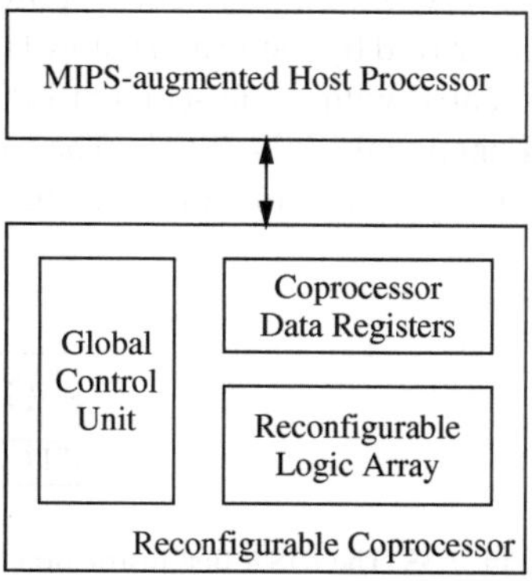

Fig. 7.34 Block diagram of a microprocessor augmented with the REMARC programmable array

- Reprogrammable Instruction Set Accelerator introduced by Trimberger with Xilinx.
- OneChip proposed by Wittig, Jacob, and Chow with University of Toronto.
- Virtual Computer designed by Steven Casselman with Virtual Computer.
- WASMII proposed by Ling with Kanagawa Institute of Technology, and Amano with Keio University, Japan.
- MorphoSys designed by H. Singh, Lu, Lee *et al.* with University of California at Irvine.

The system proposed by Trimberger [Trimberger, 1998a] consists of a host processor augmented with a multiple-context FPGA–based [Ong, 1995] referred to as *Reprogrammable Instruction Set Accelerator* (RISA). This system provides three different instruction formats:

1. The PRISC format, where the RISA-assigned instructions have the opcode field value equal to `FPGAOP` (an opcode different from `FPGAOP` specifies a hardwired instruction). An immediate data field specifies the required operation (Fig. 7.35). As in PRISC, the microcode is vertical.
2. A D/P flag specifies a Defined (that is, hardwired) or Programmed instruction (Fig. 7.36). A plurality of opcodes are assigned to FPGA. Immediate data is also provided. The microcode is also vertical.
3. The VLIW instruction format, which includes both an opcode field for the defined execution unit and an opcode field for the RISA (Fig. 7.37). As this format refers to horizontal microcode, we will dicuss it in a subsequent subsection.

Each of the above mentioned formats defines an architecture. It seems that the author claimed three architectures in the same publication. Therefore, our taxonomy provides three entries for Reprogrammable Instruction Set Accelerator: RISA', RISA", RISA"'.

Concerning the reprogramming procedure, Trimberger mentions that RISA reconfiguration is under control of a dedicated so called execution unit. However, it is not clear whether the reconfiguration is explicitly launched by the user or is automatically performed by the execution unit. Therefore, it is not obvious if an explicit SET instruction is available. Furthermore, the visibility of the temporal construction built on the multiple-context FPGA at the architecture level is also not obvious.

The *OneChip* system proposed by Wittig and Chow [Wittig and Chow, 1996] and improved by Jacob and Chow [Jacob and Chow, 1999] consists of a MIPS host augmented with a four-context FPGA–based [Jones and Lewis, 1995] Programmable Functional UNIT (PFU). The format of the PFU instruction is depicted in Fig. 7.38. The *opcode* field indicates when an FPGA-mapped operation is launched. The 4-bit

Opcode	A	B	Y	Immediate
ADD	R3	R4	R5	XXXX
FPGAOP	Rx	Ry	Rz	XXXX

Fig. 7.35 The RISA first instruction format

D/P	Opcode	A	B	Y	Immediate
D	ADD	R3	R4	R5	XXXX
P	FPGAOP(N)	Rx	Ry	Rz	XXXX

Fig. 7.36 The RISA second instruction format

Fixed Opcode	A	B	Y	Programmed Opcode	C	Immediate
ADD	R3	R4	R5	FPGAOP(N)	R22	XXXX

Fig. 7.37 The RISA third instruction format

FPGA configuration field is used to specify the required configuration. In this way, $2^4 = 16$ different PFU configurations can be encoded. According to our formalism, only a single computing unit is controlled at any given time; thus, OneChip is a vertical machine.

The PFU is granted a direct memory access. Accessing the memory space is realized by an indirect addressing mode. The starting input memory block address (R_source) and the starting output memory block address (R_dest) as well as their sizes (Source Block Size, Destination Block Size) are specified within the PFU instruction. This way, OneChip can process large chuncks of data at a rate which exceeds the performance supported by two input and one output register-based operands of the standard MIPS.

The PFU is build on a four-context FPGA. A single computing facility can be configured per context; therefore the construction is temporal. The FPGA active and idle configurations are managed by means of a so-called Reconfiguration Bits Table. The computing resources are either configured on the PFU when a configuration miss is detected, or pre-loaded by compiler directives. Since, no further details are provided, it is not obvious if a `SET` and/or `ACTIVATE` instructions are provided.

The *Virtual Computer* designed by Steve Casselman [Casselman, 1993; Thornburg and Casselman, 1994; Casselman, 1997] is an FCCM which includes a host processor and a programmable array consisting of a set of Xilinx XC4000 FPGA devices interconnected through I-Cube IQ160 *Field Programmable Interconnection Devices* (FPID) [I-Cube, 1998b]. The host processor can reconfigure each of the FPGAs and FPIDs independently.

While the FPGAs are single-context devices, FPIDs are double-context devices. In an FPID device, a configuration bit stream is shifted serially through array to reconfigure the idle context. The active context can continue to operate in parallel with idle context reconfiguration. Therefore, the construction is temporal. Moreover, as the configuration of the FPGAs may not change during a FPID configuration

Instruction format	opcode	FPGA config.	misc.	R_source	R_dest	source block size	destination block size
# of bits	6	4	2	5	5	5	5

Fig. 7.38 The OneChip FPGA instruction format

swap, we may consider that state sharing between configurations is provided. Consequently, both the SoCR and PoCR computational topologies are equally utilizable.

Unfortunately, from the description provided by the author we cannot conclude about the verticality or horizontality of the microcode. The best we can do is to assume the simplest case, i.e., a vertical microcode.

The *What A Stupid Machine It Is* (WASMII) proposed by Ling and Amano [Ling and Amano, 1993; Ling and Amano, 1995], [Takayama et al., 1999] is based on the dataflow paradigm of computation. It aims to execute programs written as large dataflows graphs that can be topologically splitted into separate subgraphs. WASMII structure is pictured on Fig. 7.39. It includes a Fujitsu's multiple-context programmable logic device (MPLD) [Yoshimi and Ikezawa, 1990; Takayama et al., 1999], a Token Router, an Input Token Registers, a Page Controller, and an external memory. Each of the subgraphs in a large graph will be sequentially configured on the MPLD raw hardware as pages, and further executed on the WASMII machine. The state information is lost on an MPLD context switching, the resources for state sharing between context being provided outside of the array. Such resources are referred to as *input token registers*. The *token router* receives values, i.e., tokens, which have been computed by the actived context and sends them to the input token registers of the appropriate idle context. A context is ready to be activated after all its input tokens have arrived. When an active context has completed its computation and dispatched all of its output tokens, one of the ready contexts is activated.

The *page controller* is responsible with preloading of pages from external memory into the multiple-context configuration memory. This task is performed according to an order decided in advance. Although the availability of an explicit SET instruction is suggested, no further details are provided. Also, we would like to mention that the process of activating and dezactivating the contexts is hidden for user.

Finally, it is worth mentioning that a spatial construction of WASMII machines can be easily built [Ling and Amano, 1993]. The resulting system is referred to as

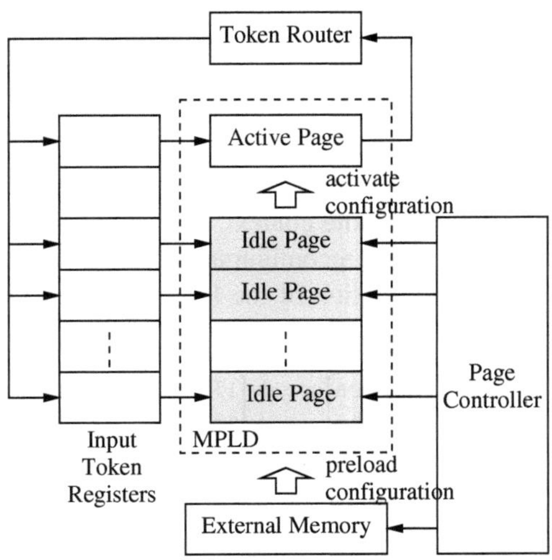

Fig. 7.39 WASMII architecture

Multi-Chip WASMII. Unfortunately, no detailed information regarding the architecture of such complex system is provided.

As depicted in Fig. 7.40, the *Morpho*ing *Sys*tem (MorphoSys) [Lee et al., 2000; Singh et al., 2000] consists of five major components: a MIPS-like processor core referred to as TinyRISC, a coarse-grain 8×8 array (RC Array) of so called reconfigurable cells, a Context Memory for the RC Array, a Frame Buffer, and a DMA Controller. Basically, each reconfigurable cell comprises an ALU, a multiplier, a shift unit, two multiplexers that select the inputs, an output register, and a register file. The RC array configuration is stored into the *Context Memory* and is broadcasted to the reconfigurable array in two modes: column-wise or row-wise. For column (row) broadcast, all eight reconfigurable cells in the same column (row) are configured by the same context word., and, therefore, perform the same operation. The context memory is organized into two blocks, one for the row mode and the other for the column mode. Up to 16 configuration planes may be stored in each block; thus, the RC array can be regarded as a 32-context reconfigurable device.

In addition to the typical RISC instructions, TinyRISC instruction set includes two classes of specific instructions for controlling the MorphoSys components: DMA instructions and RC Array instructions. The *DMA instructions* initiate data transfers between main memory and the Frame Buffer, and the load of context words from main memory into the Context Memory. The *RC Array* instructions specify the context to be activated and the broadcast mode. Since the MorphoSys engine can overlap computation with data transfers, the DMA transfers can take place concurrently with RC Array execution.

It is obvious that MorphoSys is a vertical coded machine with temporal construction. Since the user can specify what context to prefetch and what context to be executed, a SET instruction is indeed provided.

7.5.2.3 Vertical Microcoded CCMs with Spatial Constructions

In this subclass, multiple computing resources are resident on the FPGA chip as active circuits. The following machines are discussed:

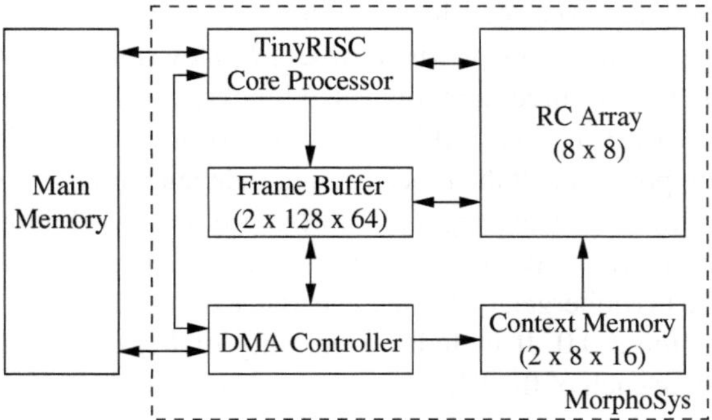

Fig. 7.40 MorphoSys

- PRISM system proposed by Athanas with Virginia Polytechnic Institute and State University, and Silverman with Brown University.
- PRISM-II/RASC proposed by Wazlowski also with Brown University.
- Multiple-RISA developed by Trimberger with Xilinx Corporation.
- T1000 designed by Zhou and Martonosi with Princeton University.
- The system proposed by Gilson.
- The Nano-Processor and the Dynamic Instruction Set Computer introduced by Wirthlin, Hutchings, *et al.* with Brigham Young University.
- CCSimP developed by Salcic *et al.* with Auckland University.
- ConCISe system proposed by Kastrup *et al.* with Philips Research.
- Chimaera developed at Northwestern University by Hauck *et al.*.
- URISC proposed by Brebner and Donlin with University of Edinburgh.
- Functional Memory developed by Halverson and Lew with University of Hawaii.
- Molen $\rho\mu$-coded processor proposed by Vassiliadis *et al.* with Delft University of Technology.
- Xputer designed by Hartenstein *et al.* with University of Kaiserslautern.
- Splash designed by Buell *et al.* with Supercomputing Research Center.
- NAPA processor designed at by Garverick, Rupp *et al.* with National Semiconductor.

PRISM (*Processor Reconfiguration Through Instruction-Set Methamorphosis*) is in fact one of the earliest proposed FCCMs [Athanas and Silverman, 1991; Athanas, 1992a]. It consists of a RISC processor and reconfigurable hardware that are coupled on an external bus. To evaluate FPGA-mapped functions, the host processor explicitly uploads the input operands to FPGA and then downloads the result back to register file. In order to handle the loading of FPGA configurations, the compiler inserts library function calls into the program stream [Athanas and Silverman, 1993]. From this description, we can conclude that a SET instruction is available.

As shown in Fig. 7.41, functions written in C language are compiled into an FPGA configuration which is further mapped into the reconfigurable hardware. A separate computing resource is generated for each FPGA-mapped function. Therefore, we classify the construction as being *spatial*. Both SoCR and PoCR computing topologies are supported.

A further development of PRISM is PRISM-II [Wazlowski et al., 1993]. As depicted in Fig. 7.42, it consists of a host augmented with a so called *Reconfigurable Architecture Superscalar Coprocessor* (RASC) [Wazlowski, 1996]. RASC contains a *spatial construction* of three interconnected blocks, where each block includes a *Reconfigurable Processing Unit* (RPU) and some local memory. Three processing configurations are possible: all the blocks acting independently; any combination of two processing blocks working together; all three blocks working together as one large block. Thus, both SoCR and PoCR computing topologies are supported.

Multiple-RISA [Trimberger, 1998b] is spatial extension of RISA already presented in Subsection 5.0.0. It includes a plurality of *Reprogrammable Execution Units* (REU), where each of the REU can be separately reconfigured. A single computing resource can be configured on each REU. Since there are multiple REUs in a multiple-RISA system, the construction is spatial. An Instruction Management

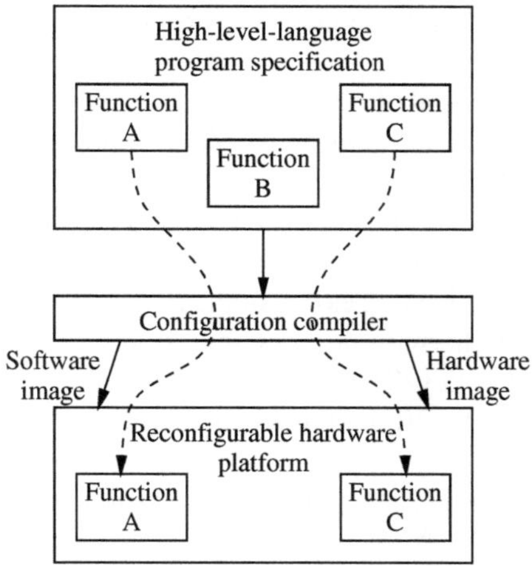

Fig. 7.41 Overview of the PRISM compiler

Logic keeps track of the computing resources which are currently configured in the REUs. If an FPGA-assigned instruction matches one of the configurations in the REUs, the decoder directs the execution to the proper computing unit. If the issued instruction does not match any of the configured computing units, a miss is indicated and an exception is raised. The processor is then stoped and a reconfiguration procedure is initiated. This way, the Multiple-RISA behaves like a spatial cache of computing units. The management of such a cache can be performed, for example, by a *least recently used* strategy.

From the above considerations, we can conclude that the user does not have a direct control the reconfiguration process. No means for pre-loading computing resources or controlling what computing resource is to be replaced are provided to the user. Consequently, an *on-demand* reconfiguration strategy is employed, i.e, an explicit SET instruction is not available. In addition, the spatial construction is not an architectural issue; it is only an implementation issue. Finally, we mention that

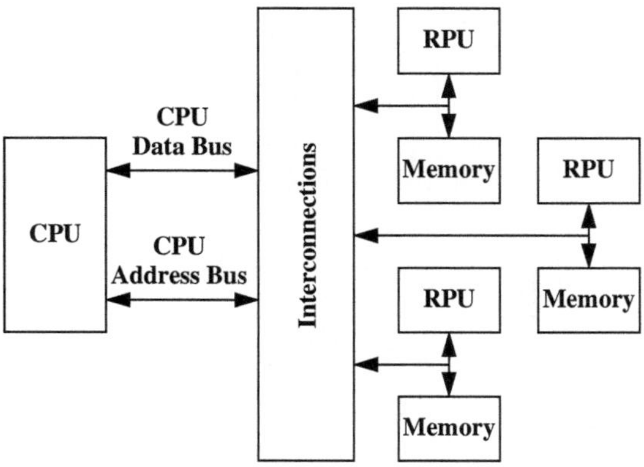

Fig. 7.42 Simplified RASC organization

the Multiple-RISA system has the same instruction set as the standard RISA system. The microcode is vertical.

About the same architectural idea with Multiple-RISA is employed by the T1000 processor proposed by Zhou and Martonosi [Zhou and Martonosi, 2000]. T1000 is a spatial extension of PRISC, and consists of a plurality of PFUs that augment a superscalar 4-issue out-of-order host. As in RISA, an explicit SET instruction is not provided. Therefore, the system automatically performs the configuration management. If the issued instruction matches one of the PFU configurations, the instruction is dispatched normally. Otherwise, a reconfiguration procedure is launched based on a *least recently used* policy. The T1000 is a vertical microcoded machine. The computational topology is SoCR, and the construction is spatial (also not visible at the architecture level).

The organization of the FCCM proposed by Gilson [Gilson, 1994b], [Gilson, 1994a] is shown in Fig. 7.43. It consists of a host processor and at least two FPGA-based *computing devices*. Each such device takes the form of a RISC engine that has a unique *reconfigurable instruction execution unit*, on which customized circuits can be implemented. Thus, construction is *spatial*. Through a so-called Host Interface, the host processor controls the reconfiguration of FPGAs by loading new configuration data into the FPGA Configuration Memory. Thus, a write into the FPGA configuration memory is in fact a SET instruction. The reconfiguration process can be performed such that when one computing device is being reconfigured, all the others may continue their execution. As the computing devices are coupled on the host bus, only one of them can be accessed at a time. Therefore, the microcode is vertical, and the computing topology is SoCR.

The *Nano-Processor* proposed by Wirthlin *et al.* [Wirthlin et al., 1994] consists of a very simple processor core augmented with a general-purpose FPGA, on which customized computing units can be instantiated at application load-time. Each custom instruction is implemented on a separate module, each module being responsible to decode its instruction opcode.

The Nano-Processor instruction format is shown in Fig. 7.44. It includes 16 bits, where the opcode takes 5 bits and the operand takes 11 bits (3 bits the page address

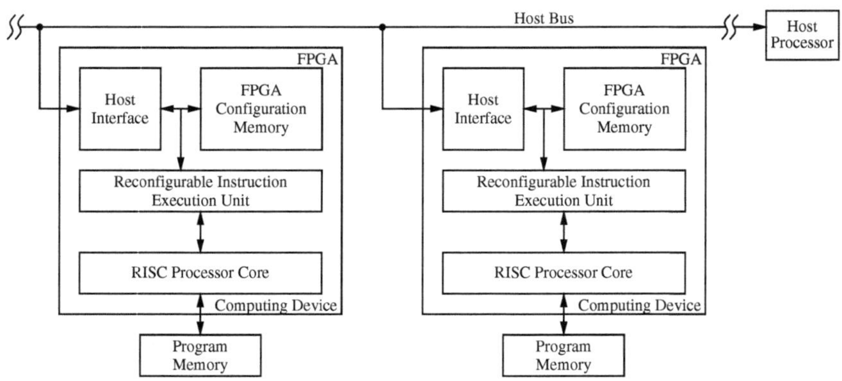

Fig. 7.43 The organization of the processor proposed by Gilson

Instruction format	Page Address	Opcode	Offset
Number of bits	3	5	8

Fig. 7.44 The Nano-Processor instruction format

and 8 bits for the offset). With the 5-bit opcode $2^5 = 32$ instructions can be encoded. Six instructions control hardwired computing resources, while the other 26 control customized computing resources.

As each instruction controls a single hardwired exclusive-or configured computing resource, the microcode is vertical. Since up to 26 computing units may simultaneously reside on the reconfigurable array, the construction is spatial. Obviously, the computing topology is SoCR. For the reconfiguration is performed at application load-time by an embedded procedure, an explicit SET instruction is not available at run-time.

Dynamic Instruction Set Computer (DISC) designed by Wirthlin and Hutchings [Wirthlin and Hutchings, 1995] is a further development of the Nano-Processor. It consists of an 8-bit processor referred to as a global controller, augmented with a programmable array with line-wise partial reconfiguration capabilites. The global controller contains a number of hardwired instructions for sequencing, status controlling, and memory transfers. The programmable array is structured on rows. A computing resource is implemented as an instruction module that streches horizontally across the entire width of the array. As depicted in Fig. 7.45, the modules are connected with the global controller through a communication network which

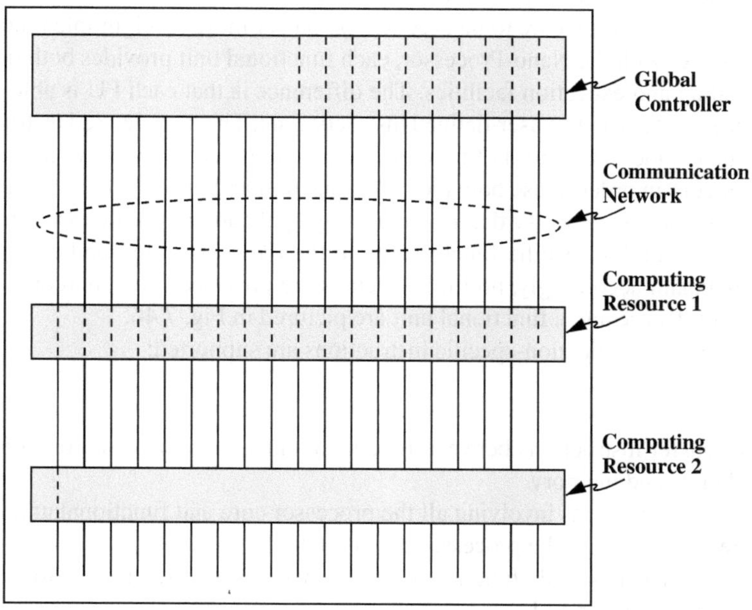

Fig. 7.45 The simplified organization of the Dynamic Instruction Set Computer

stretches vertically across the die. Each instruction module may consume an arbitrary amount of hardware just by varying its height. In addition, the instruction modules are designed to be physically relocatable, i.e., they operate properly at any vertical location.

As in the Nano Processor, every custom computing resource contains decode, data-path, and control units. By extending the decoder of the host processor, the decode unit assigns a specific op-code to the custom instructions and is responsible for acknowledging its presence to the global controller. As opposed to Nano Processor where the number of the configurable computing resources, and their associated instructions are limited by the load-time configuration strategy, the DISC computing resources use the raw hardware only when needed. This way, an arbitrary number of application-specific instructions can be implemented.

A spatial construction with SoCR computing topology can be configured on DISC. At run-time custom computing resources are configured onto the reconfigurable array until all available raw hardware is occupied. After that, new instructions can be configured onto the array only by removing old ones. Each custom decode unit compares the opcode of the issued instruction for a match against its own opcode during the first instruction cycle. On a match, the module signals the global controller that the hardware is present and instruction sequencing can continue. If a miss is detected, the global controller is stalled in order to allow for a reconfiguration process. The system performs the management of the computing resources automatically. Therefore, an explicit SET instruction is not available. Since only a single computing resource is explicitly controlled at a time, the microcode is vertical.

In the *Custom-Configurable SimP* (CCSimP) machine [Salcic and Maunder, 1996a; Danecek et al., 1995; Salcic, 1996], the instruction set of the SimP processor [Salcic and Maunder, 1996b] is augmented with user-defined (or application-specific) instructions that are to be executed on one or more configurable functional units (FU). As with the Nano-Processor, each functional unit provides both instruction decoding and execution facilities. The difference is that each FU is able to decode and execute an entire user-defined instruction set. The user-defined instructions are executed under the control of functional unit and, if necessary, they can stall the processor core in order to use both the functional unit and processor core datapaths. Although the authors specify that implementing application-specific instructions is performed by adding specific functionality in the FU control unit and configuring the FU datapath accordingly, no further details are provided. The interconnections between SimP core and a functional unit are pictured in Fig. 7.46.

Four types of application-specific instructions are supported:

1. Data transfer instructions between functional units and processor core, or functional units and memory.
2. Complex instructions involving all the processor core and functional unit datapaths resources, while the processor core is stalled.
3. Instructions that use only functional unit datapath resources, allowing for parallel execution of core and FU.
4. Instructions for core and functional unit resynchronisation.

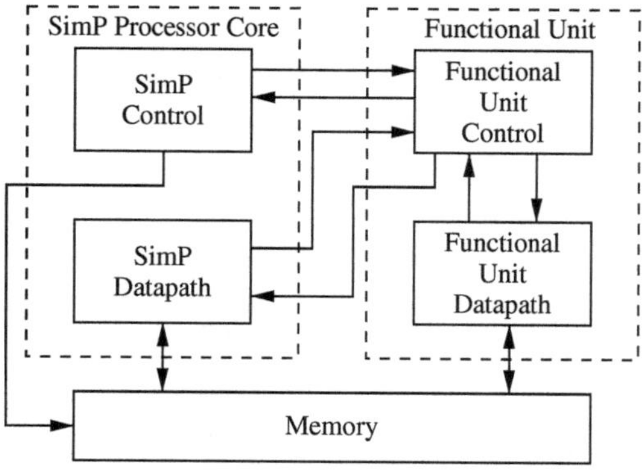

Fig. 7.46 Interconnections between SimP processor core and a functional unit

The format of an application-specific instruction is shown in Fig. 7.47. It includes three fields specifying the *instruction type*, *FU identification* (ID), and *operation code*. As it can be easily deducted, up to eight functional units, and up to 1024 application-specific operations per functional units can be encoded.

The microcode is vertical since a single FU is controlled at any given time. For the functional units are configured at application load-time by an embedded procedure, an explicit SET instruction is not available at run-time. The construction is spatial and the computing topology is SoCR.

The *ConCISe* architecture proposed by Kastrup *et al.* [Kastrup et al., 1999a] is depicted in Fig. 7.48. In order to reduce the reconfiguration overhead encountered in the PRISC engine, a spatial construction (SoCR topology) of computing units is deployed on the reconfigurable array.

As depicted in Fig. 7.49, the 11-bit FPGA configuration identifier of PRISC is splitted in two subfields. The first (11-L)-bit subfield specifies the required configuration, while the L-bit subfield is a *function identifier*. In this way, 2^{11-L} different PFU configurations can be specified by the *configuration ID* field, and 2^L functions per RFU configuration can be encoded.

Similar to PRISC, ConCISe employs an *on-demand* loading strategy, i.e., an explicit SET instruction is not available. Loading a new off-chip configuration can be performed only with low speed, as it commonly happens with a single-context FPGA. The instruction in Fig. 7.49 can control a single computing unit per time cycle; thus, ConCISe is a vertical machine.

The architecture of *Chimaera* proposed by Hauck *et al.*[Hauck et al., 1997; Ye et al., 2000], is depicted in Fig. 7.50. It consists of a host processor augmented with a *Reconfigurable Functional Unit* (RFU). The RFU gets its operands from the host's register file or a shadow register file. Up to nine input and one output operands can be used by an RFU instruction. Each RFU configuration defines from what registers

Instruction format	FU Instruction type	FU ID number	FU operation code
Number of bits	3	3	10

Fig. 7.47 CCSimP instruction format

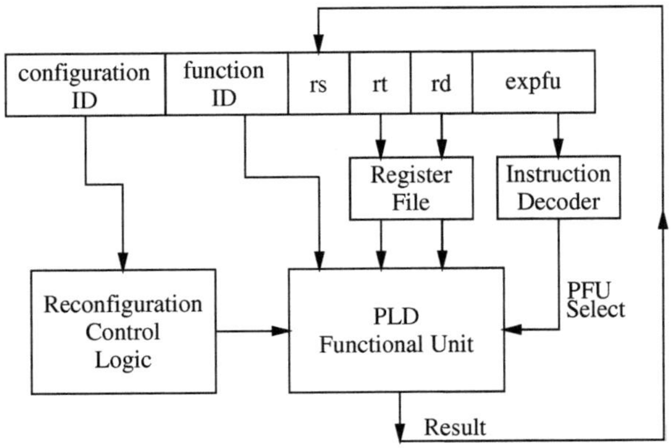

Fig. 7.48 ConCISe architecture

Instruction format	expfu	rs	rt	rd	function ID	config. ID
Number of bits	6	5	5	5	L	11-L

Fig. 7.49 The ConCISe instruction format

the reconfigurable unit reads in its operands. As such, the RFU instruction format does not provide identifiers for the input operands. It includes only the RFUOP opcode, indicating that an RFU-assigned instruction is being called, an ID operand specifying which particular reconfigurable function to call, and the destination register identifier.

The programmable array is structured on rows of active logic, where a computing resource can use one or more rows. The reconfiguration pattern is partial, the reconfigurable atom being one row. Thus, several computing resources can be

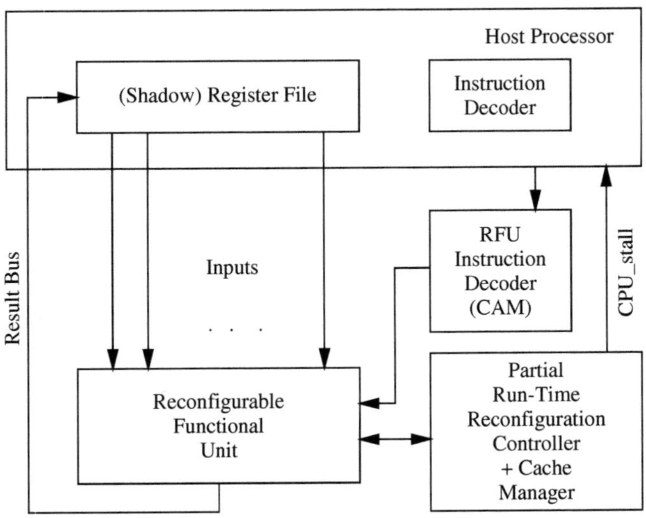

Fig. 7.50 Simplified architecture of Chimaera

active on the RFU at a time. Since each reconfigurable computing unit knows from which register to read its input operands, all mapped RFU computing resources run in parallel every machine cycle. This speculative execution model allows RFU instructions to have latencies of multiple cycles without the need for stalling the host processor. A result is written back to the (shadow) register file in just one machine cycle only when the corresponding instruction is actually called. In this way, the RFU calls act just like any other instructions, and fits into the processor's standard execution pipeline.

The RFU Instruction Decoder's Content Addressable Memory determines if the next instruction in the instruction stream has an RFUOP opcode, and, if so, whether the corresponding computing resource is currently loaded. If the resource is already loaded, the value returned by the instruction is written in the (Shadow) Register File. If the resource is not loaded, then a miss is detected, and an exception is raised. The Partial Run-Time Reconfiguration Controller stalls the processor, loads the proper computing resource from memory into the Reconfigurable Array, and launches the instruction into execution. The reconfiguration is performed on a per-row basis, the new instruction overwritting one ore more of the currently loaded instructions.

The reconfiguration process is not under the control of the user. Consequently, a dedicated `SET` instruction for reconfiguration is not provided. A spatial construction of computing resources with SoCR topology is build on the Chimaera FPGA. As only a single computing resource is controlled at any given time, the microcode is vertical.

Flexible Ultimate RISC (Flexible URISC) designed by Brebner and Donlin [Brebner and Donlin, 1998], [Donlin, 1998] contains a minimal processor core (URISC) [Jones, 1988] with a single instruction: `move memory-to-memory`. The URISC contains a set of fixed and reconfigurable units, having their input and output registers mapped in the memory space of the processor. The computation is performed by moving operands to and from these memory-mapped registers. Unconditional jumps are implemented by moving the target address into the memory-mapped *program counter*. Conditional jumps are performed by adding the contents of a memory-mapped *ALU condition code register* to a branch address and writting the result back to the program counter. This allows the destination of the jump to be offset by a truth or false value contained in the condition code register.

The configuration memory of the reconfigurable units is also mapped in the processor's memory space. Consequently, the units can be reconfigured at run-time under the command of the URISC, by the same `move` instruction. This way, a `SET` instruction is emulated. Obviously, the microcode is vertical, the construction is spatial, and the computing topology is SoCR.

A similar example is the *Functional Memory* designed by Halverson and Lew [Halverson, Jr. and Lew, 1996]. A host processor having only two instructions (`move` and `jump`) is augmented with a so-called *functional memory* that consists of an FPGA connected in parallel with a random access memory. This compound is mapped in the memory space of the host. The basic architecture of the system is depicted in Fig. 7.51.

The main processor sends configuration data to FPGA by means of `move` instructions; thus, a `SET` instruction is available. The configuration data defines both

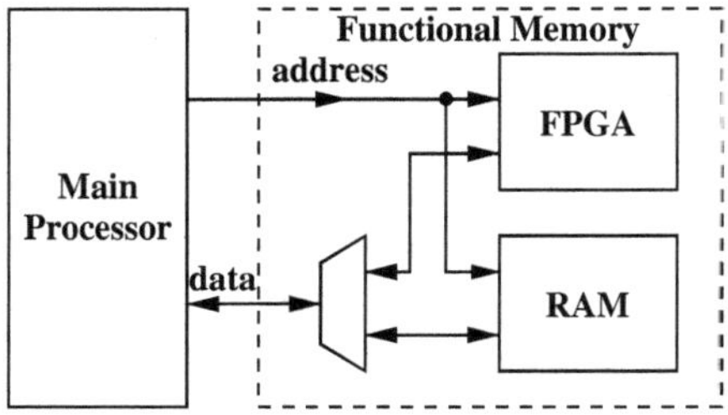

Fig. 7.51 The basic architecture of functional memory

the memory addresses the FPGA will be responsive to and the functions that the FPGA will perform. The reconfigurable unit's operands are stored into these input registers. When the main processor writes data into FPGA input registers, the data is stored also in RAM. This way, the input operands can be subsequently read by the main processor, too. When the main processor reads from FPGA mapped addresses, it downloads the values computed by FPGA.

The functional memory can be thought of as being a *spreadsheet computer*. Some locations (FPGA input registers and RAM cells) store data, while those providing the results computed by FPGA can be associated with computing formulas. Therefore, a number of memory-mapped locations assigned to the FPGA can be programmed to be the calculated result of an expression for which other memory mapped locations are the arguments. As with URISC, we classfy the microcode as vertical, the construction as spatial, and the computing topology as SoCR.

The *Molen $\rho\mu$-coded processor* has been proposed by Vassiliadis *et al.* [Vassiliadis et al., 2001; Vassiliadis et al., 2004]. In its more general form, the proposed machine organization is shown in Fig. 7.52. Instructions are fetched from the memory and stored in the instruction buffer (I_BUFFER). The ARBITER fetches instructions from the I_BUFFER and performs a partial decoding on the instructions to determine where they should be issued. Instructions that have been implemented in fixed hardware are issued to the *Core Processing* (CP) unit. The instructions entering the CP unit are further decoded and then issued to their corresponding functional units. Instructions that are to be implemented in reconfigurable hardware are issued

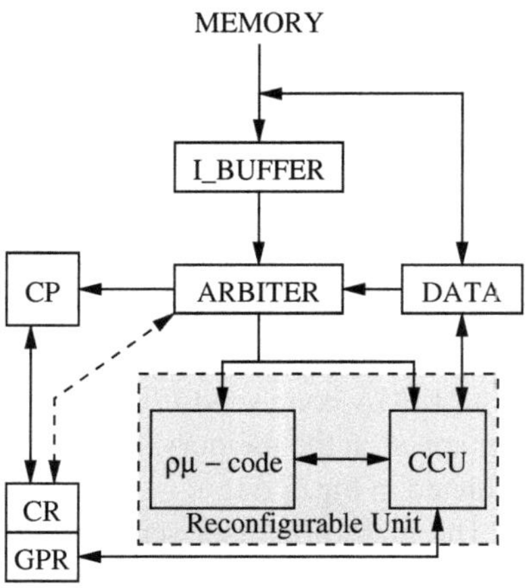

Fig. 7.52 The organization of the Molen processor

to the *Reconfigurable Unit*. The source data are fetched from the *General-Purpose Registers* (GPR) and the results are written back to the same GPRs. Other status information is stored in the *Control Registers* (CR).

The reconfigurable unit consists of a *Custom Configured Unit* (CCU) and the $\rho\mu$-code unit. An operation executed by the reconfigurable unit is divided into two distinct phases: SET and EXECUTE. The SET phase is responsible for reconfiguring the CCU raw hardware enabling the execution of the operation. Such a phase may be subdivided into two subphases, namely *partial set* (P − SET) and *complete set* (C − SET). The P − SET covers common functions of an application or set of applications. More specifically, in the P − SET phase, the CCU is *partially* configured to support these common functions. While the P − SET sub-phase can be possibly performed during the loading of a program or even at chip fabrication time, the C − SET sub-phase is performed during program execution. Furthermore, the C − SET sub-phase only partially reconfigures remaining blocks in the CCU (not covered in the P − SET sub-phase) in order to augment the functionality of the CCU; this way, other less frequent functionsare supported in reconfigurable hardware. For the reconfiguration of the CCU, reconfiguration microcode is first loaded into the $\rho\mu$-code unit and then executed to perform the actual reconfiguration. In the EXECUTE phase, the configured operation is performed by an execution microcode resident into the $\rho\mu$-code unit. In this way, different operations are performed by loading different reconfiguration microcodes and different execution microcodes, respectively.

The instruction format of the P − SET, C − SET, and EXECUTE instructions is given in Fig. 7.53. The opcode (OPC) specifies which instruction to perform. The Reconfigurable/Pageable (R/P)-bit specifies where the microcode is located and implicitly also specifies how to interpret the address fiels: as a main memory address, α, (R/P=1) or as an control store address, $\rho CS - \alpha$, for the $\rho\mu$-code unit (R/P=0). It should be noticed that the address field always points to the address of the first microcode instruction in a microroutine. That is, instead of specifying new instructions for the operations (requiring instruction opcode space), the instruction simple points to (memory) addresses.

We can conclude that Molen is a vertical microcoded machine (can also be horizontal as we will describe in a subsequent section), which has the SET instruction exposed to the user. Since more than one computing resources can be configured on the CCU at the same time, the construction is *spatial*.

The *Xputer* system has been proposed by Hartenstein *et al.* [Hartenstein et al., 1992]. As depicted in Fig. 7.54, it is composed of a *reconfigurable ALU* (rALU), a

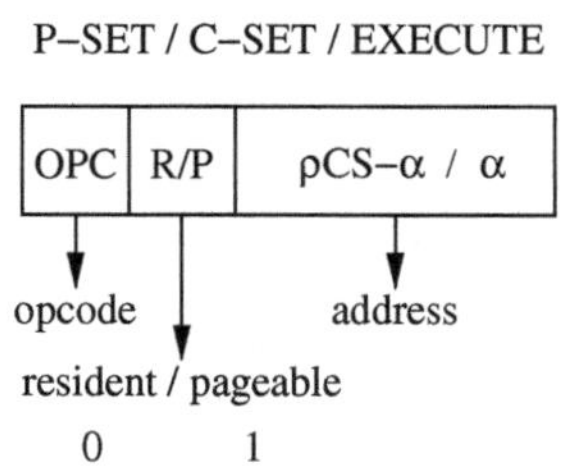

Fig. 7.53 The P − SET, C − SET, and EXECUTE instruction formats

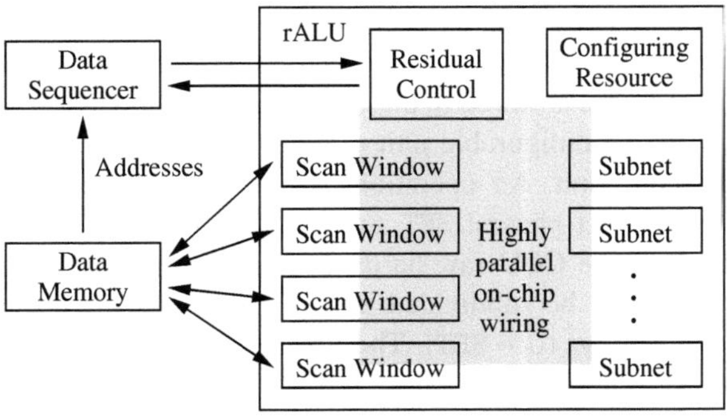

Fig. 7.54 The Xputer organization – adapted from [Hartenstein et al., 1992]

hardwired *data sequencer*, and a *data memory*. Multiple-input and multiple-output custom computing resources referred to as compound operators can be mapped on rALU. Each resource is configured as a subnet inside rALU. A set of register files which are called *scan windows* or *scan caches* defines the shape and size of windows onto the data memory array that are used to transfer data between rALU and data memory. The *data sequencer* provides a set of generic data memory address sequences that makes a scan window slide over the data memory step-by-step along a path called a *scan pattern*. A *Tagged Control Word* (TCW) is located at the end of the data sequence. When the scan cache arrives at the TCW, it changes the state of a so-called *residual control logic* in order to select further actions. A TCW decoder is configured also as a subnet within rALU.

An example of an Xputer processing procedure with 3 × 3 scan window and linear scan pattern is depicted in Fig. 7.55. An Xputer executable comprises a data map defining the order in which the data are stored in the data memory, a scan

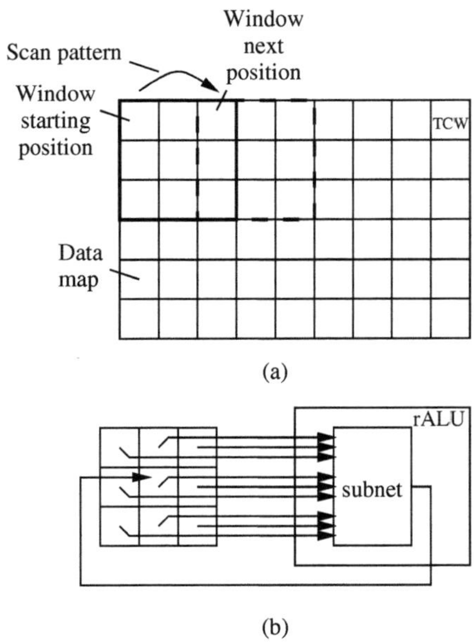

Fig. 7.55 Processing in Xputer: (a) – 3 × 3 scan window with a linear scan pattern; (b) processing within the 3 × 3 scan window

window shape specification, rALU subnets and wiring specification, a scan pattern, and a TCW. As a single computing resource (subnet) is controlled at a time, the microcode is vertical. Also, for there are multiple subnets configured at a time, the construction is spatial. The computational topology is SoCR. Finally, we mention that the subnets inside rALU are configured at application load-time. Therefore, a SET instruction is not available at run-time.

Splash system has been designed by a team with Supercomputing Research Center [Buell et al., 1996]. The first generation (Splash-1) consists of a board including a linear array of 32 Xilinx XC3090 chips, and a VME interface [Gokhale et al., 1991; Buell et al., 1996]. The second generation (Splash-2) [Arnold et al., 1992; Buell et al., 1996] is composed of up to 16 so called Array Boards, each of them containing 17 Xilinx XC4010 chips. As depicted in Fig. 7.56, Splash-2 system is connected through an Interface Board to a Sparc workstation host. The array boards can be connected in a linear or SIMD fashion. Therefore, both SoCR and PoCR computational topologies are possible. Each array board is globally reconfigurable under a host command. Thus, a SET instruction is available. Since only a single command to an array board can be issued on the SBus at a time, we classify Splash as being a vertically-coded engine.

The *National Adaptive Processing Architecture* (NAPA) introduced by Rupp *et al.* [Rupp et al., 1998; Rupp, 1998] consists of a 32-bit RISC-based core referred to as the *Fixed Instruction Processor* (FIP), which is augmentèd with a partial reconfigurable execution unit called *Adaptive Logic Processor* (ALP). The ALP is build on the CLAy FPGA from National Semiconductor's [Garverick et al., 1994; Rupp, 1995]. An interface between the FIP and ALP, which is referred to as the *Reconfigurable Pipeline Controller* (RPC), supplements the FIP instruction set with a so-called Reconfigurable Pipeline Instruction Set (RPIS). The RPC also performs configuration management. The simplified organization of the NAPA processor is depicted in Fig. 7.57.

The execution model is as follows. When the FIP reaches a point in the program where a custom instruction is to be executed, parameters are passed to the ALP together with an RPIS instruction. Then the FIP suspends execution until the ALP

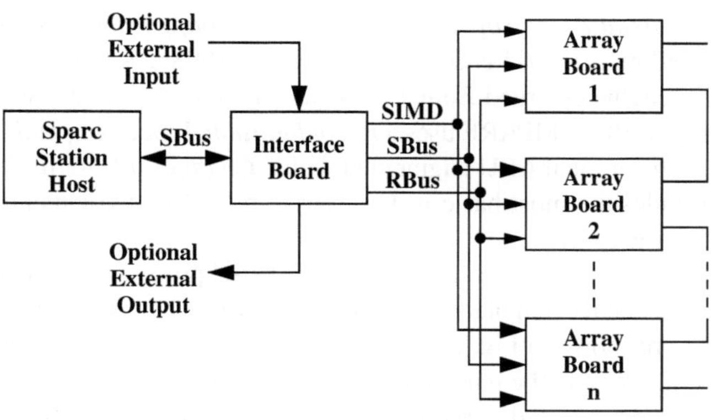

Fig. 7.56 SPLASH organization – extracted from [Buell et al., 1996]

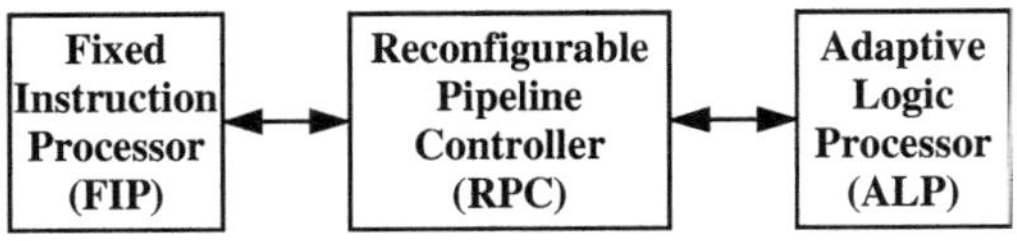

Fig. 7.57 The NAPA simplified organization – adapted from [Rupp et al., 1998; Gokhale and Stone, 1998]

completes the operation. Unfortunately, from description provided by the authors, it is not obvious if the FPGA reconfiguration is managed independently in hardware by the RPC, or the user is given an explicit SET instruction. We complete the presentation of NAPA by mentioning that layout of a custom computing unit covers a rectangular region on the FPGA that is extended over the entire length of one dimension, much like in the Dynamic Instruction Set Computer (DISC) case. Thus, we can conclude that the construction is spatial.

7.5.3 Horizontal Microcoded CCMs

The FCCMs exhibiting a horizontal microcode is the second major class of the taxonomy. For this class, two or more configuring and/or computing resources can be controlled at a given time. Three subclasses are generated by the structure of the reconfigurable computing facilities: horizontal microcoded CCMs with primordial constructions, temporal constructions and spatial constructions.

7.5.3.1 Horizontal Microcoded CCMs with Primordial Constructions

The following two FCCMs will be analyzed:

- CoMPARE system designed by Sawitzki et al. with Dresden University of Technology.
- The VLIW architecture proposed by Alippi *et al.* with Polytechnical Institute of Milano.

Common Minimal Processor Architecture with Reconfigurable Extension (CoMPARE) designed by Sawitzki *et al.* [Sawitzki et al., 1998b; Sawitzki et al., 1998a] is a horizontal engine emerged from a RISC having a reconfigurable extension. As depicted in Fig. 7.58, CoMPARE uses a *Reconfigurable Processing Unit* (RPU) that consists of a conventional ALU augmented with a *Configurable Array Unit* (CAU). The ALU provides the most basic instructions, while CAU implements additional customized instructions.

The main difference between a conventional RISC processor and CoMPARE is that the later uses the RPU in place of the standard ALU. The two instruction formats of CoMPARE are shown in Fig. 7.59. The *register (R) format* is used for arithmetic and logic operations, and the *immediate (I) format* is used for load/store operations, jump/branch operations, and CAU reconfiguring operation. In addition to the three operands encoded by the standard RISC R-format instructions, the CoMPARE RPU

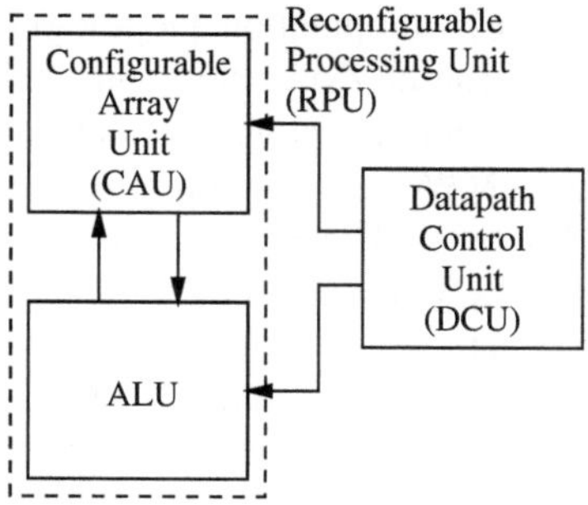

Fig. 7.58 RPU structure of CoMPARE

R Instruction format	OP-Code	Rdst	Rsrc 1	Rsrc 2
Number of bits	8	4	4	4
I Instruction format	OP-Code	Rsrc 1 or Rdst	Rsrc 2	Immediate
Number of bits	4	4	4	8

Fig. 7.59 The CoMPARE instruction formats

can process at most four input operands and produce two results in one instruction. Two out of four read operands are encoded in the instruction, while the other two are implicitly defined as being the immediate successors of the first two in the register file. Likewise, the first output operand is encoded in the instruction, while the second output operand is again the immediate successor of the first one in the register file. Hardwired instructions do not use these additional register values, only configured instructions do.

CoMPARE has four operation modes: *single ALU*, *single CAU*, *superscalar extension* (i.e., ALU and CAU in a SoCR model), and *sequential mode* (i.e., ALU and CAU in a PoCR model). These modes are shown in Fig. 7.60. Two bits in the *opcode* field of the R-format instructions are used to specify the operation mode. Three of the remaining six bits of the opcode encode 8 operations for ALU. Likewise, the other three bits of the opcode encode 8 operations for CAU.

Based on these considerations, we can conclude that CoMPARE exhibits a *spatial construction* built on CAU. The ALU and CAU can be arranged as either a *SoCR*

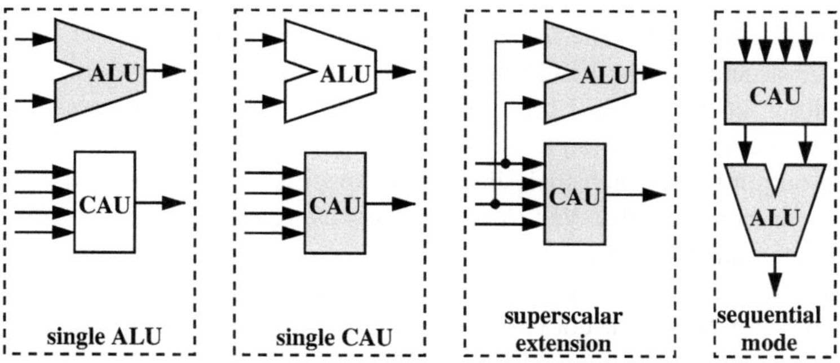

Fig. 7.60 The operation modes of CoMPARE

or *PoCR* computational topology. The microcode is horizontal, since both ALU and CAU can be explicitly controlled during a time cycle. By comparison, we would like to notice that an I-format instruction is vertical. We complete the presentation of CoMPARE by mentioning that the CoMPARE configuration data set consists of 1984 bytes. The configuration information is transferred word by word into the CAU by a LCW (load configuration word) instruction, which is actually an explicit SET instruction.

The reconfigurable VLIW processor proposed by Alippi *et al.* with Polytehnical Institute of Milano [Alippi et al., 1999] is a horizontally-coded engine that consists of a set of functional units, one of them being reconfigurable. Therefore, the construction is *spatial*, and the computing topology is SoCR. In order to keep the number of ports of the register file low, the functional units are organized in clusters, where the RFU represents a separate cluster. Special instructions that copy values between register files are provided. A dedicated *fpga-opcode* extends the initial instruction set with a user-customized instruction. Unfortunately, no further details are provided; the authors focused especially on methodologies to select the application critical parts to be implemented on the RFU. In their work, the reconfigurable VLIW architecture plays the role of a generic model rather than a fully specified architecture.

7.5.3.2 Horizontal Microcoded CCMs with Temporal Constructions

The following machines will be analyzed:

- VEGA introduced by Jones and Lewis, both with University of Toronto.
- RISA''' proposed by Trimberger with Xilinx Corporation.
- PipeRench proposed by a team with Carnegie Mellon University.

The *Virtual Element Gate Array* (VEGA) proposed by Jones and Lewis [Jones and Lewis, 1995; Jones, 1995] is composed of a 4-input LUT-based FPGA Logic Unit equiped with a large configuration memory referred to as *Logic Instruction Memory* (LIM). LIM can store up to 2048 contexts in the current implementation of VEGA. Due to the large number of contexts, a *Node Memory* (NM) is aimed to provide for state sharing between contexts. A Cache Memory is used to optimize the logic unit access to NM. The I/O Unit manages communications with the outside world. The structure of the VEGA element is given in Fig. 7.61.

Each logic instruction in LIM specifies which four LUT inputs are to be read from cache, the logic function to be evaluated by the logic unit and the location to write the output to. As each logic instruction specifies a new configuration of the LUT, we can conclude that VEGA implements as a pure temporal constructions. The computational topology is PoCR.

The data transfers between NM and cache are driven by NM control block under the command of NM instructions. Also, the data transfers between cache and I/O unit are controlled by the instructions in I/O Instruction Memory. VEGA is a three-slot horizontal machine, in which scheduling the instructions in LIM, NM, and I/O

Fig. 7.61 VEGA structure

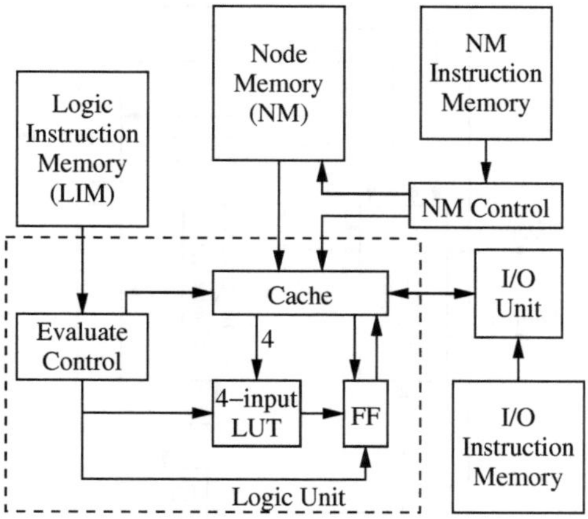

instruction memory is statically defined. For more details concerning the instruction format we refer the user to the bibliography.

The third architecture claimed by Trimberger in the RISA project [Trimberger, 1998a] will be referred to as RISA"'. Its instruction format which includes both an opcode field for the defined execution unit and an opcode field for PFU is shown in Fig. 7.62.

Therefore, RISA is a horizontal microcoded machine. As the PFU is implemented on a multiple-context FPGA, the construction is *temporal*. It is not obvious is an explicit SET instruction is provided.

The *PipeRench* coprocessor developed by a team with Carnegie Mellon University [Cadambi et al., 1998; Myers et al., 1998; Goldstein et al., 1999; Goldstein et al., 2000] is focused on implementing linear pipelines of arbitrary length. PipeRench includes a partially reconfigurable fabric that manages a "virtual pipeline" in the framework of stationary striped reconfiguration [Schmit, 1997; Cadambi et al., 1998]. PipeRench is envisioned as a coprocessor in a general-purpose computer, and has direct access to the same memory space as the host processor.

The architecture of PipeRench is depicted in Fig. 7.63. It consists of a set of identical *Stripes* which can be configured separately at run time, a *Configuration Memory* storing the stripe configurations, a *Configuration Controller*, four *Data Controllers* (DC), a *State Memory*, an *Address Translation Table* (ATT), and a *Memory Bus Controller*. PipeRench also contains a register that indicates whether it is working or idle. This register can be periodically polled by the host.

As the virtual stripes may be physically configured in any of the physical stripes in the fabric, global I/O busses stretching the fabric are provided. PipeRench

Fixed Opcode	A	B	Y	Progr. Opcode	C	Immediate
ADD	R3	R4	R5	FPGAOP(N)	R22	XXXX

Fig. 7.62 The RISA instruction format

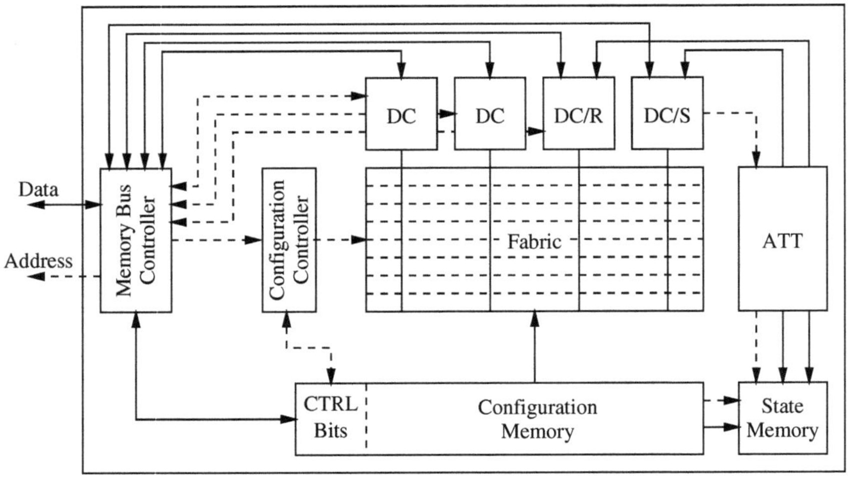

Fig. 7.63 Architecture of PipeRench (solid lines are data paths, dashed lines are address and control paths)

contains four such busses: two of these busses are dedicated for data input and output, while the other two can be used for both data input/output and storing/restoring a stripe state during hardware virtualization. At the end of each global bus is a data controller, which handles the inputs and outputs from the application. The data controllers access off-chip memory through the memory bus controller which arbitrates the access to a single external bus. Two of the data controllers (DC/R, DC/S) also can access the state memory in order to deal with saving and restoring the state of a pipeline stage.

The state information for each stripe is stored in an on-chip state memory. This memory provides one location for each corresponding location in the configuration memory. An address translation table having one entry per physical stripe keeps track of what virtual stripe is configured in a physical stripe. The virtual stripes of the application are stored into the configuration memory. A single physical stripe can be configured in one read cycle with data stored in such a memory. The configuration of a stripe takes place concurrently with execution of the other stripes.

The host processor initiates a new application by specifying the memory address of the first configuration word, the number of iterations to be performed, and the memory addresses for data input and output. The configuration controller provides the interface with the host processor. In order to map applications to a given number of physical stripes, the configuration controller must handle the time-multiplexing process of the application's stripes onto the physical fabric, schedule the stripes, as well as manage of the on-chip configuration memory. Going off-chip to fetch a new configuration word is time-consuming. To overcome this overhead, the modification of the configuration cache can take place concurrently with execution.

The PipeRench device is seen by the user as being capable of implementing pipelines of arbitrary length. A program for a PipeRench device is a chained list of configuration words, each of which includes three fields: fabric configuration bits

	Stripe config. data	Next address	First virtual stripe flag	Last virtual stripe flag	Store	Restore	Read flags	Write flags
# of bits	not obvious	8	1	1	1	1	4	4

Fig. 7.64 PipeRench configuration word format

specifying the FPGA configuration for each virtual pipeline stage of the application, a next-address field, and a set of flags used by the configuration and data controllers. A configuration word is depicted in Fig. 7.64.

The next-address field points to the next virtual stripe of the application. Thus, it implements a `JUMP TO NEXT − ADDRESS` instruction. The fields *First-* and the *Last-virtual-stripe flags* are used by configuration controller to determine the iteration count and the number of stripes in the application.

The stripe configuration data field specifies a `SET` instruction. The stripe reconfiguration can be done with or without state restoration, as stipulated by the *Restore* flag. All *Store*, *Restore*, *Read*, and *Write* flags are relevant for data controllers. The *Store* flag specifies that the stripe state should be stored into the state memory. The *Read Flags*, and *Write Flags* determine the application's memory read/write access pattern.

Once the stripe is configured, it runs automatically, without explicit control untill it is reconfigured. The configuration word controls more than one resource: the fabric by a `SET` instruction, the configuration controller by a `JUMP` instruction, and the data controllers by flags; thus, the configuration word is a horizontal instruction. As the PipeRench configuration memory is able to store more virtual stripes than physical stripes the FPGA contains, the construction is temporal. As more than one stripes are active at a time, the construction is also spatial. The computing topology is PoCR.

7.5.3.3 Horizontal Microcoded CCMs with Spatial Constructions

The following FCCMs are classified in this subclass:

- Spyder introduced by Iseli and Sanchez with Swiss Federal Institute of Technology.
- FPGA-augmented TriMedia (ρ-TriMedia) proposed by Sima *et al.*.
- RaPiD-based system proposed by Ebeling, Cronquist *et al.* with University of Washington.
- Colt proposed by Bittner and Athanas Virginia Polytechnic Institute and State University.

Iseli and Sanchez proposed *Reconfigurable Processor DEvelopment SYstem* (Spyder) [Iseli and Sanchez, 1993b; Iseli and Sanchez, 1993a; Iseli and Sanchez, 1995; Iseli, 1996; Sanchez et al., 1999]. Spyder is an FPGA-based VLIW processor

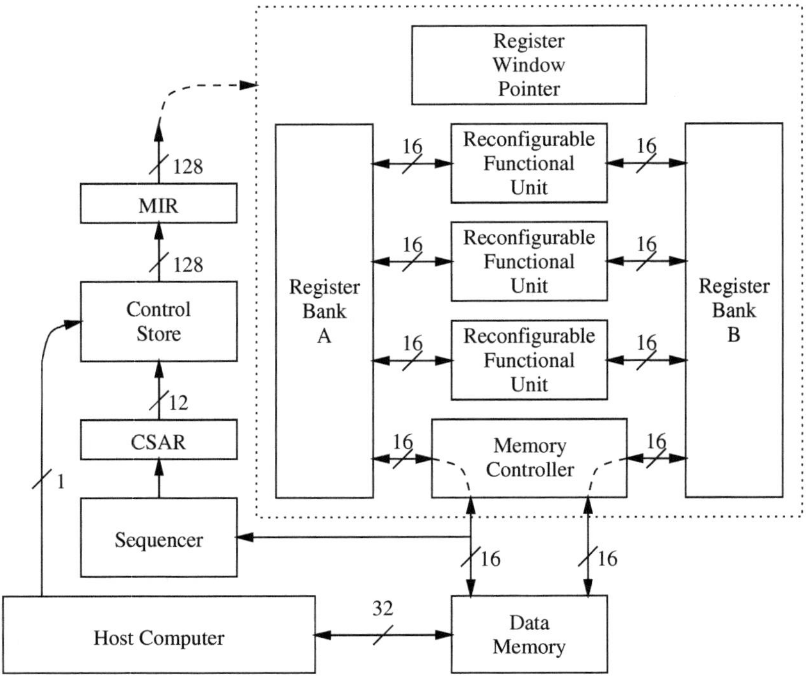

Fig. 7.65 Architecture of spyder

which consists of three reconfigurable functional units, a memory controller and a sequencer. As depicted in Fig. 7.65, the reconfigurable units and the memory controller are connected to a register file that is organized in two register banks. The connection between a unit and a register bank is supported by a bidirectional data bus.

Spyder is a `load/store` architecture, i.e., the units load and store data only from and to registers, and the data memory is accessed only by means of `load` and `store` operations. Each reconfigurable unit can read two 16-bit data words and generate two 16-bit results (one per register block). The memory controller can read from or write back to data memory two 16-bit data words (one per register block). In a new version of Spyder [Iseli and Sanchez, 1994; Iseli and Sanchez, 1995; Iseli, 1996], the execution units can be connected in a ring by two 16-bit wide data buses. This way, both SoCR and PoCR topologies are supported.

The structure of the Spyder microinstruction word is shown in Fig. 7.66. The field *Seq.* (Sequencer) specifies the operation to be performed by the sequencer. The fields *Memory Bus A*, *Memory Bus B* encode the operations for the memory controller. *Window Select*, *Register Address*, and *Write* specify the input and output operands for the memory controller and reconfigurable functional units. The *Stop* bit is used to set up breakpoints or stop the processor.

The operations performed by the three functional units are controlled by the 21 common bits of the field *RFU OP*. Their distribution and function are entirely open and depend on the configuration of the units. Therefore, the microcode is

Instr. format	Seq.	Memory Bus A	Memory Bus B	Window Select	Register Address	Write	RFU OP	Stop
# of bits	16	25	23	16	20	6	21	1

Fig. 7.66 The Spyder instruction format – from [Iseli and Sanchez, 1995], [Iseli, 1996]

horizontal, each slot driving (1) a reconfigurable unit or a memory controller, and (2) the sequencer.

Spyder is connected through a VME bus to a host workstation. The reconfiguration of the Spyder is performed under a command issued by the host. This way, the user can initiate a reconfiguration process. However, no further details are provided.

TriMedia/CPU64 is a 64-bit 5 issue-slot VLIW core, launching a long instruction every clock cycle [van Eijndhoven et al., 1999]. It has a uniform 64-bit wordsize through all functional units, the register file, load/store units, on-chip highway and external memory. Each of the five operations in a single instruction can in principle read two register arguments and write one register result every clock cycle. In addition, each operation can be guarded with an optional (4th) register for conditional execution without branch penalty. The architecture supports subword parallelism and is optimized with respect to media-processing. With the exception of floating point divide and square root, all functional units have a recovery (minimum number of clock cycles between the issue of successive operations) of 1, while their latency (clock cycles between the issue of an operation and availability of its results) varies from 1 to 4. The TriMedia/CPU64 VLIW core also supports double-slot operations, or super-operations. Such a super-operation occupies two neighboring slots in the VLIW instruction, and maps to a double-width functional unit. This way, operations with more than 2 arguments and/or more than one result are possible. The TriMedia/CPU64 organization is presented in Fig. 7.67.

In the *FPGA-augmented Trimedia/CPU64* [Sima et al., 2001; Sima et al., 2004; Sima et al., 2005] the the TriMedia-CPU64 processor is augmented with a Reconfigurable Functional Unit (RFU) which consists mainly of an FPGA core. Also, a hardwired Configuration Unit which manages the reconfiguration of the raw hardware is attached to the reconfigurable functional unit, as it is depicted in Fig. 7.68. The reconfigurable functional unit is embedded into TriMedia as any other hardwired

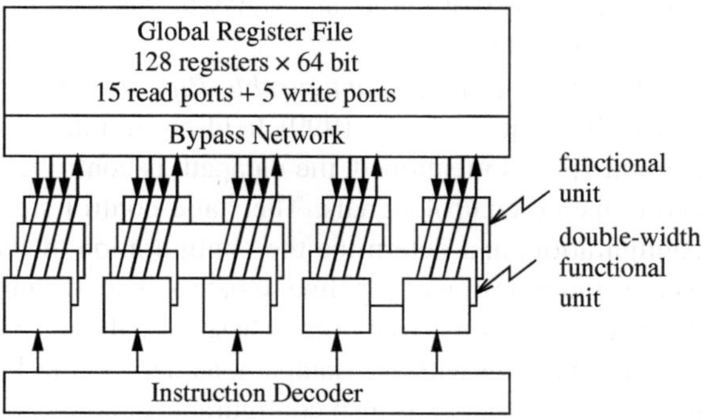

Fig. 7.67 TriMedia/CPU64 organization

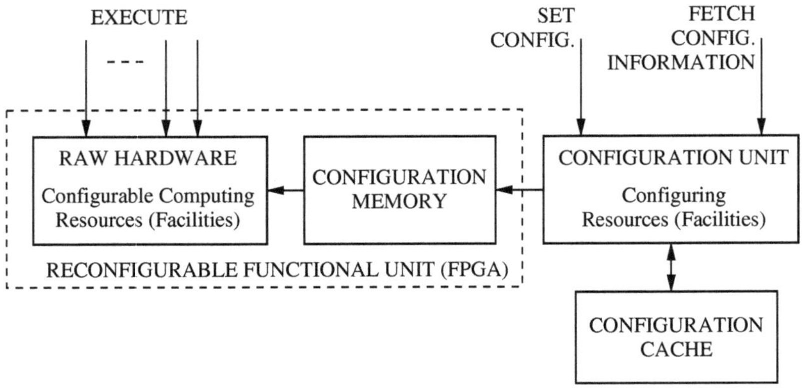

Fig. 7.68 The organization of the RFU and associated configuration unit

functional unit, i.e., it receives instructions from the instruction decoder, reads its input arguments from and writes the computed values back to the register file.

In order to use the RFU, a kernel of new instructions is needed. Loading a context information into the RFU configuration memory is performed under the command of a `SET_CONTEXT` instruction, while the `ACTIVATE_CONTEXT` instruction controls the swaping of the active configuration with one of the idle on-chip configuration. `EXECUTE` instructions launch the operations performed by the computing resources configured on the raw hardware [Sima et al., 2000]. In this way, the execution of an RFU-mapped operation requires three basic stages: `SET_CONTEXT`, `ACTIVATE_CONTEXT`, and `EXECUTE`. It should be mentioned that since the `EXECUTE` instructions are executed on the RFU without checking of the active configuration, it is still the responsibility of the user to perform the management of the active and idle configurations.

A distinct class of engines with spatial constructions is based on PoCR computational topology. The main goal of such machines is to provide reconfigurable hardware support for pipelined applications. A pipeline of computing resources configured on such machines can have a 1-D or 2-D spatial organization. In a 1-D pipeline the computing resources are chained, while in a 2-D pipeline the computing resources are connected in a mesh structure. PipeRench which belongs to 1-D pipeline class was already discussed. RaPiD which belong to the 1-D pipeline class, and Colt/Wormhole, and rDPA that belong to the 2-D pipeline class will be subsequently discussed.

The coarse-grain *Reconfigurable Pipelined Datapath* (RaPiD) [Ebeling et al., 1996], [Cronquist et al., 1999] field-programmable array has been presented in Section 7.3. As mentioned, the datapath is controlled using a combination of static and dynamic control signals. The static control signals are defined by the RaPiD configuration, and determine the configuration of the interconnection network, as well as the insertion of pipeline registers. The dynamic control signals schedule the datapath operations over time. These signals are issued by a *control path* which stretches parallel with the datapath, as depicted in Fig. 7.69. Similar with the data path, the control path is also configurable at application load-time.

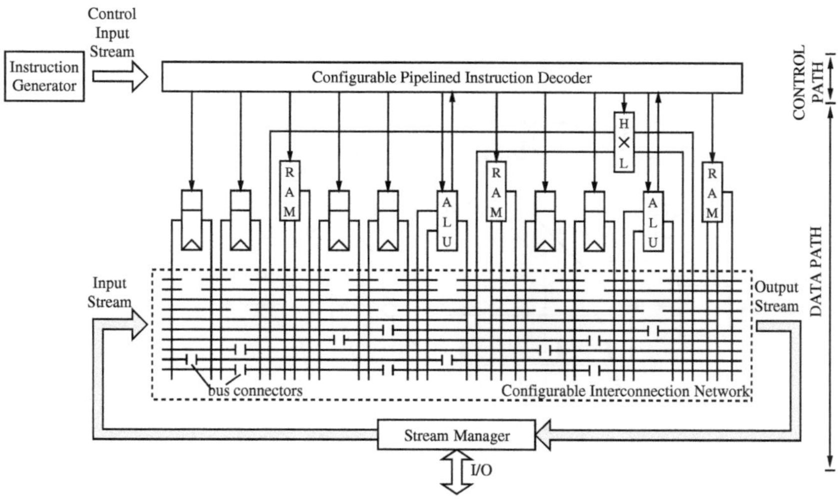

Fig. 7.69 The basic RaPiD cell

The dynamic control is managed outside of the array. The control signals are inserted at one end of the control path by an *Instruction Generator*, and are passed from stage to stage of the control path pipeline where they drive the functional units. The control data can also be generated inside the array by different functional unit condition codes, such as an ALU state or feedback from simple controllers mapped on the array, etc.

The control path pipeline plays the role of a vertical MIR with serial shifting capabilities. Therefore, we classify the microcode as horizontal. The construction is spatial and the computing topology is PoCR. The host processor supervizing the RaPiD can initiate a reconfiguration process only at the application load-time. Unfortunately, nu further details are provided.

The first FCCM implementing a 2-D pipeline we will discuss is the *Colt* system designed by Bittner *et al.* [Bittner, Jr. et al., 1996; Bittner, Jr., 1997]. Colt, which is based on the *Wormhole* FPGA reconfiguration pattern already described in Section 7.3, is depicted in Fig. 7.70. The application domain for this machine is DSP, where most computations are performed on word-wide operands.

Colt is composed of two layers. The first layer includes word-wide RPUs interconnected through a cylindrical mesh. The RPUs are identical, each of them comprising an ALU, a barrel shifter, a conditional unit, two input registers and optional delay blocks. Colt also includes a multiplier which can perform unsigned multiplication. The first layer is reconfigurable in a distributed fashion by self-steering streams, as provided by the Wormhole concept. The second layer includes a truncated crossbar, also supporting the distributed control by means of streams. Therefore, the crossbar is able to simultaneously switch multiple points under the command of the self-steering stream. The crossbar connects I/O ports (Data Ports) with the units at the edge of the second layer and with the multiplier. An I/O port cannot be connected directly to another I/O port, as such connection would lead to no computation.

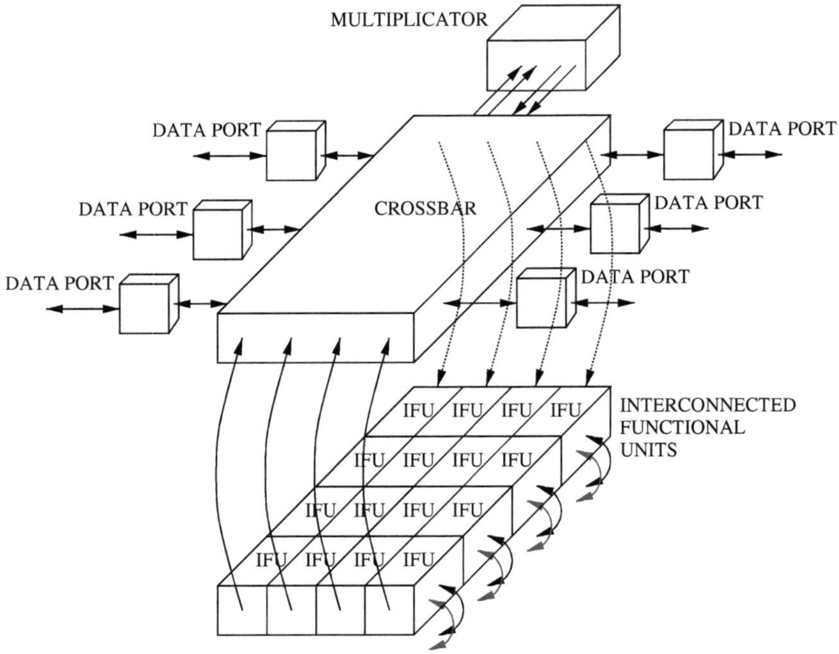

Fig. 7.70 Colt architecture

Multiple streams can enter and configure the system, each of the I/O port can be regarded as a configuring resource. Thus, the number of configuring resources is equal to 6 for the particular Colt system shown in Fig. 7.70. Also, since multiple I/O ports can simultaneously accept streams, we classify the microcode as *horizontal*. Also, it is obvious that a SET instruction is provided.

7.5.4 Taxonomy Table

Each reconfigurable machine has now been analyzed. In Table 7.3 we will summarize our taxonomy and the major bibliographical references.

7.6 Conclusions

In this chapter a taxonomy of CCMs has been presented. Several classifications on FCCMs were proposed in the past. They all rely on FCCM implementation issues. Our approach is different: we focused on architectural issues. We mainly discussed the following FCCM architectural characteristics: the architecture of the microcode, the construction class, the number of configuring resources, the computing topology, and the availability of an explicit SET instruction.

Table 7.3 The taxonomy table

Custom Computing Machine	Microcode Architecture (V/H)	Computing Construction (Pr/T/S)	Number of Configuring Facilities	Computing Topology (SoCR/PoCR/n.a.)	Explicit SET μI (Y/N)	Bibliography
PRISC	V	PrC	1	n.a.	N	[Razdan, 1994] [Razdan et al., 1994] [Razdan and Smith, 1994]
Modified PRISC	V	PrC	1	n.a.	Y	[Hauck, 1998b]
OneChip	V	PrC	not obvious	n.a.	N	[Wittig, 1995] [Wittig and Chow, 1996]
REMARC	V	PrC	1	n.a.	Y	[Miyamori and Olukotun, 1998a]
Garp	V	PrC/T ?	1	n.a.	Y	[Hauser and Wawrzynek, 1997] [Hauser, 1997]
ConCISe	V	PrC	1	n.a.	N	[Kastrup et al., 1999a] [Kastrup et al., 2000]
PAM	V ?	PrC	1	n.a.	Y	[Bertin et al., 1989b] [Bertin et al., 1989a] [Bertin et al., 1993] [Vuillemin et al., 1996]
MorphoSys	V	T	not obvious	PoCR	Y	[Lee et al., 2000] [Singh et al., 2000]
RISA'	V	T ?	not obvious	SoCR	not obvious	[Trimberger, 1998a]
RISA"	V	T ?	not obvious	SoCR	not obvious	[Trimberger, 1998a]

(continued)

Table 7.3 (continued)

Custom Computing Machine	Microcode Architecture (V/H)	Computing Construction (Pr/T/S)	Number of Configuring Facilities	Computing Topology (SoCR/PoCR/n.a.)	Explicit SET μI (Y/N)	Bibliography
OneChip-98'	V	T	not obvious	SoCR	Y	[Jacob, 1998] [Jacob and Chow, 1999]
OneChip-98"	V	T	not obvious	SoCR	Y	[Jacob, 1998] [Jacob and Chow, 1999]
Virtual Computer	V ?	T	not obvious	SoCR / PoCR	not obvious	[Casselman, 1993] [Thornburg and Casselman, 1994] [Casselman, 1997]
WASMII	V ?	T	1	PoCR	Y ?	[Ling and Amano, 1993] [Ling and Amano, 1995] [Takayama et al., 1999]
Multiple RISA	V	S	not obvious	SoCR	N	[Trimberger, 1998b]
T1000	V	S	not obvious	SoCR	N	[Zhou and Martonosi, 2000]
Gilson's CCM	V	S	1	SoCR	Y	[Gilson, 1994a]
Nano-Processor	V	S	n.a.	SoCR	N	[Wirthlin et al., 1994]
DISC	V	S	1	SoCR	N	[Wirthlin and Hutchings, 1995]
CCSimP	V	S	n.a.	SoCR	N	[Salcic and Maunder, 1996a]
URISC	V	S	1	SoCR	Y	[Brebner and Donlin, 1998] [Donlin, 1998]

Table 7.3 (continued)

Custom Computing Machine	Microcode Architecture (V/H)	Computing Construction (Pr/T/S)	Number of Configuring Facilities	Computing Topology (SoCR/PoCR/n.a.)	Explicit SET μI (Y/N)	Bibliography
Functional Memory	V	S	not obvious	SoCR	not obvious	[Lew and Halverson, Jr., 1995] [Halverson, Jr. and Lew, 1996]
PRISM	V	S	1	SoCR	Y	[Athanas, 1992a] [Athanas, 1992b] [Athanas and Silverman, 1993]
PRISM-II/RASC	V	S	1	SoCR	Y	[Wazlowski et al., 1993] [Wazlowski, 1996]
Chimaera	V	S	1	SoCR	N	[Hauck et al., 1997]
Xputer/rALU	V	S	1	SoCR	Y	[Hartenstein et al., 1992]
Splash-2	V	S	1	SoCR/PoCR	Y	[Buell et al., 1996]
AnyBoard	V	S	1	SoCR/PoCR	Y	[van den Bout et al., 1992]
NAPA	V	S	not obvious	SoCR	not obvious	[Rupp et al., 1998] [Rupp, 1998]
CM-2X	?	S	not obvious	?	not obvious	[Cuccaro and Reese, 1993]
Molen	V & H	S	1	SoCR	Y	[Vassiliadis et al., 2001] [Vassiliadis et al., 2004]
CoMPARE	H	PrC	1	SoCR / PoCR	Y	[Sawitzki et al., 1998b] [Sawitzki et al., 1998a]

(continued)

Table 7.3 (continued)

Custom Computing Machine	Microcode Architecture (V/H)	Computing Construction (Pr/T/S)	Number of Configuring Facilities	Computing Topology (SoCR/PoCR/n.a.)	Explicit SET μI (Y/N)	Bibliography
Alippi's VLIW	H	PrC	not obvious	SoCR	Y	[Alippi et al., 1999]
RISA'''	H	T	not obvious	SoCR	not obvious	[Trimberger, 1998a]
VEGA	H	T	1	PoCR	Y	[Jones and Lewis, 1995] [Jones, 1995]
PipeRench	H	S/T	1	PoCR	Y	[Goldstein et al., 1999] [Myers et al., 1998]
Spyder	H	S	not obvious	PoCR	Y	[Iseli and Sanchez, 1993b] [Iseli and Sanchez, 1994] [Sanchez et al., 1999]
ρ-TriMedia	H	S	1	SoCR	Y	[Sima et al., 2004] [Sima et al., 2005]
RaPiD	H	S	n.a.	PoCR	N	[Cronquist et al., 1999] [Cronquist et al., 1998] [Ebeling et al., 1996]
Colt/Wormhole	H	S	6	2-D PoCR	Y	[Bittner, Jr. et al., 1996] [Bittner, Jr. and Athanas, 1997b] [Bittner, Jr. and Athanas, 1997a] [Athanas and Bittner, Jr., 1998]
rDPA	H	S	perimeter/2	2-D PoCR	Y	[Hartenstein et al., 1994b] [Hartenstein et al., 1994a]

Regarding the microcode architecture, we introduced a new formalism based on microcode, in which the execution of an FPGA-dedicated instruction is performed as a microprogrammed sequence with two basic stages: a SET stage, and an EXECUTE stage. The immediate implication of this formalism is that all CCMs are microcoded machines, in which the microcode may or may not be exposed to the user.

We then turned these characteristics into classification criteria and proposed a taxonomy of CCMs. In terms of the microcode architecture, the FCCMs were classified in vertical or horizontal microcoded machines. In terms of the construction class, the set of computing resources formed a primordial, temporal or spatial construction. Concerning the computational topology, the computing resources were classified as a sea or pipeline of computing resources. The taxonomy we proposed is architectural consistent, and can be easily extended to embed other criteria.

References

Actel Corporation (1996a). ACT 1 Series FPGAs. Datasheet, Sunnyvale, California.

Actel Corporation (1996b). ACT 2 Series FPGAs. Datasheet, Sunnyvale, California.

Actel Corporation (1998a). Integrator Series FPGAs: 1200XL and 3200DX Families. Datasheet, Sunnyvale, California.

Actel Corporation (1999a). 40MX and 42MX FPGA Families. Datasheet, Sunnyvale, California.

Actel Corporation (1999b). 54SX FPGA Families. Datasheet, Sunnyvale, California.

Agrawala, Ashok K. and Rauscher, Tomlinson G. (1976). *Foundations of microprogramming; architecture, software and applications*. Academic Press, New York, New York.

Alippi, Cesare, Fornaciari, William, Pozzi, Laura, and Sami, Mariagiovanna (1999). A DAG-Based Design Approach for Reconfigurable VLIW Processors. In *IEEE Design and Test Conference in Europe*, pages 778–780, Munich, Germany.

Altera Corporation (1999c). APEX 20K Programmable Logic Device Family. Datasheet, San Jose, California.

Altera Corporation (1999d). FLEX 10K Embedded Programmable Logic Family. Datasheet, San Jose, California.

Altera Corporation (1999e). MAX 7000 Programmable Logic Device Family. Datasheet, San Jose, California.

Altera Corporation (1999f). MAX 9000 Programmable Logic Device Family. Datasheet, San Jose, California.

Amerson, R., Carter, R., Culbertson, W., Kuekes, P., and Snider, G. (1996). An FPGA for Multi-Chip Reconfigurable Logic. In *Proceedings of the IEEE Custom Integrated Circuits Conference*, pages 137–143, Santa Clara, California.

Andrews, M., editor (1980). *Principles of Firmware Engineering in Microprogram Control*. Computer Science Press, Potomac, Maryland.

Arnold, J. M., Buell, Duncan A., and Davis, E. G. (1992). Splash 2. In *Proceedings of the 4th Annual Symposium on Parallel Algorithms and Architectures*, pages 316–324, New York, NewYork.

Athanas, Peter M. (1992a). *An Adaptive Machine Architecture and Compiler for Dynamic Processor Reconfiguration*. PhD thesis, Brown University, Providence, Rhode Island.

Athanas, Peter M. (1992b). An Adaptive Machine Architecture and Compiler for Dynamic Processor Reconfiguration. Technical Report LEMS-101, Brown University, Providence, Rhode Island.

Athanas, Peter M. and Bittner, Jr., Ray A. (1998). Worm-hole run-time reconfigurable processor field programmable gate array (FPGA). U.S. Patent No. 5,828,858, October 1998.

Athanas, Peter M. and Silverman, H. F. (1991). An Adaptive Hardware Machine Architecture for Dynamic Processor Reconfiguration. In *Proceedings of the IEEE International Conference on Computer Design*, pages 397–400, Cambridge, Massachusetts.

Athanas, Peter M. and Silverman, Harvey F. (1993). Processor Reconfiguration through Instruction-Set Metamorphosis. *IEEE Computer*, 26(3):11–18.

Atmel Corporation (1999g). AT40K FPGAs with FreeRAM. Datasheet, San Jose, California.

Atmel Corporation (1999h). AT6000 Series Configuration. Application Note, San Jose, California.

Atmel Corporation (1999i). AT6000(LV) Series. Coprocessor Field Programmable Gate Arrays. Datasheet, San Jose, California.

Atmel Corporation (1999j). Configuration Compression Algorithm. Application Note, San Jose, California.

Baldwin, David R., Wilson, Malcom E., and Trevett, Neil F. (1991). Architectures for Serial or Parallel Loading of Writable Control Store. U.S. Patent No. 5,056,015, October 1991.

Bertin, P., Roncin, D., and Vuillemin, J. (1989a). *Systolic Array Processors*, chapter Introduction to Programmable Active Memories, pages 300–309. Prentice-Hall, Englewood Cliffs, New Jersey.

Bertin, Patrice, Roncin, Didier, and Vuillemin, Jean (1989b). Introduction to Programmable Active Memories. Research Report # 3, Digital Equipment, Paris Research Laboratory, Paris, France.

Bertin, Patrice, Roncin, Didier, and Vuillemin, Jean (1993). Programmable Active Memories: a Performance Assessment. Research Report # 24, Digital Equipment, Paris Research Laboratory, Paris, France.

Betz, Vaughn and Rose, Jonathan (1998). Cluster-Based Logic Blocks for FPGAs: Area-Efficiency vs. Input Sharing and Size. In *Proceedings of the IEEE Custom Integrated Circuits Conference*, pages 551–554, Santa Clara, California.

Bhat, Narasimha B. and Chaudhary, Kamal (1997). Field Programmable Logic Device with Dynamic Interconnections to a Dynamic Logic Core. U.S. Patent No. 5,596,743, January 1997.

Bittner, Jr., Ray A. (1997). *Wormhole Run-Time Reconfiguration: Conceptualization and VLSI Design of a High Performance Computing System.* PhD thesis, Virginia Tech, Blacksburg, Virginia.

Bittner, Jr., Ray A. and Athanas, Peter M. (1997a). Computing Kernels Implemented with a Wormhole RTR CCM. In Werner, Bob, Arnold, Jeffrey M., and Pocek, Kenneth L., editors, *IEEE Symposium on FPGAs for Custom Computing Machines*, pages 98–105, Napa Valley, California.

Bittner, Jr., Ray A. and Athanas, Peter M. (1997b). Wormhole Run-time Reconfiguration. In *ACM/SIGDA International Symposium on Field Programmable Gate Arrays*, pages 79–85, Monterey, California.

Bittner, Jr., Ray A., Athanas, Peter M., and Musgrove, Mark D. (1996). Colt: An Experiment in Wormhole Run-Time Reconfiguration. In Schewel, John, Athanas, Peter M., Bove, Jr., V. Michael, and Watson, John, editors, *Photonics East, Conference on High-Speed Computing, Digital Signal Processing, and Filtering Using FPGAs*, volume 2914 of *The SPIE Proceedings*, pages 187–195, Boston, Massachusetts.

Blaauw, Gerrit A. and Brooks, Jr., Frederick P. (1997). *Computer Architecture. Concepts and Evolution.* Addison-Wesley, Reading, Massachusetts.

Bolotski, Michael, DeHon, André, and Thomas F. Knight, Jr. (1994). Unifying FPGAs and SIMD Arrays. Transit Note # 95, Massachusetts Institute of Technology, Cambridge, Massachusetts.

Brebner, Gordon (1996). A Virtual Hardware Operating System for the Xilinx XC6200. In Hartenstein, Reiner W. and Glesner, Manfred, editors, *6th International Workshop on Field Programmable Logic and Applications. Field-Programmable Logic: Smart Applications, New Paradigms and Compilers*, volume 1142 of *Lecture Notes in Computer Science*, pages 327–336, Darmstadt, Germany.

Brebner, Gordon (1997). The Swappable Logic Unit: a Paradigm for Virtual Hardware. In Werner, Bob, Arnold, Jeffrey M., and Pocek, Kenneth L., editors, *IEEE Symposium on FPGAs for Custom Computing Machines*, pages 77–86, Napa Valley, California.

Brebner, Gordon (1998). Field-Programmable Logic: Catalyst for New Computing Paradigms. In Hartenstein, Reiner W. and Keevallik, Andres, editors, *8th International Workshop on Field-*

Programmable Logic and Applications. From FPGAs to Computing Paradigm, volume 1482 of *Lecture Notes in Computer Science*, pages 49–58, Tallin, Estonia.

Brebner, Gordon and Donlin, Adam (1998). Runtime Reconfigurable Routing. In *Reconfigurable Architectures Workshop*, pages 25–30, Orlando, Florida.

Brown, Stephen and Rose, Jonathan (1996). Architecture of FPGAs and CPLDs: A Tutorial. *IEEE Transactions on Design and Test of Computers*, 13(2):42–57.

Brown, Stephen D. (1994). An Overview of Technology Architecture and CAD Tools for Programmable Logic Devices. In *Proceedings of the IEEE Custom Integrated Circuits Conference*, pages 69–76, Santa Clara, California.

Brown, Stephen D., Francis, Robert J., Rose, Johnathan, and Vranesic, Zvonko (1992). *Field-Programmable Gate Arrays*. Kluwer Academic Publishers, Boston, Massachusetts.

Buell, Duncan A., Arnold, Jeffrey M., and Kleinfelder, Walter J., editors (1996). *Splash 2: FPGAs in a Custom Computing Machine*. IEEE Computer Society, Los Alamitos, California.

Buell, Duncan A. and Pocek, K. L. (1995). Custom Computing Machines: An Introduction. *Journal of Supercomputing*, 9(3):219–230.

Cadambi, Srihari, Weener, Jeffrey, Goldstein, Seth Copen, Schmit, Herman, and Thomas, Donald E. (1998). Managing Pipeline-Reconfigurable FPGAs. In *ACM/SIGDA International Symposium on Field Programmable Gate Arrays*, pages 55–64, Monterey, California.

Callahan, Timothy J., Hauser, John R., and Wawrzynek, John (2000). The Garp Architecture and C Compiler. *IEEE Computer*, 33(4):62–69.

Camarota, Rafael C., Furtek, Frederick C., Ho, Walford W., and Browder, Edward H. (1992). Programmable Logic Cell and Array. U.S. Patent No. 5,144,166, September 1992.

Camarota, Rafael C., Furtek, Frederick C., Ho, Walford W., and Browder, Edward H. (1993). Programmable Logic Cell and Array with Bus Repeaters. U.S. Patent No. 5,218,240, June 1993.

Carter, W. S., Duong, K., Freeman, R., Hseih, H.-C., Ja, J. Y., Mahoney, J. E., Hgo, L. T., and Sze, S. L. (1986). A user programmable reconfigurable logic array. In *IEEE Proceedings of Custom Integrated Circuits Conference*, pages 233–235, Rochester, New York.

Carter, William S. (1987). Configurable Logic Element. U.S. Patent No. 4,706,216, November 1987.

Casselman, Steven Mark (1993). Virtual Computing and the Virtual Computer. In Buell, Duncan A. and Pocek, Kenneth L., editors, *IEEE Workshop on FPGAs for Custom Computing Machines*, pages 43–48, Napa Valley, California.

Casselman, Steven Mark (1997). FPGA Virtual Computer for Executing a Sequence of Program Instructions by Successively Reconfiguring a Group of FPGA in Response to Those Instructions. U.S. Patent No. 5,684,980, November 1997.

Chan, Pak K. and Mourad, Samiha (1994). *Digital Design Using Field Programmable Gate Arrays*. Prentice-Hall, Englewood Cliffs, New Jersey.

Cline, B. E., editor (1981). *Microprogramming Concepts and Techniques*. PBI Books, New York.

Compton, Katherine and Hauck, Scott A. (2002). Reconfigurable Computing: A Survey of Systems and Software. *ACM Computing Surveys*, 34(2):171–210.

Cronquist, Darren C., Fisher, Chris, Figueroa, Miguel, Franklin, Paul, and Ebeling, Carl (1999). Architecture Design of Reconfigurable Pipelined Datapaths. *Advanced Research in VLSI*, pages 23–40.

Cronquist, Darren C., Franklin, Paul, Berg, S. G., and Ebeling, Carl (1998). Specifying and Compiling Applications for RaPiD. In Pocek, Kenneth L. and Arnold, Jeffrey M., editors, *IEEE Symposium on FPGAs for Custom Computing Machines*, pages 116–125, Napa Valley, California.

Cuccaro, Steven A. and Reese, Craig F. (1993). The CM-2X: A Hybrid CM-2/Xilinx Prototype. In Buell, Duncan A. and Pocek, Kenneth L., editors, *IEEE Workshop on FPGAs for Custom Computing Machines*, pages 121–130, Napa Valley, California.

Dales, Michael (1999). The Proteus Processor – A Conventional CPU with Reconfigurable Functionality. In Lysaght, Patrick, Irvine, James, and Hartenstein, Reiner W., editors, *9th International Workshop on Field-Programmable Logic and Applications*, volume 1673 of *Lecture Notes in Computer Science*, pages 431–437, Glasgow, Scotland.

Danecek, J., Drapal, F., Pluhacek, A., Salcic, Zoran, and Servit, M. (1995). A Simple Processor for Custom Computing Machines. *Journal of Microcomputing Applications*, Academic Press.

DeHon, André (1994). DPGA-Coupled Microprocessors: Commodity ICs for the Early 21st Century. In Buell, Duncan A. and Pocek, Kenneth L., editors, *IEEE Workshop on FPGAs for Custom Computing Machines*, pages 31–39, Napa Valley, California.

DeHon, André (1995). Notes on Coupling Processors with Reconfigurable Logic. Transit Note # 118, Massachusetts Institute of Technology, Cambridge, Massachusetts.

DeHon, André (1996a). DPGA Utilization and Application. In *ACM/SIGDA International Symposium on Field Programmable Gate Arrays*, pages 115–121, Monterey, California.

DeHon, André (1996b). DPGA Utilization and Application. Transit Note # 129, Massachusetts Institute of Technology, Cambridge, Massachusetts.

DeHon, André (1996c). Dynamically Programmable Gate Arrays: A Step Toward Increased Computational Density. *Fourth Canadian Workshop on Field-Programmable Devices*, Toronto, Canada.

DeHon, André (1996d). Reconfigurable Architectures for General-Purpose Computing. A. I. 1586, Massachusetts Institute of Technology, Cambridge, Massachusetts.

DeHon, André, Bolotski, Michael, and Knight, Jr., Thomas F. (2000). DPGA-Coupled Microprocessors. U.S. Patent No. 6,052,773, April 2000.

DeHon, André, T. Knight, Jr., Tau, E., Bolotski, Michael, Eslick, I., Chen, D., and Brown, J. (1998). Dynamically Programmable Gate Array with Multiple Context. U.S. Patent No. 5,742,180, April 1998.

Donlin, Adam (1998). Self Modifying Circuitry - A Platform for Tractable Virtual Circuitry. In Hartenstein, Reiner W. and Keevallik, Andres, editors, *8th International Workshop on Field-Programmable Logic and Applications). From FPGAs to Computing Paradigm*, volume 1482 of *Lecture Notes in Computer Science*, pages 199–208, Tallin, Estonia.

Ebeling, Carl, Cronquist, Darren C., and Franklin, Paul (1996). RaPiD – Reconfigurable Pipelined Datapath. In Hartenstein, Reiner W. and Glesner, Manfred, editors, *6th International Workshop on Field Programmable Logic and Applications. Field-Programmable Logic: Smart Applications, New Paradigms and Compilers*, volume 1142 of *Lecture Notes in Computer Science*, pages 126–135, Darmstadt, Germany.

Furtek, Frederick C., Mason, Martin T., and Luking, Robert B. (2000a). Field Programmable Gate Array Having Access to Orthogonal and Diagonal Adjacent Neighboring Cells. U.S. Patent No. 6,014,509, January 2000.

Furtek, Frederick C., Mason, Martin T., and Luking, Robert B. (2000b). FPGA Logic Cell Internal Structure Including Pair of Look-Up Tables. U.S. Patent No. 6,026,227, February 2000.

Garverick, Tim, Sutherland, Jim, Popli, Sanjav, Alturi, Venkata, Smith, Jr., Arthur, Pickett, Scott, Hawley, David, Chen, Shao-Pin, Moni, Shankar, Ting, Benjamin S., Camarota, Rafael C., Day, Shin-Mann, and Furtek, Frederick (1994). Versatile and Efficient Cell-to-Local Bus Interface in a Configurable Logic Array. U.S. Patent No. 5,298,805, March 1994.

Gilson, Kent L. (1994a). Integrated Circuit Computing Device Comprising a Dynamically Configurable Gate Array Having a Microprocessor and Reconfigurable Instruction Execution Means and Method Therefor. U.S. Patent No. 5,361,373, November 1994.

Gilson, Kent L. (1994b). Integrated Circuit Computing Device Comprising a Dynamically Configurable Gate Array Having a Reconfigurable Execution Means. WO Patent No. 94/14123, June 1994.

Gokhale, Maya, Holmes, W., Kopser, A., Lucas, S., Minnich, R., Sweely, D., and Lopresti, D. (1991). Building and Using a Highly Parallel Programmable Logic Array. *Computer*, 24(1):81–89.

Gokhale, Maya B. and Stone, Janice M. (1998). NAPA C: Compiling for a Hybrid RISC/FPGA Architecture. In Pocek, Kenneth L. and Arnold, Jeffrey M., editors, *6th IEEE Symposium on FPGAs for Custom Computing Machines*, pages 126–135, Napa Valley, California.

Goldstein, Seth Copen, Schmit, Herman, Budiu, Mihai, Cadambi, Srihari, Moe, Matt, and Taylor, R. Reed (2000). PipeRench: A Reconfigurable Architecture and Compiler. *IEEE Computer*, 33(4):70–77.

Goldstein, Seth Copen, Schmit, Herman, Moe, Matthew, Budiu, Mihai, Cadambi, Srihari, Taylor, R. Reed, and Laufer, Ronald (1999). PipeRench: A Coprocessor for Streaming Multimedia Acceleration. In *The 26th International Symposium on Computer Architecture*, pages 28–39, Atlanta, Georgia.

Gray, J. P. and Kean, T. A. (1989). Configurable Hardware: A New Paradigm for Computation. In *Proceedings of the Decennial Caltech Conference*, pages 279–295, Pasadena, California.

Guccione, Steven Anthony (1995). *Programming Fine-Grained Reconfigurable Architectures*. PhD thesis, University of Texas, Austin, Texas.

Guccione, Steven Anthony and Gonzales, Mario J. (1995). Classification and Performance of Reconfigurable Architectures. In Moore, Will and Luk, Wayne, editors, *5th International Workshop on Field-Programmable Logic and Applications*, pages 439–448, Oxford, United Kingdom.

Halverson, Jr., Richard P. and Lew, Art Y. (1996). Computer System and Method Using Functional Memory. U.S. Patent No. 5,574,930, November 1996.

Hartenstein, Reiner, Becker, Jürgen, and Kress, Reiner (1996). Custom Computing Machines versus Hardware/Software Co-Design: From a Globalized Point of View. In Hartenstein, Reiner W. and Glesner, Manfred, editors, *6th International Workshop on Field Programmable Logic and Applications. Field-Programmable Logic: Smart Applications, New Paradigms and Compilers*, volume 1142 of *Lecture Notes in Computer Science*, pages 65–76, Darmstadt, Germany.

Hartenstein, Reiner W., Hirschbiel, A. G., Schmidt, K., and Weber, M. (1991/1992). A Novel Paradigm of Parallel Computation and its Use to Implement Simple High-Performance Hardware. *Future Generation Computer Systems*, 7(2-3):181–198.

Hartenstein, Reiner W., Kress, Rainer, and Reinig, Helmut (1994a). A New FPGA Architecture for Word-Oriented Datapaths. In Hartenstein, Reiner W. and Servít, Michal Z., editors, *4th International Workshop on Field-Programmable Logic and Applications. Field-Programmable Logic: Architectures, Synthesis and Applications*, volume 849 of *Lecture Notes in Computer Science*, pages 144–155, Prague, Czech Republic.

Hartenstein, Reiner W., Kress, Rainer, and Reinig, Helmut (1994b). An FPGA Architecture for Word-Oriented Datapaths. In *Proceedings of the Second Canadian Workshop on Field-Programmable Devices*, Kingston, Ontario, Canada.

Hauck, Scott (1998a). The Future of Reconfigurable Systems. In *Proceedings of the 5th Canadian Conference on Field-Programmable Devices*, Montreal, Canada.

Hauck, Scott A. (1998b). Configuration Prefetch for Single Context Reconfigurable Coprocessors. In *6th ACM/SIGDA International Symposium on Field Programmable Gate Arrays*, pages 65–74, Monterey, California.

Hauck, Scott A. (1998c). The Roles of FPGA's in Reprogrammable Systems. *Proceedings of the IEEE*, 86(4):615–638.

Hauck, Scott A., Fry, Thomas W., Hosler, Matthew M., and Kao, Jeffrey P. (1997). The Chimaera Reconfigurable Functional Unit. In Werner, Bob, Arnold, Jeffrey M., and Pocek, Kenneth L., editors, *5th IEEE Symposium on FPGAs for Custom Computing Machines*, pages 87–96, Napa Valley, California.

Hauser, John R. (1997). The Garp Architecture. Technical Report, University of California at Berkeley, Berkeley, California. http://www.cs.berkeley.edu/projects/brass/ documents/GarpArchitecture.ps

Hauser, John R. and Wawrzynek, John (1997). Garp: A MIPS Processor with a Reconfigurable Coprocessor. In Werner, Bob, Arnold, Jeffrey M., and Pocek, Kenneth L., editors, *5th IEEE Symposium on FPGAs for Custom Computing Machines*, pages 12–21, Napa Valley, California.

Hennessy, John (1989). RISC Architecture: A Perspective on the Past and Future. In *Proceedings of the Decennial Caltech Conference*, pages 37–42, Pasadena, California.

Hudson, Rhett D., Lehn, David, Hess, Jason, Atwell, James, Moye, David, Shiring, Ken, and Athanas, Peter M. (1998). Spatio-Temporal Partitioning of Computational Structures onto Configurable Computing Machines. In Schewel, John, editor, *Reconfigurable Computing Conference, Configurable Computing: Technology & Applications*, volume 3526 of *The SPIE Proceedings*, pages 62–71, Boston, Massachusetts.

I-Cube Corporation (1998b). IQ Family Register Programming User's Reference. Datasheet, Campbell, California.

Iseli, Christian (1996). *Spyder: A Reconfigurable Processor Development System.* PhD thesis, Swiss Federal Institute of Technology, Lausanne, Switzerland. Thesis no. 1476.

Iseli, Christian and Sanchez, Eduardo (1993a). Beyond Superscalar Using FPGAs. In *IEEE International Conference on Computer Design*, pages 486–490, Los Alamitos, California.

Iseli, Christian and Sanchez, Eduardo (1993b). Spyder: A Reconfigurable VLIW Processor using FPGAs. In Buell, Duncan A. and Pocek, Kenneth L., editors, *IEEE Workshop on FPGAs for Custom Computing Machines*, pages 17–24, Napa Valley, California.

Iseli, Christian and Sanchez, Eduardo (1994). A Superscalar and Reconfigurable Processor. In Hartenstein, Reiner W. and Servít, Michal Z., editors, *4th International Workshop on Field-Programmable Logic and Applications. Field-Programmable Logic: Architectures, Synthesis and Applications*, volume 849 of *Lecture Notes in Computer Science*, pages 168–174, Prague, Czech Republic.

Iseli, Christian and Sanchez, Eduardo (1995). Spyder: A SURE (SUperscalar and REconfigurable) processor. *The Journal of Supercomputing*, 9(3):231–252.

Jacob, Jeffrey A. (1998). Memory Interfacing for the OneChip Reconfigurable Processor. Master's thesis, University of Toronto, Toronto, Canada.

Jacob, Jeffrey A. and Chow, Paul (1999). Memory Interfacing and Instruction Specification for Reconfigurable Processors. In *7th ACM/SIGDA International Symposium on Field Programmable Gate Arrays*, pages 145–154, Monterey, California.

Jenkins, Jesse H. (1994). *Designing with FPGAs and CPLDs.* Prentice-Hall, Englewood Cliffs, New Jersey.

Johnson, David C., Fuller, Douglas A., Engelbrecht, Kenneth L., Marlan, Gregory A., Arnold, Ronald G., and Fagerness, Gerald G. (1998). Method and Apparatus for Performing Microcode Paging During Instruction Execution in an Instruction Processor. U.S. Patent No. 5,796,972, August 1998.

Johnson, William Mike (1991). *Superscalar Microprocessor Design.* Prentice Hall, Englewood Cliffs, New Jersey.

Jones, D. W. (1988). The Ultimate RISC. *Computer Architecture News*, 16(3):48–555.

Jones, David (1995). A Time-Multiplexed FPGA Architecture for Logic Emulation. Master's thesis, University of Toronto, Toronto, Canada.

Jones, David and Lewis, David M. (1995). A Time-Multiplexed FPGA Architecture for Logic Emulation. In *Proceedings of the IEEE 1995 Custom Integrated Circuits Conference*, pages 487–494, Santa Clara, California.

Kastrup, Bernardo, Bink, Arjan, and Hoogerbrugge, Jan (1999a). ConCISe: A Compiler-Driven CPLD-Based Instruction Set Accelerator. In Pocek, Kenneth L. and Arnold, Jeffrey M., editors, *7th IEEE Symposium on FPGAs for Custom Computing Machines*, pages 92–100, Napa Valley, California.

Kastrup, Bernardo, Trum, Jeroen, Moreira, Orlando, Hoogerbrugge, Jan, and van Meerbergen, Jef (2000). Compiling Applications for ConCISe: An Example of Automatic HW.SW Partitioning and Synthesis. In Hartenstein, Reiner W. and Grünbacher, Herbert, editors, *10th International Conference on Field-Programmable Logic and Applications. The Roadmap to Reconfigurable Computing*, volume 1896 of *Lecture Notes in Computer Science*, pages 695–706, Villach, Austria.

Kastrup, Bernardo, van Meerbergen, Jef, and Nowak, Katarzyna (1999b). Seeking (the right) Problems for the Solutions of Reconfigurable Computing. In Lysaght, Patrick, Irvine, James, and Hartenstein, Reiner W., editors, *9th International Workshop on Field-Programmable Logic and Applications*, volume 1673 of *Lecture Notes in Computer Science*, pages 520–525, Glasgow, Scotland.

Lee, Ming-Hau, Singh, Hartej, Lu, Guangming, Bagherzadeh, Nader, Kurdahi, Fadi J., Filho, Eliseu M.C., and Alves, Vladimir Castro (2000). Design and Implementation of the MorphoSys Reconfigurable Computing Processor. *Journal of VLSI and Signal Processing-Systems for Signal, Image and Video Technology*, 24(2-3):147-164.

Lew, Art and Halverson, Jr., Richard (1995). A FCCM for Dataflow (Spreadsheet) Programs. In Athanas, Peter M. and Pocek, Kenneth L., editors, *3rd IEEE Symposium on FPGAs for Custom Computing Machines*, pages 2–10, Napa Valley, California.

Ling, X. P. and Amano, H. (1993). WASMII: a Data Driven Computer on a Virtual Hardware. In Buell, Duncan A. and Pocek, Kenneth L., editors, *IEEE Workshop on FPGAs for Custom Computing Machines*, pages 33–42, Napa Valley, California.

Ling, Xiao-Ping and Amano, Hideharu (1995). WASMII: An MPLD with Data-Driven Control on a Virtual Hardware. *Journal of Supercomputing*, 9(3):253–276.

Liu, Philip S. and Mowle, Frederic J. (1978). Techniques of Program Execution with a Writable Control Memory. *IEEE Transactions on Computers*, C-27(9):816–827.

Lucent Technologies (1999a). ORCA Series 2 Field-Programmable Gate Arrays. Datasheet, Allentown, Pennsylvania.

Lucent Technologies (1999b). ORCA Series 3C and 3T Field-Programmable Gate Arrays. Datasheet, Allentown, Pennsylvania.

Mangione-Smith, W. H. and Hutchings, Brad L. (1997). Reconfigurable Architectures: The Road Ahead. In *Reconfigurable Architectures Workshop*, pages 81–96, Geneva, Switzerland.

Mangione-Smith, W. H., Hutchings, Brad L., Andrews, D., DeHon, André, Ebeling, Carl, Hartenstein, R., Mencer, Oscar, Morris, J., Palem, K., Prasanna, V. K., and Spaanenburg, H. A. E. (1997). Seeking Solutions in Configurable Computing. *IEEE Computer*, 30(12):38–43.

Marquardt, Alexander (Sandy), Betz, Vaughn, and Rose, Jonathan (1999). Using Cluster-Based Logic Blocks and Timing-Driven Packing to Improve FPGA Speed and Density. In *7th ACM/SIGDA International Symposium on Field Programmable Gate Arrays*, pages 37–46, Monterey, California.

Mason, Martin T., Evans, Scott C., and Aranake, Sandeep S. (1999). Method and System for Configuring an Array of Logic Devices. U.S. Patent No. 5,946,219, August 1999.

Miyamori, Takashi and Olukotun, Kunle (1998a). A Quantitative Analysis of Reconfigurable Coprocessors for Multimedia Applications. In Pocek, Kenneth L. and Arnold, Jeffrey M., editors, *6th IEEE Symposium on FPGAs for Custom Computing Machines*, pages 2–11, Napa Valley, California.

Miyamori, Takashi and Olukotun, Kunle (1998b). REMARC: Reconfigurable Multimedia Array Coprocessor (Abstract). In *6th ACM/SIGDA International Symposium on Field Programmable Gate Arrays*, page 261, Monterey, California.

Miyamori, Takashi and Olukotun, Kunle (1999). REMARC: Reconfigurable Multimedia Array Coprocessor. *IEEE Transactions on Information and Systems*, E82-D(2):389–397.

Miyazaki, Toshiaki (1998). Reconfigurable Systems: A Survey (Embedded Tutorial). In *Proceedings of the Asian and South Pacific Design Automation Conference, ASP-DAC '98*, pages 447–452, Pacifico Yokohama, Yokohama, Japan.

Miyazaki, Toshiaki, Yamada, Kazuhisa, Tsutsui, Akihiro, Nakada, Hiroshi, and Ohta, Naohisa (1995). Telecommunication-oriented FPGA and Dedicated CAD System. In Moore, Will and Luk, Wayne, editors, *5th International Workshop on Field-Programmable Logic and Applications*, pages 54–67, Oxford, United Kingdom.

Myers, Matthew, Jaget, Kevin, Cadambi, Srihari, Weener, Jeffrey, Moe, Matthew, Schmit, Herman, Goldstein, Seth Copen, and Bowersox, Dan (1998). *PipeRench Manual*. Carnegie Mellon University, Pittsburgh, Pennsylvania. http://www.ece.cmu.edu/research/piperench/.

Oldfield, John V. and Dorf, Richard C. (1995). *Field-Programmable Gate Arrays: Reconfigurable Logic for Rapid Prototyping and Implementation of Digital Systems*. John Wiley & Sons.

Ong, R. (1995). Programmable Logic Device Which Stores More Than One Configuration and Means for Switching Configurations. U.S. Patent No. 5,426,378, June 1995.

Patterson, David A. and Hennessy, John L. (1996). *Computer Architecture. A Quantitative Approach*. Morgan Kaufmann, San Francisco, California, second edition.

Patterson, David A. and Sequin, Carlo H. (1981). RISC I: A Reduced Instruction Set VLSI Computer. In *Proceedings of the Eighth Annual Symposium on Computer Architecture*, pages 216–230, Minneapolis, Minnesota.

Radunovi¢, B. and Milutinovi¢, V. (1998). A Survey of Reconfigurable Computing Architectures. In Hartenstein, Reiner W. and Keevallik, Andres, editors, *8th International Workshop on*

Field-Programmable Logic and Applications. From FPGAs to Computing Paradigm, volume 1482 of *Lecture Notes in Computer Science*, pages 376–385, Tallin, Estonia.

Ramakrishna Rau, B. and Fisher, Joseph A. (1993). *Instruction-Level Parallelism*, chapter Instruction-Level Parallelism: History, overview, and perspectives, pages 9–50. Kluwer Academic Publishers, Boston, Massachusetts.

Rauscher, Tomlinson G. and Adams, Phillip M. (1980). Microprogramming: A Tutorial and Survey of Recent Developments. *IEEE Transactions on Computers*, C-29(1):2–20.

Rauscher, Tomlinson G. and Agrawala, Ashok K. (1978). Dynamic problem-oriented redefinition of computer architecture via microprogramming. *IEEE Transactions on Computers*, C-27(11):1006–1014.

Razdan, Rahul (1994). *PRISC: Programmable Reduced Instruction Set Computers*. PhD thesis, Harvard University, Cambridge, Massachusetts.

Razdan, Rahul, Brace, Karl, and Smith, Michael D. (1994). PRISC Software Acceleration Techniques. In Werner, Bob, editor, *IEEE International Conference on Computer Design*, pages 145–149, Los Alamitos.

Razdan, Rahul and Smith, Michael D. (1994). A High Performance Microarchitecture with Hardware-Programmable Functional Units. In Proceedings of the *27th Annual International Symposium on Microarchitecture – MICRO-27*, pages 172–180, San Jose, California.

Rupp, Charlé R. (1995). *CLAyFun Reference Manual*. National Semiconductor Corp., Santa Clara, California.

Rupp, Charlé R. (1998). Reconfigurable Computer Architecture for Use in Signal Processing Applications. U.S. Patent No. 5,784,636, June 1998.

Rupp, Charlé R., Landguth, Mark, Garverick, Tim, Gomersall, Edson, and Holt, Harry (1998). The NAPA Adaptive Processing Architecture. In Pocek, Kenneth L. and Arnold, Jeffrey M., editors, *6th IEEE Symposium on FPGAs for Custom Computing Machines*, pages 28–37, Napa Valley, California.

Salcic, Zoran (1996). SimP – A Simple Custom-Configurable Processor Implemented in FPGA. Technical Report, Auckland University, Department of EEE, Auckland, New Zealand.

Salcic, Zoran and Maunder, Bruce (1996a). CCSimP – An Instruction-Level Custom-Configurable Processor for FPLDs. In Hartenstein, Reiner W. and Glesner, Manfred, editors, *6th International Workshop on Field Programmable Logic and Applications. Field-Programmable Logic: Smart Applications, New Paradigms and Compilers*, volume 1142 of *Lecture Notes in Computer Science*, pages 280–289, Darmstadt, Germany.

Salcic, Zoran and Maunder, Bruce (1996b). SimP – a Core for FPLD-based Custom Configurable Processors. In *ASIC Conference – ASICON*, pages 197–201, Shanghai, China.

Salcic, Zoran and Smailagic, Asim (1997). *Digital Systems Design and Prototyping Using Field Programmable Logic*. Kluwer Academic Publishers, Boston, Massachusetts.

Sanchez, Eduardo, Sipper, Moshe, Haenni, Jacques-Olivier, Beuchat, Jean-Luc, Stauffer, André, and Perez-Uribe, Andrés (1999). Static and Dynamic Configurable Systems. *IEEE Transactions on Computers*, 48(6):556–564.

Sawitzki, Sergej, Gratz, Achim, and Spallek, Rainer G. (1998a). CoMPARE: A Simple Reconfigurable Processor Architecture Exploiting Instruction Level Parallelism. In Hawick, K. A. and Heath, J. A., editors, *Proceedings of the 5th Australasian Conference on Parallel and Real-Time Systems*, pages 213–224, Adelaide, Australia.

Sawitzki, Sergej, Gratz, Achim, and Spallek, Rainer G. (1998b). Increasing Microprocessor Performance with Tightly-Coupled Reconfigurable Logic Arrays. In Hartenstein, Reiner W. and Keevallik, Andres, editors, *8th International Workshop on Field-Programmable Logic and Applications. From FPGAs to Computing Paradigm*, volume 1482 of *Lecture Notes in Computer Science*, pages 411–415, Tallin, Estonia.

Scalera, Stephen M. and Vázquez, Jóse R. (1998). The Design and Implementation of a Context Switching FPGA. In Pocek, Kenneth L. and Arnold, Jeffrey M., editors, *6th IEEE Symposium on FPGAs for Custom Computing Machines*, pages 78–85, Napa Valley, California.

Schmit, Herman (1997). Incremental Reconfiguration for Pipelined Applications. In Werner, Bob, Arnold, Jeffrey M., and Pocek, Kenneth L., editors, *5th IEEE Symposium on FPGAs for Custom Computing Machines*, pages 47–55, Napa Valley, California.

Sima, Mihai, Cotofana, Sorin, van Eijndhoven, Jos T.J., Vassiliadis, Stamatis, and Vissers, Kees (2001). 8 × 8 IDCT Implementation on an FPGA-augmented TriMedia. In Pocek, Kenneth L. and Arnold, Jeffrey M., editors, *9th Annual IEEE Symposium on Field-Programmable Custom Computing Machines*, Rohnert Park, California.

Sima, Mihai, Cotofana, Sorin D., van Eijndhoven, Jos T.J., Vassiliadis, Stamatis, and Vissers, Kees A. (2005). IEEE-compliant IDCT on FPGA-augmented TriMedia. *Journal of VLSI Signal Processing Systems for Signal, Image, and Video Technology*, 39(3): 195–212.

Sima, Mihai, Cotofana, Sorin D., Vassiliadis, Stamatis, van Eijndhoven, Jos T.J., and Vissers, Kees A. (2004). Pel Reconstruction on FPGA-augmented TriMedia. *IEEE Transactions on VLSI Systems*, 12(6):622–635.

Sima, Mihai, Vassiliadis, Stamatis, Cotofana, Sorin, van Eijndhoven, Jos T.J., and Vissers, Kees (2000). A Taxonomy of Custom Computing Machines. In Veen, Jean Pierre, editor, *Proceedings of the First PROGRESS Workshop on Embedded Systems*, pages 87–93, Utrecht, The Netherlands.

Singh, Hartej, Lee, Ming-Hau, Lu, Guangming, Kurdahi, Fadi J., Bagherzadeh, Nader, and Filho, Eliseu M. Chaves (2000). MorphoSys: An Integrated Reconfigurable System for Data-Parallel and Computation-Intensive Application. *IEEE Transactions on Computers*, 49(5):465–481.

Smith, Michael John Sebastian, editor (1997). *Application-Specific Integrated Circuits*. Addison-Wesley, Reading, Massachusetts.

Takayama, Atsushi, Shibata, Yuichiro, Iwai, Keisuke, Miyazaki, H., Higure, K., Ling, X.-P., and Amano, Hideharu (1999). Implementation and Evaluation of the Compiler for WASMII, a Virtual Hardware System. In *Proceedings of the 1999 International Conference on Parallel Processing*, pages 346–351, Aizu-Wakamatsu, Fukushima, Japan.

Tau, Edward, Chen, Derrick, Eslick, Ian, Brown, Jeremy, and DeHon, André (1995a). A First Generation DPGA Implementation. In *The Third Canadian Workshop on Field-Programmable Devices*, Montreal, Canada.

Tau, Edward, Chen, Derrick, Eslick, Ian, Brown, Jeremy, and DeHon, André (1995b). A First Generation DPGA Implementation. Transit Note # 114, Massachusetts Institute of Technology, Cambridge, Massachusetts.

Teich, Paul R., Asghar, Saf, and Lee, Sherman (1999). Retargetable VLIW Computer Architecture and Method of Executing a Program Corresponding to the Architecture. U.S. Patent No. 5,943,493, August 1999.

Thornburg, Mike and Casselman, Steve (1994). Transformable Computers. *The 8th International Parallel Processing Symposium*, pages 674–679, Cancun, Mexico.

Trimberger, Stephen M. (1994). *Field-Programmable Gate Array Technology*. Kluwer Academic Publishers, Boston, Massachusetts.

Trimberger, Stephen M. (1995). Microprocessor-based FPGA. International Patent Application under P.C.T., No. WO 95/04402, February 1995.

Trimberger, Stephen M. (1998a). Reprogrammable Instruction Set Accelerator. U.S. Patent No. 5,737,631, April 1998.

Trimberger, Stephen M. (1998b). Reprogrammable Instruction Set Accelerator Using a Plurality of Programmable Execution Units and an Instruction Page Table. U.S. Patent No. 5,748,979, May 1998.

Trimberger, Stephen M., Carberry, Richard A., Johnson, Robert Anders, and Wong, Jennifer (1998). Programmable Logic Device Including Configuration Data or User Data Memory Slices. U.S. Patent No. 5,784,313, July 1998.

Trimberger, Stephen M., Carberry, Richard A., Johnson, Robert Anders, and Wong, Jennifer (1999). Method of Time Multiplexing a Programmable Logic Device. U.S. Patent No. 5,978,260, November 1999.

Trimberger, Steve, Carberry, Dean, Johnson, Anders, and Wong, Jennifer (1997). A Time-Multiplexed FPGA. In Werner, Bob, Arnold, Jeffrey M., and Pocek, Kenneth L., editors, *5th IEEE Symposium on FPGAs for Custom Computing Machines*, pages 22–28, Napa Valley, California.

van den Bout, David E., Morris, Joseph H., Thomae, Douglas, Labrozzi, Scott, Wingo, Scot, and Hallman, Dean (1992). AnyBoard: An FPGA-based reconfigurable system. *IEEE Design and Test of Computers*, 9:21–30.

van Eijndhoven, Jos T. J., Sijstermans, F. W., Vissers, Kees A., Pol, E.-J. D., Tromp, M. J. A., Struik, P., Bloks, R. H. J., van der Wolf, Pieter, Pimentel, A. D., and Vranken, Harald P.E. (1999). TriMedia CPU64 Architecture. In *Proceedings of International Conference on Computer Design*, pages 586–592, Austin, Texas.

Vassiliadis, Stamatis, Wong, Stephan, and Cotofana, Sorin (2001). The MOLEN $\rho\mu$-coded Processor. In Brebner, Gordon and Woods, Roger, editors, *11th International Conference on Field-Programmable Logic and Applications*, volume 2147 of *Lecture Notes in Computer Science*, pages 275–285, Belfast, Northern Ireland, United Kingdom.

Vassiliadis, Stamatis, Wong, Stephen, and Cotofana, Sorin D. (2003). Microcode Processing: Positioning and Directions. *IEEE Micro*, 23(4):21–30.

Vassiliadis, Stamatis, Wong, Stephen, and Georgi N. Gaydadjiev, Koen Bertels, Georgi K. Kuzmanov Elena Moscu Panainte (2004). The Molen Polymorphic Processor. *IEEE Transactions on Computers*, 53(11):1363–1375.

Villasenor, John and Mangione-Smith, William H. (1997). Configurable Computing. *Scientific American*, pages 55–59. http://www.sciam.com/0697issue/0697villasenor.html.

Vuillemin, Jean E., Bertin, Patrice, Roncin, Didier, Shand, Mark, Touati, Hervé H., and Boucard, Philippe (1996). Programmable active memories: Reconfigurable systems come of age. *IEEE Transactions on VLSI*, 4(1):56–69.

Wazlowski, M., Agarwal, L., Lee, T., Smith, A., Lam, E., Athanas, P., Silverman, Harvey, and Ghosh, S. (1993). PRISM-II Compiler and Architecture. In Buell, Duncan A. and Pocek, Kenneth L., editors, *IEEE Workshop on FPGAs for Custom Computing Machines*, pages 9–16, Napa Valley, California.

Wazlowski, Michael Edward (1996). *A Reconfigurable Architecture Superscalar Coprocesor*. PhD thesis, Brown University, Providence, Rhode Island.

Wilkes, Maurice V. (1951). The best way to design an automatic calculating machine. In *Report of the Manchester University Computer Inaugural Conference*, pages 16–18. Electrical Engineering Department of Manchester University.

Wirthlin, Michael J. (1997). *Improving Functional Density through Run-Time Circuit Reconfiguration*. PhD thesis, Brigham Young University, Provo, Utah.

Wirthlin, Michael J. and Hutchings, Brad L. (1995). A Dynamic Instruction Set Computer. In Athanas, Peter M. and Pocek, Kenneth L., editors, *3rd IEEE Symposium on FPGAs for Custom Computing Machines*, pages 99–109, Napa Valley, California.

Wirthlin, Michael J., Hutchings, Brad L., and Gilson, Kent L. (1994). The Nano Processor: A Low Resource Reconfigurable Processor. In Buell, Duncan A. and Pocek, Kenneth L., editors, *2nd IEEE Workshop on FPGAs for Custom Computing Machines*, pages 23–30, Napa Valley, California.

Wittig, R. (1995). OneChip: An FPGA Processor with Reconfigurable Logic. Master's thesis, University of Toronto, Toronto, Canada.

Wittig, Ralph D. and Chow, Paul (1996). OneChip: An FPGA Processor With Reconfigurable Logic. In Pocek, Kenneth L. and Arnold, Jeffrey M., editors, *4th IEEE Symposium on FPGAs for Custom Computing Machines*, pages 126–135, Napa Valley, California.

Xilinx Corporation (1996c). XC6200 Field Programmable Gate Arrays. Datasheet, San Jose, California.

Xilinx Corporation (1999). *The Programmable Logic Data Book*. Xilinx Corp., San Jose, California.

Ye, Zhi Alex, Moshovos, Andreas, Hauck, Scott, and Banerjee, Prithviraj (2000). CHIMAERA: A High Performance Architecture with a Tightly–Coupled Reconfigurable Functional Unit. *33rd Annual International Symposium on Microarchitecture*, pages 225–235, Monterey, California.

Yoshimi, Masahisa and Ikezawa, Toshi (1990). Multifunction Programmable Logic Device. Japan Patent Abstract No. 2,130,023, May 18, 1990.

Zhou, Xianfeng and Martonosi, Margaret (2000). Augmenting Modern Superscalar Architectures with Configurable Extended Instructions. In *International Parallel Processing Symposium*, pages 941–950, Cancun, Mexico.

Index

MIX
Papier aus verantwortungsvollen Quellen
Paper from responsible sources
FSC® C105338

If you have any concerns about our products, you can contact us on
ProductSafety@springernature.com

In case Publisher is established outside the EU, the EU authorized representative is:
Springer Nature Customer Service Center GmbH
Europaplatz 3, 69115 Heidelberg, Germany

Printed by Libri Plureos GmbH
in Hamburg, Germany